AF348939

Creation:
Theory and Theology

by

Richard R. Wilkinson

Science & Religion; One Truth!

Copyright

Contents

0: About this Book

What triggered the creation debate?

The Scopes Monkey Trial, as it came to be known, was a farce from the beginning. The ACLU advertised in Tennessee newspapers for a willing teacher to be the defendant in a contrived case that would test the constitutionality of the Butler act, recently passed by the state legislature to ban the teaching of evolution in the public schools. Local physics teacher John Thomas Scopes, who only substituted in biology class, answered the ad as urged by several Dayton townspeople who anticipated that the court proceedings and publicity would stimulate local business.

The scheme worked. In 1925, the State of Tennessee charged Scopes with teaching human evolution to high school students in violation of the Butler act. Scopes did not remember ever teaching evolution but agreed to serve as defendant and allowed himself to be "arrested" by his friend Sue Hicks, the inspiration for Johnny Cash's *Boy Named Sue* hit. Sue was named after his mother who died from complications giving birth to him.

As foreseen, thousands of journalists accompanied a steady stream of trial participants and spectators to Dayton. The trial got national attention and was featured on the front page of numerous newspapers across the country. Part of the theatrics was a dressed-up chimp named Joe whom trial promoters gleefully welcomed to town and exploited for publicity purposes. After eight days in court, the jury deliberated only nine minutes before finding Scopes guilty.

The judge fined him $100 (equivalent to about $1,500 in 2020). But, since the law said that fines greater than $50 must be levied by the jury, Scopes got off on the technicality, an outcome that the judge probably intended. Since the appeals court judge characterized the matter as undeserving of further legal attention, the judicial theatrics stopped there. But the monkey business had served its intended purpose. Dayton was national news for many weeks.

More important than the benefits to the town were the increased public awareness of the religion-versus-evolution controversy and the accompanying constitutionality questions, paving the path for modern science to be taught in schools unfettered by conflicting traditional beliefs.

Attitudes have swung dramatically toward science since then, especially among youngsters. The early 21st century saw the number of adults in the United States (US) who accept human evolution increase to 40 percent and the numbers in Europe rise even further. About 25 percent of Americans believe in divinely-guided evolution.

It was not Darwin's intent to elbow God aside with the book that transformed the world's views about life on Earth. He used the word "Creator" nine times and the word "God" twice in his final edition of *Origin of Species*. He even offered some advice that was so religious it cannot be taught in public schools. His second page says, "Let no man out of a weak conceit of sobriety, or an ill-applied moderation, think or maintain, that a man can search too far or be too well studied in the book of *God's word*, or in the book of *God's works*." (*Origin of Species*, p2) Examining both of God's records, the one observed in nature and the one contained in scripture, is the best approach to this day.

But later, even Darwin had to concede his failure to reconcile the two records, writing, "An Agnostic would be the more correct description of my state of mind." (*Life and Letters*, 1879 letter) This book will do what Darwin did not.

Darwin is not alone. The apparent contradictions between God's word and His works seem unresolvable to almost everyone. Preachers, pastors, prophets, popes, pundits, and pretenders, whose work would benefit most from harmony with science, only complicate the situation by attacking evolution and/or stubbornly interpreting scripture in ways that are irreconcilable with science.

So, why the trend toward trusting scientists over seers? Parents of today are more and more comfortable allowing a child to make up its own mind about religion. So, public schools, which must include evolution in the curricula, have much more influence on the views of the rising generation than parents do. And virtually all children in the USA attend 180 days of school every year whereas very few children attend church even once per week.

The relative ignorance of average laypersons in science and religion makes them vulnerable to one-sided sales pitches. Scientists say what will garner more grants. Preachers say what will garner more gifts. And authors write what will sell more books.

Of tens of thousands of distinct Christian, Jewish, and Islamic sects, none offer an official explanation of how the Genesis meshes with science on the topic of creation. Furthermore, it is impossible to find any two priests, rabbis, or imams in their own sect that have identical opinions on creation versus science, much less any two congregants. If the Old Testament (OT) is God's word, then this embarrassment of disagreements says more about mankind's capacity to understand than it does about His ability to communicate!

Likewise, scientists have an embarrassment of contradictory views. Most famously, relativity theory and quantum theory clash when dealing with very short distances. And, scientists even disagree on what gravity is, some thinking it is a force, others seeing it is a feature of spacetime.

String theory appears to solve all the problems but, after half a dozen decades of tweaking, it now conflicts with nothing but can't predict anything. And, if one asks any two physicists for a descriptive definition of time, one will likely get two different answers.

Secularization of rising generations has trended upwards for the last half century. The Pew Research Center found that the percentage of people 18 and older who identify as Christians dropped 8% in the seven years ending with 2014 while those identifying as atheists and agnostics increased 6.7% and those identifying as non-Christians rose 1.2%.

(pewforum.org/2015/05/12/americas-changing-religious-landscape, Downloaded (DL) 28 Jun 2017) In 2020, church membership in the USA fell below 50% for the first time ever. (news.gallup.com/poll/341963/church-membership-falls-below-majority-first-time.aspx, DL 31 Mar 2021)

Clergy faith is also waning. For example, "According to an investigation of 860 pastors in seven Dutch Protestant denominations, 1 in 6 clergy are either agnostic or atheist." (bbc.com/news/world-europe-14417362, Downloaded (DL) 1 Feb 2019)

For many people, especially youth, the biggest obstacles to faith are the apparent conflicts with science and Genesis. Consequently, many never get past Genesis to the most important parts of the Bible, such as the condescension of God, His exemplary life, His atoning sacrifice, His resurrection, and the happiness He offers to His followers.

The object of this book is to model at least one possible story or view of science and religion in which theory and theology do not conflict. This book borrows heavily from religious sources whose ideas are the most comprehensive and from scientists whose ideas best fit the approach taken.

Does curiosity kill the cat?

It's true that cats sometimes poke their noses into dark places to their own detriment, but the ones that find mice are better off. Besides, the risk-reward ratio is low, especially if they have nine lives. Curiosity is a valuable attribute. Perhaps that is why someone added a fix to the phrase in the early 20th century; "But, satisfaction brought it back."

Satisfaction is the purpose of this book for those whose curiosity brought them here. Enoch, an example of a hyper-righteous person, "was one of the curious ones: 'I raised my eyes and contemplated… the sky with its glittering stars, the sun and the moon, … the earth and all that is in it… How do they operate? How do they keep going? Who can explain to me the alterations of dawn and dusk, day and night, moon and stars…?' And prostrating himself he prayed for enlightenment." (Hugh Nibley, April Ensign, 1977, *A Strange Thing in the Land, Part II*)

And, according to apocryphal writings, Enoch's curiosity was richly rewarded with divine knowledge of the creation, astronomy, ethics, and eschatology; "And I wrote 360 books. The Lord set me on the path of understanding and wisdom." (2 En 10:7) God loves curiosity and will reward it in people whose priorities remain in line with His.

What unusual punctuation is used?

End-of-sentence punctuation (".", "?", "!", & "...") will precede any parenthetical source citations associated with that sentence for consistency. For sentences that end with quotation marks, the end-of-sentence punctuation is inside the close-quote mark if a full standalone sentence was quoted independent and after the close quote mark in all other cases. There will be no space between a date and its BC/AD designation.

What sources are cited?

Bible references will be from the King James Version (KJV) unless preceded by the initialization of another version in the citation. Not all translations are created equal but some passages in non-KJV bibles are translated in ways that better serve the purpose of this book. The Greek and Hebrew Concordance referenced is James Strong's as shown on biblehub.com. His numbering system for terms will be formatted as "Str#" followed immediately by the number. In the context of the Old Testament, the number will refer to a Hebrew term and in the context of the New Testament to a Greek term. And, simple citation formatting has been chosen that, though standard for this book, has not been adopted by any other work nor any organization. The abbreviations citation details will be immediately followed by the relevant number without a space in between. Though citation formats are my own standard, any source can easily be found using the information provided as search criteria in modern query technology. All emphasis (underlining and italicizing) in quotes is by me.

Ancient religious literature is often cited, not because it should be considered scripture, but because it reveals ideas in the minds of worshipers in times past. Writers of spiritual literature seldom wove tales out of thin air. They borrowed and embellished existing written and oral tradition.

Who is the target Audience?

This book is for all who are fascinated or bewildered by the science-vs-religion debate. Since it draws heavily from the extensive creation theology of The Church of JESUS CHRIST of Latter-day Saints, some of its adherents are likely to be receptive. For anyone embarking on a proselyting mission, reading this entire book beforehand will help to steer conversations away from doctrinal pitfalls. Atheists and agnostics who enjoy reading religious philosophies will find this book entertaining, particularly if they see conflicts between religion and science as the primary obstacle to faith. This book is especially intended for anyone who is on the cusp of having their faith destroyed by perceived discrepancies between the Bible and Science.

What if you already know it all?

This book starts at the ground floor and builds layer by layer, simple at first to support the advanced ideas later on. The order in which topics are presented is a chronological journey of discovery from the fundamental concepts to the loftier ones, each idea adding to the prior. The early pages lay the groundwork while the remainder of the book deals with creation history in roughly chronological order. Even savants such as yourselves should find plenty of juicy tidbits of information and logic from the beginning that may be useful to you in discussions with your friends and families. The later pages in the book are heavier. The weight is worth the wait!

Pretend you're starting fresh. Judge not this book by the first few thoughts that fail to align with your views. The big picture reveals itself one brush stroke at a time. God designed the creation to be unfathomable. So, everyone should approach the subject with humility. "He has made everything beautiful… [and] set eternity in the human heart; yet *no one can fathom what God has done from beginning to end*." (*New International Version (NIV)*, Ec 3:11)

What if you're a professional critic?

This book is a mix of theory and theology, as per the subtitle. It doesn't purport to be completely true in every tiny detail. The breadth and depth of topics makes that impossible. However, if science is the knowledge of the creation and religion is knowledge of the Creator, there must be a way to harmonize these realms of truth. As this is a difficult task, the book pulls from sources far and wide, from Jewish lore, Islamic literature, pseudepigrapha, restorationist doctrine, apocryphal books, ancient philosophers, early Christian theologians, professors of scripture, near-death experiences, prophets of modern movements, et cetera. These sources are not of equal weight. Scripture is of course the most authoritative. But please don't dismiss viewpoints out of hand simply because you are a stranger to a source.

What if you're new to this topic?

Many of you expect answers before pondering the questions. This book generally ponders the questions and their context first before giving the answers. For the topics at hand, one-line answers seldom serve satisfactorily.

Enjoy! Have fun trying out some of the ideas on your kith and kin of include them in your social media posts.

1: Conundrums and Conflicts

Are the conundrums inscrutable?

In the beginning, Genesis confused me. The more I studied, the more questions came to mind, most of which never came up in Sunday school, nor seminary, nor college religion classes. Early on I observed that authors on the topic frequently forced pieces together that didn't fit and glossed over spaces where no pieces were available. At length I resolved to formulate a contradiction-free description of creation myself, one in which all the puzzle pieces from science and religion fit nicely together.

When I attempted on occasion to discuss ideas with others, I found that attitudes make very effective barriers. Once, while at the public library looking for books on the origins of life, earth, and the universe, I happened to meet a fellow congregant and, in the conversation that followed, mentioned that I was researching the creation. He confided that he had once sought many of the same answers but gave up when he found himself so confused that he began to question his own beliefs. "It's a dangerous quest," he said, and warned me that my faith might falter if I ventured too far.

Shortly after, while browsing at a Christian bookstore, I asked a sales clerk whether there were any works by credible authors that harmonized science and scripture on the subject of creation. After searching in vain she said with a sigh, "Maybe there are some questions we aren't supposed to know the answer to." She is not the only believer with that view.

I figured God might not be in the business of stitching His religion to the science of men. And, without divine revelation, only someone familiar with both science and religion would have a shot at piecing together a good creation picture. I was in no special circumstance to expect a divine revelation anyway, but I reasoned that if Enoch, Abraham, and Moses could have their curiosity about the stars rewarded in stunning 3-D visions, then surely God would not deem it an evil endeavor for me to seek similar knowledge in the context of contemporary science. So, I began collecting sources, reading voraciously, and making notes, and am quite happy with the results – this book.

The sections in this chapter are designed to pique the reader's curiosity by asking some of the tough questions. The answers will be discussed at sufficient length in later chapters.

Shouldn't truth agree?

One Sunday evening, years ago, as I was bidding my little daughter goodnight, she asked, "Daddy, how could Adam have named the dinosaurs if they had already died long before?" Her elementary school teacher had taught that the extinction of dinosaurs occurred millions of years before humans arrived on the scene but her Sunday school teacher had taught that Adam had named all the animals.

What was I to tell her? I didn't have an answer sufficiently simple for a 5-year-old, brilliant though she was. So, I stumbled through several theories known to me at the time and thought that I had at least done a fine job of illustrating alternative ways to resolve the discrepancy. Many scholars much smarter than me had no single simple answer either. Her reaction? "You don't really know, do you Daddy?"

Ouch! Knowing stuff was Daddy's job! A child can ask a simple question that even a sage finds impossible to answer. If I had it to do over again, I'd quote the scripture, "Adam gave names to all cattle, and to the fowl of the air, and to every beast of the field." (Gn 2:20) Then I'd say, "Dinosaurs were not cattle, nor fowls, nor beasts of the field, so they went without names until scientists began to dig up their bones." Perhaps that would have satisfied her long enough for me to formulate a simpler and more legitimate answer.

Early scientists had to tread carefully to avoid the wrath of the Catholic Church. When Copernicus wrote about heliocentrism (sun at center) in 1543AD, he did so in hypothetical terms to preserve his good relationship with the church, merely postulating, "*If* the earth were in motion *then* the observed phenomenon would result." (*On the Revolutions of the Heavenly Bodies*) There is nothing in the Bible that rules out heliocentrism other than colloquial perspectives of the sun rising and setting that are still pervasive in modern times despite widespread knowledge of Earth's daily rotation.

In 1593, astronomer Giordano Bruno was tried on charges of heresy for suggesting that myriad earth-like planets might orbit countless stars throughout the universe. The Roman Catholic Inquisition burned him at the stake for not recanting his ideas, a practice that tended to discourage faithful friars from sharing novel notions. Bruno suffered one of the most excruciating deaths for stating what would later be proven true. Thousands of exoplanets have been discovered around stars in recent decades, some of which appear quite earth-like, and scientists assume the universe is full of such star systems yet none of these discoveries have caused a single religious leader to protest.

When Galileo publicly shared his observations in the early 1600's that the planets circled the sun, he was accused of heresy by the church. His heliocentric solar system model was called "foolish and absurd in philosophy, and formally heretical since it explicitly contradicts in many places the sense of Holy Scripture". (*Assessment made at the Holy Office*, 24 Feb 1616, Rome)

So, the church issued an injunction against Galileo to stop talking about heliocentrism. Seventeen years later, the church again charged Galileo with heresy and interrogated him under threat of torture. His prison sentence was later commuted to house arrest for the remainder of his days and his books were banned along with those written by Copernicus.

What did the Catholic Church's commission of theologians use as Biblical proof Galileo was wrong? They cited the example where Joshua commanded the sun to stop. If it could stop, they reasoned, it had to be moving. Therefore, it was the sun that moved and not the earth. Therefore, Galileo was contradicting scripture. (Jsh 10:12)

However, in Old Testament times, the "Hebrew word translated 'stand still' has the same root as a Babylonian word used in ancient astronomical texts to describe eclipses." (cosmosmagazine.com/space/biblical-miracle-turns-out-to-be-solar-eclipse-study-finds, DL 31 Oct 2017) The following verse in the Hebrew says the moon also "took a stand". (Strg#5975) If an eclipse changed the outcome of a battle, there's no issue with science and no need to explain why other ancient writings of that time say nothing of the sun and moon stopping in their tracks for a whole day.

As proof that the earth doesn't move, the commission also cited the psalmist's verse that says God "laid the foundations of the earth, that it should not be removed for ever". (Ps 104:5) But, such poetry cannot be taken as literal astronomical fact any more than Carol King's lyrics to, *I Feel the Earth Move*. The ancient Hebrew conception of Earth and skies is occasionally reflected in Biblical verbiage. For example, the Bible speaks of the "pillars of the earth". (1 Sm 2:8) Could the commission's theologians have also taken that term literally almost 100 years after Magellan circumnavigated the globe? Why did they cherry-pick what to take literally when a scientist's life was at stake? Perhaps church leaders chafed at being upstaged by a scientist.

After seeing what had happened to Bruno and Galileo, the great Descartes abandoned plans to publish his now-famous treatise, *The World*, the result of brilliant research on space, particles, light, and the laws of motion. Descartes credited a "divine spirit" for much of his new ideas and tried to stay on the church's good side, but to no avail. His books were added to the *Index of Prohibited Books* that no catholic should read. (sparknotes.com/philosophy/descartes/context.html, DL 4 Sep 2016)

Over time, as a preponderance of evidence mounted for science that was once called heresy by the Roman Catholic Church, bishops and popes became more careful, waiting nearly a century after Darwin published *On the Origin of Species* before commenting. As of now, most mainstream religions are taking reasonably science-friendly positions but a number of fundamentalist Christian groups still treat evolution and old-Earth science like boogeymen. One such group spent $27 million to build a Creation Museum in 2007 and $160 million to build an Ark Encounter theme park in 2016. Both projects aimed to promote a young-earth model along with alternatives to evolution. Exhibits even depict humans and dinosaurs living alongside each other.

It's tough for students of faith to sit through science classes, study material that conflicts with their core beliefs, and provide answers they feel are false on exams. Referring to aspects of science that seemed to disagree with Genesis, a student in my biology class once asked, "How can scientists be so stupid?" I was afraid of repercussions if I discussed religion in a public school, so I simply responded that I had managed to forge an understanding of science and religion that eliminated contradictions and recommended he try to do the same.

Rather than resolve contradictions, some people either mentally compartmentalize religion and science or avoid learning enough to see any conflicts. More analytical types of people enjoy pointing out perceived contradictions between scientific and religious versions of creation as they understand them, either hoping for interesting and helpful responses from others, or trying to persuade others that religion is nonsense. Here's a sampling of alleged religion-science contradictions based on misunderstandings and/or ignorance of facts:

♦ *How old is the earth?* The Bible says six thousand years, but science says 4.5 billion years.

♦ *How long did creation take?* The Bible says six days, but science says 13.8 billion years.

♦ *How was man created?* The Bible says from the dust of the earth in one day by God, but science says from lower orders of animals over billions of years by evolution.

♦ *How was life started?* The Bible says by the word of God instantaneously, but science says by the natural formation of replicating molecules near deep sea vents or warm ponds over billions of years.

♦ *Whence came human diversity?* The Bible says all people come from Adam and Eve, but science says skin, eye, and hair color varieties arose from genetic mutations over tens of thousands of years.

♦ *What was created first?* The Bible says plants were created before the sun and other stars, but science says the sun was created even before the earth and any of its lifeforms.

A comprehensive list would be much longer. So, it's no surprise that, as of 2019, most Americans, 59%, believed there were contradictions between science and religion generally while 30% believed *their* religion conflicted with science. Obviously, people's perceptions are heavily biased by their own religious views and their understanding of science. Logic dictates that true science and true religion must be in harmony.

Chemist, geologist, and apostle James Talmage wrote, "The Creator has made a record in the rocks for man to decipher; but he has also spoken directly regarding the main stages of progress by which the earth has been brought to be what it is. The accounts cannot be fundamentally opposed; one cannot contradict the other; though man's interpretation of either may be at fault." (*Juvenile Instructor*, Dec 1965, p475)

Does Genesis contradict itself?

Commonly asked questions about internal contradictions in Genesis include, but are not limited to, the following:

♦ Why did God create light on Day One, but the sun on Day Four? (Gn 1:3, 16)

♦ How could the phrase "evening and the morning" mark the time of each day before the sun was created and why doesn't it circumscribe a 24-hour cycle? (Gn 1:5, 8, 13)

♦ If each creation day lasted many years, how did plants survive without the sun from the second to the third day? (Gn 1:11-17)

♦ Why does the creation sequence in Genesis chapter two contradict the creation sequence in chapter one?

♦ If God created everything from nothing (creatio ex nihilo) as most of Christianity teaches, why did He start with dust when creating Adam and with a rib when creating Eve?

♦ If God is perfect, why did many of His creations turn to evil?

♦ If they were commanded to multiply, why didn't Adam and Eve have children in the garden?

♦ If Adam named all creatures, and a whale was worthy of note, why weren't other massive monsters such as mastodons, pterodactyls and titanosaurs worthy of note? (Gn 1:21)

♦ If there were no death in Eden, how did Adam eat without crushing living fruit cells and walk without crushing insects?

♦ If the Garden of Eden brought forth nothing but pleasant plants spontaneously, why did God require them to dress and keep it? (Gn 2:16)

♦ If God wanted man to become more like Himself, why didn't He create humans wise from the beginning instead of staging the Tree of Knowledge drama?

♦ Why did God provide a Tree of Life if He planned to bar them from it as soon as they needed it? (Gn 3:22)

♦ If God didn't want Adam and Eve to eat of the Tree of Knowledge why did He plant it right in their yard? (Gn 2:9)

♦ If God is good, why did He create the bad guy Satan and set him loose in the Garden of Eden? (Gn 3:1)

Not every rational reader of Genesis spends time pondering such questions as these but, for those who do, the reaction is often a rejection of the Bible and all it represents.

What harm can eisegesis do?

Virtually all of the previously mentioned creation conundrums arise from eisegesis, meaning the interpreter injects his own ideas into the text or proposes meanings that are poorly supported by the text. Unfortunately, such interpretations have hardened into traditional beliefs over the centuries, making the Genesis creation story much more difficult to swallow for the average person.

Perhaps the most insidious idea injected into the story of Adam and Eve is that their transgression makes every infant sinful from the moment of conception. If that idea came from the text, why wasn't it taught during New Testament (NT) times? The doctrine of original guilt was heavily influenced by heathen and heretics and was formulated after Christ's inspired apostles were gone.

The sins of any parents are often visited upon their children, and sometimes subsequent generations in the form of consequences. The same is true in the case of Adam's transgression. But that doesn't mean babies are born sinful. The old notion that if you reap a consequence you must have sown a sin is from the devil.

Early theologians reasoned that, since babies are born sinful, and baptism is a prerequisite to heaven, any babies that die without baptism die in sin and go to hell. That problem was "fixed" by initiating the baptism of infants. But paedobaptism is in direct conflict with Christ's declaration of toddler innocence when He said, "Let the little children come to Me, and do not forbid them; for of such is the kingdom of God." (Lk 18:16)

When paedobaptism became widespread tradition, it was salt in the wound of countless bereft mothers who were forced to watch as their unbaptized babies were buried at the fringes of church cemeteries, outside hallowed ground, where murderers, rapists, robbers, and others considered unfit for heaven were interred. And preachers were known to use the original sin doctrine to flagellate parents when pounding out hellfire and damnation from the austerity of the pulpit.

When the question of Jesus inheriting original sin arose, theologians solved the problem by concocting the doctrine of Mary's Immaculate Conception, the idea that she was spared the curse of Adam's sin through the merits of her future Son.

But wait! Couldn't Mary have sinned after she was born, reacquired her sinful state, and passed *that* sinfulness on to Jesus? Something had to be done. So, doctrinal experts proposed that Mary was born of an immaculate conception and was so full of grace that she remained without sin. Thus, Jesus could be born to a mother who had no sin of any kind! (*CCC (Catechism of the Catholic Church)*, 1994, p491) The teaching that Jesus was not the only human who never sinned is at best wrong and at worst blasphemy! Such a doctrine suggests that God unfairly gave Mary an advantage that has been withheld from everyone else.

Another example of harmful misinterpretation is the idea that people who die having never heard of Jesus are doomed. This leaves the vast majority of Earth's inhabitants headed for hell without hope.

Misinterpretations can also set believers up for disillusionment as in cases where denominations have determined specific dates from the Bible on which Jesus would return, only to have those dates come and go without Him showing. It was especially difficult for believers who had disposed of their property, dressed themselves in white, and waited to be raptured up to heaven, only to find themselves earthbound and penniless after the date of the expected parousia (return of the Christ) had passed.

For example, a Baptist preacher named Miller announced that Jesus would return in 1844. Out of the disappointment that followed, several new sects were born, among them the Seventh-day Adventist church.

Bad interpretations of Genesis have led thousands of young people to reject religion entirely and, in some cases, join atheist activists in mocking religion.

Are fossils a problem for Genesis?

Scientists have found hundreds of thousands of fossils that are millions and billions of years older than the circa 6,000 years of Earth's history presented by the Bible. Yet a 2014 Gallup poll showed that about 40% of Americans, or about half the Christian population, believe God created modern humans less than 10,000 years ago. (livescience.com/46123-many-americans-creationists.html, DL 25 Apr 2020)

So, Christianity in the USA is split right down the middle on the question of humanity's age. Christian apologists have tried to resolve the issue with various creation scenarios, one being that the Biblical creation story is only the final round of a series creation sagas, each of which ended with mass extinctions of which the fossils are evidence. However, such theories strongly conflict with the fossil record that shows all mass extinctions occurred in a continuous multibillion-year struggle for survival that prove the *Jurassic Park* quote true, "Life finds a way."

Others have suggested that the fossils got here from outer space, arriving on pieces of exploded worlds where dinosaurs and other ancient extinct creatures once lived. One problem with this idea is that all the space rocks were melted by extreme heat when they accreted to form Earth. It took millions of years for molten Earth too cool sufficiently for a hard crust to form. If any alien fossils had survived the accretion energies, they would have been lost anyway in the molten sea of lava.

If fossils had survived the accretion energies and the molten ball stage of our planet, they would be older than our 4.6 billion-year-old Earth. But every fossil found thus far is much younger than Earth. For example, the frozen and thus fresh flesh of some Siberian mammoths dates radiometrically to 40 TYA (thousand years ago), the same age as the ice in which they were found and about 4.6 billion years *after* Earth was formed.

Could paleontologists be faking their fossil finds? When the word "fossil" is mentioned, you might picture a dinosaur skeleton in a museum pinned together in a lifelike posture. But you don't need to go to a museum to find fossil stuff. It's all around you. As I write this, the light in my office is partially powered by fossil fuel. Two thirds of the US's electricity come from coal, a fossil fuel created from dead plants in a millions-of-years process similar to that of oil, which I pour into my car periodically in the form of gasoline and lubricant.

As I hammer on my computer keyboard, I notice the faint smell of the hot plastic computer casing. It, along with my phone, my drinking cup, and my wastebasket, are all made from processed crude oil. I'm wearing some fossil-based nylon clothing, I'm sitting on a fossil-based chair with my feet resting on fossil-based flooring. And, my desk is surfaced with fossil-based laminate.

All of these materials come from the oil that was created by geologic processes involving many layers of dead plants. Millions of years were necessary for the accumulation. And millions more years of heat and pressure were necessary for conversion to oil. It is difficult to fathom how anyone in a modern society, surrounded by myriad objects made from a fossil resource, can cling to young-earth creationism!

The foundation of my house is concrete, a material that hardens due to the action of calcium, which is taken from limestone (calcium carbonate). Limestone calcium comes from the skeletal remains of marine organisms, such as coral, deposited in layers over timespans of millions of years and subjected to intense heat.

Chalk is also a type of limestone, formed from the decomposition, sedimentation and compression of plankton skeletons over timespans of millions of years. So, it's ironic when a religion teacher uses a chalkboard to teach that the earth is only 6,000 years old!

Some young-Earth creationists propose that the dinosaur fossils were formed when Noah's flood killed them all off. However, if all the extinct creatures represented by the fossil record had lived in the short span of 1600 years between Adam's fall and Noah's flood, the entire face of the planet would have been teaming so densely with hungry creatures that Noah would have found no trees with which to build his ark.

Christians are often flabbergasted that atheists can deny the reality of a Creator while surrounded by the evidence of

His creation while atheists are often flabbergasted that Christians can deny the age of the earth while surrounded by the evidence of its ancientness. Conflicting facts should prompt rational people to revisit their assumptions.

How scientific is the Bible?

Are there any passages in the Bible that may have contributed to science? Well, there's one that says God "hangs the earth upon nothing". (Jb 26:7) Job knew circa 2200BC (Before Christ) that Earth was not sitting on the back of Atlas, or on a giant turtle, or on pillars, or on a solid foundation. Copernicus gets second-place credit for stating around 1500AD (*Anno Domino*: Year of the Lord) that Earth floats free in space, but there is no evidence he got the idea from the Bible.

The cleansing rituals given to the Hebrews in Moses' day hinted at the importance of hygiene thousands of years before germs were discovered, which were first observed under Leuwenhoek's microscope in the 1670's. God appears to have designed spiritual cleansing rituals with physical microbes in mind.

But, no. The Bible is not a science text. More than half of the scientists in the US believe in God but none of them are known to use the Bible as a handbook for their research. As Galileo said, "The Bible tells us how to go to heaven, not how the heavens go." (astronomytrek.com/20-interesting-astronomy-quotes, DL 25 Apr 2020) So, while the creation days of Genesis matter, the message is not about the science.

The Bible doesn't mention DNA (Deoxyribonucleic Acid: double-helix genetic code), microorganisms, planetary motion, galaxies, time dilation, superclusters, particle physics, quantum mechanics or the expansion of the universe. If God wanted to share science through the Bible, he would have had his prophets include a fundamental equation or two, such as $E=mc^2$ or $F=ma$. Or perhaps He would have revealed the secrets of refrigeration to the children of Israel.

Although modern life is enriched and highly accessorized by myriad technologies, not a whisper about them can be found in scripture. If Adam's family learned how to control fire and create wheeled conveyances from God, it was not something the prophets deemed appropriate for inclusion in the Bible.

But wouldn't it be nice if the Bible had at least provided enough clarifications to answer the questions about conflicts with science? For example:

♦ If Genesis describes the creation of all things, why doesn't it begin with the Big Bang rather than with a barren Earth? (Gn 1:1)

♦ If Earth first appeared with waters already on its face, as per Genesis 1:1, what prevented all them from freezing in the darkness while awaiting the creation of the Sun on Day Four?

♦ If God created the sun on Day Four, was the earth not orbiting anything until then? (Gn 1:16)

♦ If God created the moon to "rule the night" then why are half the nights without it? (Gn 1:16)

♦ If the creation is only 6,000 years old, how do telescopes pick up incoming light that is billions of years old?

♦ If humans didn't come from evolution, why do chimps have 98.8% of their DNA in common with humans? (Cosmos Magazine, 2017, Issue 73, *Evolving a Human Brain*)

♦ If Adam and Eve were created directly from dust, why does Judeo-Christian art show them with navels?

♦ If Adam and Eve were created immortal, how did the forbidden fruit change them?

♦ What caused all the life on Earth which did not eat the forbidden fruit to become mortal?

♦ If the forbidden fruit gave Adam and Eve knowledge of good and evil, why didn't God preserve it in orchards to be given to ignorant little children?

♦ If all humanity came from one couple, why don't more ancient cultures with writing systems earlier than 3000BC, such as the Chinese, Sumerians, or Akkadians, have an Adam and Eve story?

♦ How was a snake without lips or larynx able to speak to Eve? (Gn 3:1)

♦ If God cursed that one snake to go on its belly, why did all the snakes in the world also lose their legs? (Gn 3:14)

♦ If God is perfect, why did He create animals with unnecessary features, such as lizards with subcutaneous vestigial legs that don't ambulate, birds with wings that can't fly, and mole rats with eye spots that can't see?

♦ Why did early Bible people live for hundreds of years but today there are few centenarians?

♦ If God has always existed, why did he wait an eternity before creating anything?

♦ Why would God be so improvident as to create a universe at least 94 billion light-years across when only 100 thousand light-years of space are needed for our entire galaxy?

The fact that the Bible provides no clear answers to these science-flavored questions is evidence that such topics are outside the scope of scripture. It would seem a nice thing, especially for the youth, if the answers were handed to humanity from heaven on a silver platter. God could do that. He is the Master Engineer with a knowledge of all science. But He knows that man's hunger for happiness would not be satiated solely by science which, though much less perfect than if it came from God, is already severely upstaging faith.

What mysteries can science solve?

We live in an age of miracles and science gets all of the credit. For industrialized nations, the magic of electricity puts radios, televisions, lights, refrigerators, air conditioners, phones, and computers into the hands of every commoner. The global information superhighway is at virtually everyone's fingertips. Despite a burgeoning population, 216 million fewer people are hungry than 25 years ago. Medicine gives limbs to the maimed, hearing to the deaf, and sight to the blind. Smallpox and rinderpest have been wiped off the face of the earth and polio will likely soon follow.

In the realm of cosmogony, astronomers have mapped the microwave echo of the Big Bang and measured the expansion rate of the universe. In the realm of microphysics, scientists can cause a single particle to be in two places at once.

Brilliant Nobel Prize winning Richard Feynman told how, as a boy, he asked his father, "When I pull the wagon, the ball rolls to the back of the wagon. And when I'm pulling it along and I suddenly stop, the ball rolls to the front of the wagon. Why is that?" His father answered, "Nobody knows. The general principle is that things that are moving try to keep moving, and things that are standing still tend to stay still, unless you push them. This tendency is called inertia, but nobody knows *why* it's true."

Without an explanation from God, *why* creation behaves the way it does will always be a mystery in its fundamentals. Despite the exponential success in discovering *how* things in the universe work, science is still unable to answer *why* they work that way.

Oh sure, a scientist can try to pass off a law of physics as the *why*. But every such law is *inferred* from observed facts for which science has no *why*. For example, one might explain that Earth moves through space the way it does because of the law of gravity. But science can't say why every particle attracts every other particle in the universe with a force directly proportional to the product of their masses and inversely proportional to the square of the distance between their centers of mass. A person of faith might simply answer, "Because God wants it that way!"

Even Isaac Newton, the genius who devised the classical equation for gravitational attraction, credited intervention by God for things he couldn't explain, such as why the stars don't all fall in on each other. Since then, science has shown more forbearance, deferring tough conundrums indefinitely, hoping that new investigative tools and better mathematical models will eventually enable them to solve the mysteries without invoking God.

Why is that taboo? Because, the more scientists assume the solutions must come from God, the less likely they are to solve the mysteries themselves. Since God gave mankind intelligence and curiosity, it's obvious He expects us to use them.

There will always be limits on science at the edge of discovery. For example, what happened at the very start of the Big Bang will likely remain beyond experiment. Likewise, humans will never see what lies beyond this universe, if anything. And, since equipment large enough to sense a graviton would be so huge that it would collapse into a black hole under the gravitational force of its own mass, scientists will never directly prove it exists.

It's likely that string theorists will never devise an experiment that will test their math. And, physicists might never discover the nature of dark matter or dark energy. After decades of research, the mystery persists as to why antimatter is so rare. Light from around black holes can be detected by pairs of radio telescopes positioned far apart across the planet and deciphered by computers, but the inside of their event horizons can never be probed by man. And, simultaneous knowledge of a quantum particle's position and momentum will always be impossible.

But knowing they can never know something does not discourage theorists from spinning up possible new models. The Big Bang theory has been around for many decades, but that hasn't stopped theorists from coming up with alternative versions such as the Big Bounce theory and the Oscillating Universe theory.

Sexy new theories garner more contributions from government, academia, and philanthropy which funds more research resulting in new sexy discoveries and theories. And the cycle repeats.

For Einstein, the mind of God was the ultimate mystery. "I want to know God's thoughts," he said, "The rest are mere details." Einstein responded to success at solving a mystery with a modicum of humility, once declaring, "When the solution is simple, God is answering."

It would be unfair to demonize scientists for what appears to be worship of nature. But it would not be a terrible thing if scientists avoided arrogantly denying God. Perhaps it's for publicity. Stephen Hawking made global news when he declared, "The simplest explanation is; there is no God. No one created our universe." This he did without any proof of God's absence, a very unscientific thing to do. God is a mystery that science is unequipped to crack.

Science continues to scoop up fields of inquiry, the result being that society's focus is shifted ever more away from traditional beliefs. No matter what the topic, some scientist inevitably chimes in. One tells us how to raise our children, another how to succeed in marriage, and yet another how to overcome addiction. Knowledge is power. But science is also a business. Books, articles, grants, counseling, advisory services, consultancies, and other science-based goods and services represent cash flows that reward knowledge.

But don't motives matter? Much of human behavior that was once guided by unmonetized religious conviction has become the purview of one money-hungry branch of science or another. In the heyday of Newton, most cutting-edge science was done by devoutly religious men living lives of self-imposed poverty. But science today is no longer such a quest to understand God through deciphering His creation.

Some scientists are even claiming to have identified the biological cause of religious belief, an idea that undermines God's role in the human journey. Such natural tendencies were likely designed by God to impel humans to seek Him but, again, science must stop short of invoking God.

Most importantly, science is ill-equipped to comment on life's purpose or on morality. There are no equations for physical laws governing ethics. And, there is no man-made technology that can extract spiritual truths from the universe. The job of science will always be to model laws that predict *how* things behave. The job of explaining *why* will always fall to religion.

What problems can religion solve?

Without religion, the future of humanity would be bleak and brief. It's a picture of earthlings crawling inexorably through life toward death's door, chasing temporal goals, and hoping as they expire that the world was a better place because of them. A brilliant few might take comfort that they contributed to humanity's lasting book of collective knowledge. But, without religion, that book would be burned, along with the entire world, when the sun swells red and engulfs the earth.

Without divine intervention, everything in the universe will eventually wind down, spiral into black holes, and slowly evaporate into the cold, dark, final state of massless thermal equilibrium, all potential expended. The existence of life will have been a mere blip in time, without meaning or memory.

A much brighter picture is offered by theistic religions through faith in a life after death and the possibility of an eternal future unimaginably brighter and more pleasant than the present.

Aside from offering hope eternal, religion has a long history of explicating the inexplicable, evolving with the times to assuage new fears and answer new questions. Religion's very pervasiveness from antiquity throughout the cultures of the world underscores its historic value. It has been the bedrock of culture for tens of thousands of years.

In modern times, churches make immeasurable philanthropic contributions, assisting the world's poor and distributing essential food, water, and clothing in times of crisis. Earth angels, moved by religious conviction, spread messages from the Bible to promote selflessness, service, charity, and education. Faith-based hospitals, schools, and humanitarian relief agencies provide invaluable services that would otherwise be scarce in many areas.

But the times are changing. As the trend away from religion continues, many are questioning its value, especially in view of the problems they see. If religion is true, why does the Bible disagree with science? Why does the Bible contradict itself? Why are there tens of thousands of disparate Bible-based religions? Why do the major Christian sects have so much blood and violence in their histories? Why are there so many church scandals? Why has "Christianity" ever persecuted scientists who spoke the truth? If God is at the helm, why have churches been wrong so often?

What can religion not do?

The more educated people of today look back across the ages and see how often domineering churches got the science wrong. Copernicus, Galileo, and Kepler were right about planetary orbits, and the Catholic Church was eventually forced to concede that the earth doesn't need to be in the center of the solar system and planetary orbits don't need to scribe perfect circles!

And, to say that the tenets of Christianity and Judaism have been more static than the knowledge being amassed by science is a gross understatement. In fact, it is still the official position of most mainstream Christian churches that the heavens have been closed to divine revelation since the death of Christ's apostles! In other words, the Bible is it! So, what are the odds that such deniers will ever receive fresh apostolic or prophetic revelation?

It is obvious from the extreme diversity of doctrine and disagreement among Christian denominations that some course corrections are desperately needed to unify the faith. But do they receive revelation from God to remedy this? No. Instead, they vie with each other for attendees and donors, even going so far as to mock one another's doctrines, sacred rituals, and holy vestments in books, seminars, sermons, videos and even full-length movies.

For professional religionists, the fact that faith is their financial support compels them to be sensational in their sermons and profound in their pamphlets. Success depends on drawing people to their pulpit and buyers to their books. Thus, the pressure to hone their writing and speaking skills and to promote their priestly career often distracts them from the focus on their own spiritual purity, rendering them unworthy of inspiration. Religion does not guarantee righteous motives for its purveyors. Priestcraft is evil.

Many priests and pastors jazz up their Sunday services with coffee, donuts, pop music, and charismatic mummery to attract the crowds. And, to avoid conflict with the secular world, many churches adjust their theologies to accommodate social and scientific trends. For example, when the Big Bang theory became popular, Pope Pius XII responded by welcoming the evidence of a beginning, "Creation took place. We say: therefore, there is a Creator." (Meeting of the Pontifical Academy of Sciences, 22 Nov 1951) Ironically, the Big Bang theory is a model of the evolution of the universe, not an explanation of how it began!

Religion cannot ensure its adherents won't stumble. Even its most popular clerics are wont to wallow in wickedness. Jim Bakker, perhaps the most popular televangelist ever, was accused of drugging and raping his secretary and was sentenced to 45 years in prison for accounting fraud. After railing against sexual sin in all its forms, the great Ted Haggard was found to have been consorting with a male prostitute for many years. Televangelist Jimmy Swaggart, who frequently condemned sexual impurity in all its forms, was found to be a pornography addict and to have cheated on his wife with a prostitute. The list is far too extensive to include here. The list of child and youth abuse cases filed against churches is even longer!

Religious doctrines have not been leveraged much into modern science, although here and there a godly scientist credits personal revelation for a breakthrough. It took lawsuits against cigarette makers to motivate scientists to establish as fact what God told Joseph Smith a century earlier, that tobacco was "not good for man". (DC 89:8)

Despite the controversy surrounding the Biblical and scientific versions of creation, neither prophets, preachers, popes nor priests have received any revelation by vision or voice that added to the book of science knowledge. Nor have the most devoutly religious scientists such as Bacon, Galileo, Pascal, Boyle, Newton, Kepler, Faraday, and Maxwell ever credited divine revelation as the source for their scientific breakthroughs. So, when it comes to sustained discovery of truth about the creation, religion is at a disadvantage, at least for now, while science finds more pieces of the creation puzzle every day.

Three and a half centuries after Galileo was put under house arrest by the church for publishing scientific truth, the pope apologized – after a fashion, "The Church's experience during the Galileo affair and after it, has led to a more mature attitude… It is through research that man attains to Truth." (*Cosmos, Bios, Theos*, Margenau & Varghese, 1992, p96)

Apparently, not even the science mistakes of the Catholic Church serve as lessons to everyone. Robert Sungenis recently used scripture to promote geocentrism in his book *Galileo was Wrong*. He offered $1,000 to anyone who could prove heliocentrism. Of course, it's fairly simple for young students to do so by comparing images of the sky six months apart to see the stellar aberrations, parallax, and Doppler shifts that prove; "It is the earth that moveth and not the sun." (*BOM (The Book of Mormon)*, Hel 12:15) But, since Robert is the judge, none of the proofs submitted has yet merited the payout.

Although none of the religious doctrines or myths from around the world have contributed directly to scientific knowledge, the recent explosion of new insights into the creation is surely part of God's plan.

Does science support creationism?

Attempts to use science to bolster belief in the Bible usually fall flat on close inspection. For example, in the video *Science Confirms the Bible*, creationist Ken Ham claims that the discovery of Earth's spherical shape gave credence to the prophetic status of Isaiah, who wrote that God "sitteth upon the circle of the earth". (Is 40:22) However, a look at the original Hebrew suggests that Isaiah meant God sits *above* (Str#5921) the *horizon* (Str#2329) of the *land* (Str#776). After all, God's throne is in <u>heaven</u>. Furthermore, the earth is not actually spherical. It's spheroidal. It's oblate. If it had been God's intent to describe the shape of our entire planet through the mouth of Isaiah, he would surely have been more accurate.

A number of fundamentalists say that the expansion of the universe validates the Biblical passages that say God "made the earth" and "stretched out the heavens". (Jer 51:15; Is 45:12) However, it's much more likely that those prophets had a one-time creation event in mind rather than the Big Bang expansion of the universe that is *still underway* today. Otherwise the prophets would probably have said that God *is* stretching out the heavens.

The beloved Bible has the honor of being the longest-standing widely published book. It is therefore also the most scrutinized, and science is the standard against which today's students judge it. The fact that upcoming science-savvy generations are much more skeptical of the Bible is evidence that science tends to be fairly useless in proving its validity and may actually be dissuading youth from trusting it. As science progresses, the case against traditional interpretations of the biblical creation story strengthens.

The older and more religious generations tend to stick science in one mental box and religion in another and never the twain shall mix. This need to compartmentalize is evidence that science does more to undermine faith in the Bible than it does to support it. Many people of faith simply dismiss any geologic or anthropologic evidence that threatens their beliefs, deeming them part of an evil plot against religion.

To be sure, science that seems to contradict religion should be taste-tested carefully before being swallowed. Conversely, any theology that seems to contradict science should be reevaluated for potential reinterpretation. If both are true, then there must logically exist at least one interpretation for each that will resolve the contradictions. God and Nature do not disagree.

The failure of science to support creationism does not mean scientists are all atheists. For example, Michio Kaku, one of the most prominent popularizers of physics, shared his stance on creationism thusly, "We are in a world made by rules created by an intelligence… It is clear that we exist in a plan which is governed by rules that were created, shaped by a universal intelligence and not by chance." (express.co.uk/news/science/742567/PROOF-of-God-real-Michio-Kaku, DL 20 Mar 2019)

Does science contradict itself?

People won't tolerate lies, especially ones told by representatives of God who, being holy and inspired, are expected to get things right the first time around. On the other hand, science is allowed a great deal of leniency in the matter of truth because, as they say, scientific discovery is a process and it's okay that matters of consensus today can be disproved tomorrow. That's progress!

Yes, science is self-correcting. New discoveries have often contradicted previous "facts" and that's a wonderful thing! For example, the universe was said to be static in the early 1900s. But within a couple of decades, it was discovered to be expanding. And, until the end of the 1900s, scientists believed the expansion was being decelerated by gravity. But then it was discovered that just the opposite was true. And scientists, scrambling for an explanation, cooked up the idea of a ubiquitous repulsive force, dubbed *dark energy*. And, perhaps some other explanation will soon be found that aligns better with observations! Who knows?

These bout-faces were never considered scandalous. Nobody got irate and swore off science. There were no public protests. Publicity surrounding the discoveries was sensational and positive. It was celebrated as major scientific progress! As Richard Feynman said to a 1966 gathering of High School science teachers, "Science is the belief in the ignorance of experts."

A Nobel Prize awaits whomever solves the most famous science contradiction of our era, the clash between general relativity (GR) and quantum mechanics (QM) where gravity is concerned. Quantum physicists theorize that an impossible-to-detect particle called a graviton carries the attraction between objects whereas relativity physicists say that gravity is a curvature in spacetime that tells objects with mass to move toward each other.

This clash of genuinely incompatible theories is proof that GR or QM or both are flawed models of reality. A Grand Unified Theory (GUT) or Theory of Everything (TOE) is needed that seamlessly describes the behavior of all things large and small. It is the holy grail of physics.

The general public primarily hears about the most successful theories while countless less successful theories remain in the shadows or are scrapped. But truth is not determined by popular vote. A popular theory could be scrapped at any moment. And science is never settled. "Public policy depends on consensus—majority rules. Science, on the other hand, depends on just one investigator who happens to come up with the most reliable conclusions that are verifiable by new and often more precise observations." (scienceisneversettled.com, DL 26 Apr 2020)

Einstein became the most famous scientist of all time (pun intended) by zeroing in on the contradictions, the paradoxes, and the holes. That's what drives discovery. Science is a methodology for discovering truth and it employs the best practices and tools it can find to describe physical realities. As practices and tools improve, better science eclipses older science.

Automobile mechanics often make several wrong guesses before successfully diagnosing a problem. God knows what's wrong right off. But if a mechanic blamed your engine trouble on the lack of a car blessing from God, or on a curse from Satan, or on pernicious pixies, the repair shop would soon

32

lose its credibility and its customers. Likewise, scientists must take the stance of methodological materialism, assuming that no magic from angels or demons is whimsically at work in natural phenomena. In other words, the scientific method assumes God isn't interfering.

However, scientists who go beyond methodological materialism and adopt ontological materialism, the belief that nothing exists beyond the physical realm, are making a scientifically untestable faith claim.

Science is a dynamic discovery process that involves self-correction. Controversies and contradictions catalyze progress.

How much should humans know?

Per the *Book of Mormon*, "There are many mysteries which are kept, that no one knoweth them save God himself." (*BOM*, Alma 40:3) Hey, why is God withholding? He knows everything! Why doesn't He do a comprehensive data dump and help us reconcile the truths of science and religion?

"It is given unto many to know the mysteries of God; nevertheless they are laid under a strict command that they shall not impart only according to the portion of his word which he doth grant unto the children of men, according to the heed and diligence which they give unto him." (Alma 12:9)

The Lord explained to Joseph Smith and Oliver Cowdery; "Great and marvelous are the… mysteries of his kingdom… which surpass all understanding in glory… Which he commanded us we should not write… and are not a lawful for man to utter; Neither is man capable to make them known." (DC 76:114-116)

The Lord is more interested in character development than in brain expansion. He expects accountable people of every mental capacity to acquire faith and persevere in following Him. He is quite hands-off on other matters because they are less important to the plan of happiness for His children. His program is for people to acquire wisdom, not by high-speed downloads from heaven, but rather "precept upon precept, line upon line; here a little, and there a little". (Is 28:10)

But properly prioritized curiosity is definitely a good thing. Abraham did some stargazing and wondered about what he saw in prayer. In response, God showed him a vision of the cosmos. But, in doing so, God conveyed a spiritual lesson rather than an astronomical one.

God revealed visions of the future to Daniel but then told him to seal up the transcription of it "until the time of the end". (Dn 12:4) Why not share it all immediately? Because the people were not ready. Daniel lived in an era when rejecting prophets, murdering messengers, and destroying sacred records was commonplace.

In vision, the apostle John heard "thunders" speak and would have written it down, but a voice from heaven commanded, "Seal up those things which the seven thunders uttered, and write them not." (Rv 10:4) And, not even the angels know the day that Christ will return in glory. (Mt 24:36)

Joseph Smith said, "I could explain a hundredfold more than I ever have of the glories of the kingdoms manifested to me in the vision, *were I permitted, and were the people prepared to receive them.*" (*HC (History of the Church)*, bk5, p402)

Joseph Smith was murdered for his faith and people still tend to reject prophets. So, who can blame God for holding back? But the day will come, after the wicked have been swept away and "the Lord shall come, he shall reveal all things— Things which have passed, and hidden things which no man knew, things of the earth, by which it was made, and the purpose and the end thereof— Things most precious, things that are above, and things that are beneath, things that are in the earth, and upon the earth, and in heaven". (DC (*Doctrine and Covenants*) 101:32-34) When that day of knowing arrives, religion and science will merge into one whole truth.

Would everybody embrace the Bible if God had omitted the scientifically controversial parts? Of course not! Genesis is a puzzle without a picture on the box. People who want to stop feeling conflicted about science and scripture are responsible for solving the discrepancies to their own satisfaction.

Is it okay to pursue knowledge beyond what is taught by scripture and in church? Joseph Smith said, "I never hear of a man being damned for believing too much." (*HC*, bk6, p477)

The glorified Lord spent quite some time with His apostles; "After his suffering, he showed himself to these men and… over a period of forty days… spoke about the kingdom of God." (Acts 1:3) Why does the Bible include so much more of Christ's teachings as a mortal than of His teachings as a glorified God? Were the apostles too distracted to write? Or did the risen Lord simply rehash teachings from His mortal ministry? Not likely! "There are also many other things which Jesus did, the which, if they should be written every one, I suppose that even the world itself could not contain the books." (Jn 21:25) Many of His teachings and doings during that post-resurrection period were for the apostles' ears only!

During the risen Lord's visit to a tribe in ancient America, He commanded that all of His sayings be written *except certain ones*. "If it shall so be that they [the gentiles in the last days] shall believe these things [the stuff okay to be written] then shall the *greater things* be made manifest unto them." (*BOM*, 3 Ne 26:9)

There are things angels should not know and things that the man Jesus was not supposed to know, such as the hour of His coming in glory; "Of that day and that hour knoweth no man, no, not the angels which are in heaven, *neither the Son*." (Mk 13:32)

When three Nephites were translated to not see death until Christ's second coming, God would not allow Mormon to even write their names in the records, let alone record the experience they had while carried away into heaven. (*BOM* (*Book of Mormon*), 3 Ne 28:13-14, 25) Some things are so sacred that they must remain secret; "Secret things belong unto the LORD our God." (Dt 29:29) As Jesus commanded, "Give not that which is holy unto the dogs, neither cast ye your pearls before swine, lest they trample them under their feet, and turn again and rend you." (Mt 7:6)

Consider how Christians balk at the extended theology of The Church of JESUS CHRIST of Latter-day Saints (LDS). Traditions blind people to the possibility of additional revealed truth. They have their Bibles and refuse to believe that God would have talked to anyone but the Jews in ages past. God seldom injects precious insights into closed minds.

God has even designed the universe to keep secrets. There are certain limitations on how far and how deep scientists can delve. When they "play" the movie of an expanding universe backward mathematically, their equations break down before the beginning, the primal point of creation, as if the Creator's initial touch was cut from the film.

Even the term "observable universe" is astronomy's admission that light from a certain distance away will never reach us. The expanding universe makes it forever impossible to see what's beyond that horizon.

To know everything about the natural world one would need to crack into the spiritual realm upon which all reality is based. Science can't do that. Quantum Mechanics (QM) scientists are stymied by the spooky behavior of what they believe are the most fundamental bits of reality where the line between the physical stuff and the spirit stuff begins to blur. Heisenberg's Uncertainty Principle is one of the many admissions by physicists that some things are unknowable.

String theorists hope to explain everything in terms of fundamental rubber-band-like thingies. Their description of quantum strings borders on paranormal talk and perhaps that's the reason they've hit a brick wall in their research. Maybe there's a line at the boundary of the physical world that they shouldn't and can't cross.

But it is right for science to try every door it finds. And hopefully, on the other side of most doors, myriad other doors will appear. But they sometimes encounter a locked door that they cannot open.

For example, while scientists were struggling to understand the behavior of "normal" matter and energy, they stumbled into "exotic" matter and energy. These two phenomena are total mysteries and have been described as "dark". Nothing is known of their fundamental particles or their quantum fields or the laws of physics that determine their behaviors and interactions. All they know, after decades of chasing these elusive evil fraternal twins is that together they make up 96% of the mass of the universe. Wow! That's a whopping unknown! Perhaps God is administering a huge dose of humility to the soul of science.

Given all the remaining enigmas on the frontier of discovery, there is no justification for arrogance in science. In the words of one agnostic, "Modern science has been a voyage into the unknown, with a lesson in humility waiting at every stop." (*Pale Blue Dot*, Carl Sagan, 1994, p23) And in the words of a 12th-century Jewish philosopher, "Teach thy tongue to say 'I do not know' and you will progress." (Maimonides)

Einstein spent the last few decades of his life pursuing a Grand Unified Theory (GUT) but died unsuccessful. Likewise, humanity will run out of time before it unravels all the mysteries of the natural world. Each remaining problem is exponentially tougher to solve than the prior one. It's a side-effect of being human. "We see through a glass darkly." (1 Cor 13:12)

Must religion and science conflict?

A theory is an attempt to explain a set of facts or phenomena. Theology is the study of religious questions. As long as there are missing facts, faulty interpretations, or false assumptions in religion and science, their theories and theologies will never completely agree.

Religions are more vulnerable to criticism because of disagreements within their own ranks. For example, some

Baptist preachers teach that once people are saved, they can never be unsaved (OSAS = once saved always saved – no possibility of a fall from grace) while other Baptist preachers insist that people can lose salvation if they return to evil. (Mt 10:22)

Here's another example of an internal clash on doctrine. Within each of several evangelical sects, there is disagreement on the timing of the rapture. Some preach that the righteous will be snatched to the clouds (raptured) before the final tribulation and thus be spared the horror of the apocalyptic end of days, as depicted in the movie *Left Behind*, while others teach that the rapture will happen after the troubles are over as Jesus descends.

So, although the Catholic and Protestant sects all preach from the Bible, the disagreements between and within Christian denominations make disagreements with science inevitable. Christians even disagree on what to disagree with science about. To make matters worse, Christianity can't even agree on which version of the Word of God to use. Due to dissatisfaction with the accuracy of available Bible translations, repeated efforts to render a correct version have resulted in over 900 different complete or partial translations.

As for science, there are disagreements even within the same field. For example, some physicists claim information is destroyed when it falls into a black hole while others argue that physical information cannot be destroyed, just unknowable. There are no two leading scientists who agree on all the issues. But disagreements in science are all taken in stride. There are no evil theories, only questionable ones. Nobody's eternal happiness is at stake.

A believer's fear of science is a sign he or she needs stronger faith and more knowledge. Scientists who fear or despise religion are straying from the scientific method. Just because scientists and religionists debate each other does not mean they should hate each other. Society benefits from both the secular and spiritual contributions. Einstein said, "Science without religion is lame; religion without science is blind." (*Reader's Digest*, Nov 1973)

In science, new ideas and corrections to old ones are celebrated. Each religion and person of faith should at least respect rather than reflexively criticize doctrine that differs from their own.

But, if it's not in the Bible, Christianity generally shuns it, sometimes citing the words at the end of the Bible, "If any man shall take away from the words of the book of this prophecy, God shall take away his part out of the book of life." (Rv 22:19) Little do most who quote this verse know that "this book of prophecy" meant John's Revelation, not the whole Bible, which was compiled into a set of books centuries after John's vision. John himself wrote his first and second books of John *after* Revelation. So, obviously, his warning applied only to Revelation.

The Greeks figured lightning bolts were projectiles thrown by Zeus. Socrates said they were vortexes of air. But Benjamin Franklin used a kite and a key to prove lightning was electricity and Tesla invented a coil that created lightning bolts. So now religions all agree with science on the nature of lightning. This kind of progress should ultimately lead to a point where there is no distinction between science and religion. All truth is part of one great whole. Science that threatens faith may be scary but we'll never see the whole picture unless we keep our eyes open!

The chapters that follow propose a set of perspectives that eliminate the science-versus-religion creation conflicts. This is where the milk stops and the meat starts. What follows doesn't purport to be correct in all the details, but it should demonstrate the *possibility* of harmony between science and religion.

2: From Chaos to First Cause

What came first?

OK, we're finished with the appetizers and ready to explore answers. But where should we begin? Let's take a theme from Maria in the *Sound of Music* and start "at the very beginning." But where is that? Whenever someone proposes an absolute starting point for all existence, somebody else inevitably pipes up and asks, "What came before that?" With the chicken-and-egg question, it's easy to conclude the egg came first because chickens come from oviparous dinosaurs. But then it merely becomes a dinosaur-and-egg question, which doesn't move the needle much.

How far back can the chain of causation be traced? Maria also sang, "Nothing comes from nothing!" And at least some early Christian theologians agreed with her.

Clement of Alexandria (120-215AD) said that God created the earth "out of a confused heap". (*Hymn to the Paedagogus*)

Origen Adamantius (184-254AD) quoted the first verse of Genesis thusly, "In the beginning God made heaven and earth, and the earth was invisible and not arranged," and then reasoned, "By the words 'invisible and not arranged' Moses would seem to mean nothing else than shapeless matter." (*De Principiis*, bk4, sec33)

Modern scientists also say that Earth was made from chaotic debris from stellar explosions, as sung by Moby in his 2002 hit *We Are All Made of Stars*. Indeed, all objects in the solar system were made of elements forged in the bosoms of exploded stars – chaotic matter! And of what are stars made? They formed from hydrogen and helium which, in turn, formed from energy in the crucible of The Big Bang. At that point, we arrive at the really tough question, "What came before *that*?"

More and more scientists are open to the possibility that the Big Bang was not the beginning of everything.

40

(scienceline.org/2008/07/physics-heger-bigbang, DL 17 Jul 2020)

Some physicists speculate that the Big Bang popped out of a quantum field fluctuation. The laws of physics dictate that quantum fields must always be a little bit restless. When an unimaginably large universe is the result, that's some strong restlessness.

Others say the universe sprang from a metastable false vacuum. Huh, what? A vacuum fluctuated that was only pretending to a vacuum because real vacuums couldn't exist before the Big Bang brought space into existence? Something like that.

Still others are considering the possibility that our universe sprang from a multiverse mother, one of countless child universes. If that's true, then "universe" (all things in one) is a misnomer for our cosmos.

When physicists spout these kinds of quasi-gibberish ideas, they're actually saying, "We don't really know what existed before the Big Bang." None of these ideas are testable. Even farther beyond reach are answers to questions about the multiverse; Are there more than one? And, what came before that?

Does everything have a cause? If the universe was caused by God, who or what caused God? If he simply "exists" without creation, couldn't the universe do the same? The answers are in this book. Read on!

Has there ever been nothing?

"Dear God, Nice job on the creation of the universe!"

"Ah, it was nothing."

Yes, mainstream Christianity teaches that God created everything from nothing; God + Nothing = Everything.

In the first several centuries of Catholicism, a large dose of Hellenism crept into the church, including the doctrine of *creatio ex nihilo* (creation from nothing). Around 400AD, St.

Augustine wrote, "Although the world has been made of some material, that very same material must have been made out of nothing." (*Warfare of Science with Theology*) The Catholic Church recently reaffirmed this, "To create means to call into existence from nothing." (*Fifth Encyclical Letter of Pope John Paul II*, 30 May 1986, p60)

Of course, no dictionary on earth includes the "from nothing" bit in the definition of *creation*. Paul wrote, "Through faith we understand that the worlds were framed by the word of God, so that things which are seen were not made of things which do appear." (Hb 11:3) Science agrees with Paul that visible stuff is made of invisible stuff.

The late great Stephen Hawking said, "Because there is a law such as gravity, the Universe can and will create itself from nothing... It is not necessary to invoke God." (*The Grand Design*, Hawking & Mlodinow, 2010) Make up your mind Stephen! Did the universe come from nothing or did it come from gravity? How did gravity come to exist?

In March of 2013, physicists and philosophers convened. "It was all much ado about nothing… Empty space with nothing in it was quickly agreed not to be nothing as well as impossible. In our universe, even a dark, empty void of space, absent of all particles, is still something." (livescience.com/28132-what-is-nothing-physicists-debate.html, DL 18 Jul 2020) Our kind of space has quantum stuff everywhere.

How close to nothingness is empty space? Not close at all! Quantum physicists say that even in the coldest, emptiest, remotest, and darkest pockets of the universe, space is a bubbling soup of quantum activity where energy levels wiggle above and below zero. They can't help it. Natural law requires wiggling by all quantum fields – the gravitational field, the electron field, the neutrino field, the electromagnetic field, and several other fields permeate the entire universe.

We humans may be able to imagine the absence of *anything* but not the absence of *everything*. The absolute absence of everything could not exist, as redundant as that may sound! Anything that exists is something. Yet *creatio ex nihilo* says that *nothingness* obeyed God's command and became something! Even if nothingness could exist, wouldn't God have had to create it before giving it a command?

Let's look at God's usual *modus operandi* for creation. To create babies, He starts with mothers. To create plants, He starts with seeds. To create Adam, he began with the dust of the earth. To create Eve, He started with a rib. (Gn 2:7-22) To create wine at a wedding, He started with water. To create food for hungry crowds, He began with fishes and loaves. To create animals, he commanded; "Let the *earth* bring forth grass… Let the *waters* bring forth abundantly… Let the *earth* bring forth the living creature." (Gn 1:11, 20, 24) So, everything-from-something is the Biblical creation pattern!

Although it's illogical and inconsistent with scripture, *creatio ex nihilo* is what preachers like Pastor Piper propound; "God addresses his command to nothingness, and even nothingness obeys." (desiringgod.org/messages/he-commanded-and-they-were-created, DL 13 Sep 2018)

If God's method had been to create everything from nothing with no intermediate states, the first and only verse of Genesis could have been, "Let there be everything!" Done! Why drag it out across multiple "days" or billions of years?

Here's a look at the Hebrew for the first verse in Genesis; "The word create came from the word *baurau*, which does not mean to create out of nothing; it means to organize--the same as a man would organize materials and build a ship… God had materials to organize the world out of chaos." (*Journal of Discourses (JD)*, Joseph Smith, bk6, pp4-6)

Baurau/bârâ' is also used later for God's creation of the people of Israel who came from biological parents. (Is 43:15) The same word was used for His creation of a clean heart from an unclean heart. (Ps 51:12) Something from something.

The primary meaning of the Hebrew word from which "create" is translated in the first verse of Genesis is *shape*. (Strong's #1254, בָּרָא) For example, in 1 Samuel 2:29, the Lord asks, "Wherefore kick ye at my sacrifice and at mine offering, which I have commanded in my habitation; and honourest thy sons above me, to *make* yourselves fat with the chiefest of all the offerings of Israel my people?" The word "make" comes from the same word, yet it's obvious that the priests were getting fat by eating the choicest morsels from the animal sacrifices they performed — not *from nothing*!

Some religious apologist's use a bit of science to argue that, since the sum of all positive energy in the universe equals the sum of all negative energy in the universe, the net of everything is zero and therefore the universe could have come from nothing. Fallacious! A net value of zero is not equivalent to null. For example, the charges on an electron and a positron net to zero but when you combine the two particles the mutual annihilation results in a total energy blast of 2.022 Mega-Electronvolts (MeV). Hardly nothing!

One prominent physicist pithily expressed his doubts *creatio ex nihilo*; "This unthinkable void converts itself into the plenum of existence -- a necessary consequence of physical laws. Where are these laws written into that void? What 'tells' the void that it is pregnant with a possible universe?" (*Perfect Symmetry*, Heinz Pagels, 1985, p365)

Why is there something?

If creation was ordered from chaos, what kind of disordered stuff was it? And, why was there something rather than nothing? Leibniz goes down in history as the first to ask a question like this, but every great physics genius since his day has pondered it. (*Leibniz Selections*, Wiener, 1951, p527) Today, the question is asked more abstractly, "Why should existence of any kind be the normal state?" Conversely, can there even be a "state" if there is no existence?

Physicist Frank Wilczyk said, "The reason that there is *something* rather than *nothing* is that *nothing* is unstable." Does that clarify anything? No — perhaps that's why it was funny for other physicists. And, as a brain bomb bonus, there's a tantalizing paradox in the question; the only known attribute of nothing is that it can't exist in physical reality, leaving existence as the only tenable alternative.

Antitheist Lawrence Krauss wrote a book called *A Universe From Nothing* in which he recycles the best attempts of past physicists to explain existence in purely scientific terms, ascribing the Big Bang to a quantum field fluctuation. But philosopher and quantum theory specialist David Albert challenged the book's theme by asking, "Where, for starters, are the laws of quantum mechanics themselves supposed to have come from?" (nytimes.com/2012/03/25/books/review/a-universe-from-nothing-by-lawrence-m-krauss.html, DL 27 Apr 2020)

True nothingness would preclude the existence of God. Even as far back as the 5th century, Parmenides argued it's illogical to think *nothing* could have existence. Chaos is a more sensible initial state. (parmenides.me, DL 27 Apr 2020)

In summary, if the above-quoted scientists/philosophers are correct that nothing can't exist or was unstable, there must have always been something, even it was merely the quantum mechanical laws that govern the state transition from nothingness to somethingness.

Whence came Chaos?

Did God create a primeval chaos so He would have something to organize? If He's anything like humans, He prefers order. Even newborn babies cry when immersed in chaos. "God is not the author of disorder." (*NIV (New International Version)*, 1 Cor 14:33)

So, how did chaos come to be? Greek mythology says Chaos came into being on its own. Chaos was personified it as a god. Chaos then somehow produced Gaia, the personification of Earth, who bore a son, Uranus, whom she then married. Yuck! Their offspring became the gods of the Greek pantheon. So, Chaos gets the credit for creating itself before anything else. And Gaia is lame for *not* creating *herself* first. How personifications transmogrified into physical realities is part of the mysterious magic of myth.

Genesis says nothing about chaos. According to scripture, there was just an empty earth at the start; "The earth was without form, and void." (Gn 1:1) That's disorder or chaos. But, unlike many creation myths featuring chaos, the Genesis version has no violence among gods —no murders of kin, no dismemberments, no swallowing of children, and no fights amongst brothers.

Stars explode, but new elements are created in the process. Planets and star systems perish, but from the debris God makes new ones. He said, "As one earth shall pass away, and the heavens thereof, even so shall another come; and there is no end to my works." (*PGP (Pearl of Great Price)*, Mo 1:38) Destruction and recycling, though chaotic, are important parts of creation. Chaos is an avoidable state between stages of order. Chaos is also the uncreated default state, the realm of disorder that persists until touched by the hand of a Creator.

Did chaos have time?

Stephen Hawking wrote, "Time itself, had a beginning in the Big Bang." (hawking.org.uk/the-beginning-of-time.html, DL 1 Jan 2018) But, if time was created, as important as it is for the accomplishment of any work, wouldn't Genesis have mentioned it at the top of Day One? Genesis ticks off the evening and morning of each day without saying anything about the creation of time itself.

Was there ever a time when time did not exist? Oops, that's an oxymoronic question. Let's reword that. Has time always existed? That's not much better because "always" implies an infinite amount of time. Let's try again. Did time have a beginning?

To answer that, let's get a definition of time. Per Merriam-Webster online, time is "a nonspatial continuum that is measured in terms of events which succeed one another from past through present to future." So, was God's act of creating the time continuum the first event? No, His decision to create something was the first event on the continuum and His command that creation start was the second event. The absence of a time interval between any two events in a sequence would be a paradoxical simultaneity, at least in the physical world.

In the context of creation studies, the appropriate definition of chaos is, "The disordered state of unformed matter and infinite space supposed in some cosmogonic views to have existed *before the ordered universe.*" (Wordnik, DL 27 Apr 2020) So, by definition, sequence is implied if chaos reigned before our cosmos was organized. And sequence implies time!

If chaos involved matter, it would have had mass which cannot exist without time. And if chaos had no beginning, then neither did time. Even when events are random, as is the case in chaos, some events still precede and are the cause of other events. That is the very definition of time. So, yes. Time is a characteristic of chaos.

Did chaos have space?

Did God invent space at the beginning of creation? Genesis doesn't mention it. If there had been no space, Genesis should have started differently; "In the beginning, God created space. And the space was without form and void, and darkness was everywhere…"

As stated in the previous section, chaos, by definition, contained matter, perhaps not the baryonic matter familiar to earthling scientists, but matter nevertheless. And space is a characteristic of matter. Without space, God would have had no place to do things, and nothing to govern. As He said, "there is no kingdom in which there is no space." (DC 88:37)

How was order imposed?

Until about 1,500 years ago, most people didn't talk much about physical laws, attributing natural events and behaviors to divine whim or happenstance. A few people noticed patterns in nature's behavior, such as "What goes up must come down," or "The further it falls, the faster it falls," or "The bigger they are, the harder they fall." But, for most people, the fact that things tended to stay down was as much a part of life on land as floating was for fish in water and therefore too obvious for comment.

Then, about 400 years ago, René Descartes decided that logic was the path to all truth while Galileo tortured the truth from nature using observation and experiment. Standing on their shoulders, Isaac Newton employed math to describe gravity; $F=Gm_1m_2/d^2$. Thus was born the Universal Law of Gravitation. *Any two masses attract each other with a force directly proportional to the product of their masses and inversely proportional to the square of the distance between them.* Other brilliant minds began to wonder what else in nature followed laws that could be described by math. One small discovery for Newton, one giant epiphany for mankind!

The Book of Abraham has God announcing the plan to assemble the components of creation; "We will... organize them; and behold, they shall be very obedient." (*PGP*, Abr 4:31) Obedient to what? Physical laws! God laid down the laws once and all nature has obeyed since. A physical law is an inference that something always happens under certain conditions without exceptions. Thousands of physical laws make life possible – pinning down the exact number is somewhat like finding the end of a rainbow as there will always be more to discover.

Laws impose order on nature even better than they do on society because, unlike people, things in nature never break its laws – fortunately! If just one fundamental physical law, such as the law of gravity, suddenly switched off, it would be a very bad day for the universe. Correction, the day would not finish! The atmosphere would be blown away by the final moments by solar wind. Earth's rotation speed of 1,000 miles per hour (MPH) at the equator, combined with its surface curvature of 8 inches per mile, would cause the immediate outward flinging of everything not attached to deep-rooted rock – people, cars, rocks, gravel, houses, oceans, etc. Then Earth would rip apart. Its molten interior would spin outward as glowing hot splatter. All the debris that was previous planets and moons would head toward outer space. The explosion of the sun would overtake the fleeing planets and

48

obliterate them. Stars everywhere would likewise explode. Black holes would unravel in gargantuan flashes of energy. Galaxies would fly apart. And the entire universe would expand hyper-accelerated toward its ultimate demise.

Providentially, gravity is very obedient – and constant. Life would be very different if gravity only changed in strength unpredictably. If just one of the fundamental physical laws had gone rogue at creation, we would not be around to complain about it!

Did chaos have a choice?

As was taught in the modern School of the Prophets, "God spake, chaos heard, and worlds came into order." (*LOF (Lectures on Faith)*, 1:23) Abraham wrote that God then "watched those things which they had ordered *until they obeyed*." (*PGP*, Abr 4:18)

So, did all the little quantum and sub-quantum thingies of which the universe is made have a choice of whether to follow the rules? Yes, otherwise God would not have had to watch them. Were any disobedient? If so, they were cycled out of creation just as God cycled Satan out of heaven.

What's in it for every little thingy to obey? The answer to that will be discussed in a subsequent section. ☺

Did creation need God?

Stephen Hawking said, "M-theory predicts that a great many universes were created out of nothing. Their creation does not require the intervention of some supernatural being or god. Rather, these multiple universes arise naturally from physical law." (*The Grand Design*, Hawking, 2010, ch1)

Hawking contradicted himself when he said universes are created *out of nothing from physical laws*. Which is it? Out of nothing or from physical laws? And, if nothingness somehow harbored physical laws in advance of creation, what or who programmed it to have physical laws before there was anything physical?

M-theory says nothing about *why* or *how* such physical laws began to exist. The fact that they did is a matter of faith for the few physicists who agree with Hawking. Physicists say all kinds of silly things, like, "Just twist space the right way, you might create an entirely new universe… It could be that this universe is merely the science fair project of a kid in another universe," said senior Search for Extraterrestrial Intelligence (SETI) astronomer Seth Shostak. (space.com, DL 19 Aug 2017) If creating our universe was child's play, it would be even easier for an extremely advanced alien who wants us to call Him God.

Did the cosmos self-cause?

Could the universe have caused its own beginning? This calls to mind the famous 1948 *Drawing Hands* lithograph by the paradox-loving artist M. C. Escher in which two hands draw each other into existence on paper. The analogy would be better if Escher had made one hand draw itself into existence. But either way, it's hard to imagine how the sketch got started before any part of a pencil appeared.

Could the universe have travelled back in time and created itself, like a scientist travelling back in time and giving his earlier self the secrets of time travel that allowed him to invent the machine in which he travelled back? That's another causal loop paradox, a type of bootstrap paradox.

Self-creation is a bootstrap paradox, lacking a point of origin. And there is no way to flip the arrow of time or the direction of causation back and forth to resolve such paradoxes. That is to say, the requirement that a cause must exist for every effect is true regardless of the direction of time and the sequencing rules. How could non-existence get the urge to exist?

So, the next time you run out of gasoline in your automobile, don't wait for fuel to create itself in your tank.

Self-creating fuel might not know in advance of existing what the proper octane is for your engine.

What caused the cause?

The famous First Cause Argument or Cosmological Argument, batted about by ancient Greek philosophers, went something like this:

1) *Everything that begins to exist must have a cause.* This is the normal human experience and scientists have never seen anything spring into existence without a cause. The assumption that absolutely everything has a cause has proved to be a key success factor in scientific progress.

2) *An infinite regress of causes is impossible.* An infinite or indefinite chain of causes is deemed fallacious when proposing the cause of a thing if the cause is just as much in need of an explanation as the thing in question.

3) *Causal loops are impossible.* A thing can't bring itself into existence.

4) *Therefore, a first cause kicked everything off.* This doesn't necessarily follow since assumption #2 is unproven.

5) *The eternal first cause is a god.* This is a leap of faith. Any cause, even a small one, would suffice if existence is assumed to have evolved.

Smart guy Aristotle, who gets credit for starting the First Cause Argument, made two naked assertions, claiming that an infinite chain of causality is impossible and that an uncaused causer is the only candidate for first cause. Neither of these assumptions can be proven. There is no reason why an *uncaused cause* should be a better anchor for a causal chain than an *infinite regress*. In other words, an infinite cause is no better than an infinite chain of causality.

The strong case for cause is: All observations indicate everything in existence has a cause. So, it is a law of nature. Since the universe exists, the universe had a cause.

Desperate to know what causes universes, we now turn to the experience of Mellen-Thomas Benedict who was miraculously healed from terminal brain cancer after 90 minutes of clinical death during which his spirit saw big bangs happening for the creation of countless other universes from the void, which is chock full of energy and exists simultaneously inside and outside of all things. Universes are created and destroyed in recycling rounds but the void persists between cycles and throughout it all. (ndestories.org/mellen-thomas-benedict, DL 24 May 2018)

Scientists have given the realm in which universes cycle various names such as the multiverse, superverse, metaverse, uberverse, megaverse, and hyperverse, since "universe" is already taken and they don't want anyone to conflate our universe with the infinite realm from which it popped. As of now, it seems "multiverse" is the most popular.

The idea of a multiverse has long been used in science fiction when writers need characters to travel vast distances without suffering death from old age before arriving at their destinations. It's also a handy mechanism for avoiding time dilation problems from faster-than-light journeys. After all, only a few hard-core sci-fi fans enjoy reading that a traveler returns home to find all his family and friends dead of old age. The solution is usually a shortcut through hyperspace, subspace, or jump-space. Curiously, travelers never dawdle in this alternate realm and, if they look outside in transit, they either see stars streaking by, which is totally against everything an astrophysicist would suggest, or they see a very boring blurry glow of some sort.

How do *creatio ex nihilo* believers react to the multiverse idea? They say the multiverse is just one more way that science is taking the Creator out of the story. But scientists are actually unhappy they have to invoke the multiverse to explain the origins of our cosmos. Physics professor Graham Ross lamented, "It's pretty unsatisfactory to use the multiverse hypothesis to explain only things we don't understand." (arxiv.org/pdf/1504.00464.pdf, DL 7 Jan 2018) Time will tell whether science will come to shun the multiverse as a cause like it shunned God.

If a multiverse caused the event that launched the Big Bang, the question is, "What or who created the multiverse?" The logical answer is, a parent multiverse.

If God caused the event that launched the Big Bang, the question is, "What or who caused God?" Again, the answer would logically be that it would take a god to create a god, unless it's a feckless golden calf or the like. And, in fact, there are many who read the Bible and believe that Jesus is a god and that he came from His Father who is also a god. There you have a one-generation pattern for how gods are created! More on this later.

Is Genesis the real beginning?

Genesis means "origin" but the Hebrew name for the *First Book of Moses* in the Torah is "Bərēšīṯ" – literally "First" or, as English translations render verse 1, "In the Beginning".

Genesis starts with Elohim "preparing the heavens and the earth". (*YLT (Young's Literal Translation)*, Gn 1:1) The Hebrew *shamayim*, from which *heavens* is translated, comes from Akkadian *samu*, meaning *sky*, and Hebrew *mayim*, meaning waters. So, Genesis begins with the earth and its sky-waters rather than with the cosmos.

Elohim improved Earth by separating the *mayim* from the *mayim* or "waters from the waters" by an "expanse". (*YLT*, Gn 1:6) This is a good representation of how the physical earth's atmosphere evolved. After 2 billion years of toxic volcanic steam enshrouding the molten earth, the cooling atmosphere began to condense into thick clouds that rained for millions of years to create the first ocean on the surface – the water above divided from the water below by air.

So, the cosmos was already more than 9 billion years old by the point in time that "the Spirit of God moved upon the face of the waters" enshrouding planet Earth. (Gn 1:2)

Even the New Testament (NT) disagrees with Genesis as to the first thing in creation; "In the beginning was the Word, and the Word was with God." (Jn. 1:1) The Word, of course, refers to the premortal Jesus. For John, Christ's premortal appointment to the role of Creator was the true beginning.

Paul agreed; "God… hath in these last days *spoken unto us by his Son*, whom he hath appointed heir of all things, *by whom also he made the worlds.*" (Hb 1:1-2) John called the Son the "Word" because the Father appointed Him to be His Divine Voice in all things – to order the chaos, to manage all creations, and to instruct all creatures.

John's creation story is not the only one that begins in Jesus' premortal realm. Abraham said, "Intelligences… were organized before the world was… And God saw these souls that they were good… and he said unto those who were with him: We will go down, for there is space there, and we will take of these materials, and we will make an earth." (*PGP*, Abr 3:22-24) Like John, Abraham thought it important to start where God the Father and His premortal Son were planning the creation.

What did the early church teach?

In the first century AD, Philo of Alexandria wrote opinions in the context of *creation from preexisting materials*. Later, church father Justin Martyr (110-165AD) said, "We have been taught that He in the beginning did of His goodness, for man's sake, create all things *out of unformed matter.*" (*First Apology of Justin*, bk1, ch10)

Saint Clement of Alexandria (150-215AD), said, "*Out of a confused heap* who didst create this ordered sphere, and *from the shapeless mass of matter* didst the universe adorn." (logoslibrary.org/clement/instructor/313.html, DL 18 Jul 2020)

So, whence came the doctrine of *creatio ex nihilo*? It was first advanced by the gnostic heretic theologian Basilides in the 2nd century and was soon popularized by other influential writers and teachers of his day. So, by the end of that century, "Christian philosophers had discarded the idea of creation from chaos in favor of *creatio ex nihilo.*" (*Restoring the Ancient Church*, Barry Bickmore, 1999, p100)

Also, some experts, even Hebrew linguists, argue that the Hebrew word *bara*, from which the word "created" is translated in Genesis 1:1, implies *creatio ex nihilo*. However, *bara* is used in a significant number of other places in the OT in contexts that are not creation from nothing. For example, "Why do you… *make* yourselves fat…" (*NKJV*, 1 Sm 2:29, Str#1254) Obviously, the temple priests were not making themselves fat from nothing — they started with their bodies and added the "best of all the offerings"!

Catholicism began teaching *creatio ex nihilo* without resistance from the laity who couldn't read the Latin Bibles and tended to dose off during Latin mass. Martin Luther taught that solely the will of God was involved in the creation. Almost all Christian sects today hold firmly to *creatio ex nihilo* despite admitting that the Bible doesn't explicitly teach it. So, ironically, the doctrine of creating from nothing was created from nothing!

Can we know the beginning?

Scientists are using particle colliders to get a peek at the conditions that begat the universe. News articles occasionally have headlines such as, "Physicists Recreate Conditions of the Big Bang". The reader is then disappointed because the best that was achieved was the creation of quark soup or some other type of stuff that may have been present at the onset of the Big Bang. (livescience.com/6128-big-bang-conditions-created-lab.html, DL 29 Apr 2020)

When scientists try to trace the expansion of the universe back to its beginning, they arrive at a very hot and dense point where their modelling math begins to go crazy, producing lots of infinities. There is where the trail disappears into a cosmic mystery.

Early in their research, scientists thought the entire universe had been crammed into a singularity of infinite force, energy, and density. But science began abandoning the idea that singularities exist in the middle of black holes and the theory of a dimensionless and infinitely hot "cosmic egg" birthing the baby universe has also lost popularity. A singularity just seems impossible. Max Tegmark, astrophysicist at Massachusetts Institute of Technology (MIT), said, "This is the single most embarrassing problem of physics." (scienceline.org/2008/07/physics-heger-bigbang, DL 7 Jan 2018)

Would a peek at the instant of the Big Bang reveal the hand of God in it? It depends on the mindset of the observer. To people of faith, all creation is evidence of God. But to a scientist, even a peek at the moment just before the Big Bang is unlikely to expose God's role in creation. It is likely that the environment that gave rise to the Big Bang is, like our universe, governed by a set of reliable natural laws. God is supernatural and cannot be seen. So, science would continue to say that physical laws were the sole cause of the Big Bang.

Would a peek outside the universe help track down the First Cause? In fact, physicists are hoping for just such a peek. There might be some kind of extremely intense and non-interactive radiation from the multiverse that survived the hyper-density of the Big Bang. Patterns in that radiation, if it can be detected, might reveal something about the conditions that caused the Big Bang.

Some physicists think that gravity from the multiverse or from another universe might bleed into our universe. If it could be detected and analyzed, it might tell scientists something about the cause of the Big Bang. If our universe came from a parent universe, there might a grandparent universe, sibling universes, and cousin universes. But, even if these are somehow detected, the question would always remain, what caused the very first parent universe to exist?

Many feel that the universe is unlikely to reveal the secrets of its beginnings to scientists. As C. S. Lewis put it, "I felt in my bones that this universe does not explain itself." (*Fundamentals of the Faith*, As quoted by Peter Kreeft, 1990)

Is God the First Cause?

Whether they believe the Big Bang is 13.8 billion years old or that the earth is only 6,000 years ago, Bible-based religionists generally think God is older than it all, that He is from eternity. If the Big Bang happened, then God either pulled the trigger or was the author of the natural laws that made it happen.

Unless they're up against a Gish galloper (one who dominates the debate by overwhelming the opponent with a steady flood of argumentum verbosium), scientists win disputes about First Cause based on evidentiary grounds. Lacking credible hard evidence to prove God is the First Cause, Christian debaters usually try to poke holes in scientific theories instead of proving their own positions.

Proving that God caused the Big Bang is as impossible for a Christian as proving that a multiverse caused the Big Bang is for a scientist.

Claims made by the Bible are not counted as evidence in debates by godless scientists. If neither a young-Earth Christian nor an old-Earth Christian can prove their position using the Bible, then certainly they will fail to make their case for the Intelligent Design of the universe when up against a scientist.

Even if it could be proven that God caused the Big Bang, that doesn't necessarily mean that He is the First Cause. The fact that God created a Son opens the door to the possibility that gods have created other gods in an infinite lineage of gods culminating in our God.

3: God's Existence

Is God's existence logical?

Most people of faith believe like the ancient prophet Alma did who argued to an antichrist, "The scriptures are laid before thee, yea, and all things denote there is a God; yea, even the earth, and all things that are upon the face of it, yea, and its motion, yea, and also all the planets which move in their regular form." (*BOM*, Alma 30:44)

The 11th-century Saint Anselm of Canterbury is famous for what is now known as the ontological argument for God, which goes something like this: If we can agree on the definition of God as a being greater than anything or anyone imaginable, then God must exist because one who exists is greater than one who does not exist. (*Proslogion*, chs2-3)

The first weakness of the argument is, of course, the huge "if" at the beginning — there is no consensus on the definition of God. The second weakness is that consensus wouldn't make the definition necessarily correct. No amount of consensus can call something into being from nonexistence!

If there's a *reductio ad absurdum* argument for God, it is this; Atheists say God doesn't exist because He can't be detected while believing, "Nothing created everything", a more absurd position than "God is the Creator!"

Is God needed?

In the old days, God was the reason and explanation for everything. He made the sun rise and the rain fall. He sent bumper crops in good years and the famines in bad years. Plagues were punishments and bounties were blessings. Then science rose up and took the mystery out of the weather, the biome, and the stars.

Without divine wisdom, is it even possible to invent workable moral codes? Can a godless society unify on a single set of values, set goals accordingly, track what behaviors support or undermine those goals, and then code a book of laws for their citizens to follow. Could such a society establish a culture that would evoke the desired behaviors?

Those are difficult questions to answer because there are so few religion-free societies to study. One example, however, is the Pirahã people, a remote hunter-gatherer tribe in Brazil whose members do not believe in any kind of deity, creator, or spirits. They have no sense of right or wrong, nor does their language even support a conversation on the subject of morality. Their lack of a written language prevents them from locking down any system of justice. To them, a deed may be appealing or repellent but never right or wrong. They live only in the current moment. And, their decisions are made solely on the basis of anticipated short-term outcomes. In one case, a group of them agreed to kill a visiting researcher in exchange for some whiskey.

Perhaps the most famous question best answered by God is, "What is the meaning of life?" It is asked most often in jest, as if it were impossible to answer. But it is a serious question that can be broken down into the three questions that people have been pondering for centuries:

1. *Whence came I?* Did I exist in some form before I was born or am I merely a chance collection of molecules?

2. *Why am I here?* Is there a master plan, destiny, or divine purpose to which I should look for guidance in life, or any reason for my existence beyond mere happenstance?

3. *Whither go I?* Is there an existence after death that should inform my choices in life or is this life all there is?

Science can't answer such questions. God has already answered them and this book will review each in turn.

God's word still defines the purpose of life and the universal moral truths. Most nonbelievers are unaware how much current law has its roots in religious codes of conduct. Without God, the law of the land becomes the boundary between right and wrong in rule-of-law countries, a very low standard of ethics indeed.

Most importantly; "If there is no God we are not, neither the earth; for there could have been no creation of things, neither to act nor to be acted upon; wherefore, all things must have vanished away." (*BOM*, 2 Ne 2:13) Yes, we need Him.

Did man create God?

The earliest religions are thought to have been invented to make sense of the world, to explain evil, and to provide hope in the face of hardship or death. A year of drought or a destructive flood could be attributed to the anger of a god. And people invented rituals and offerings hoping to appease divine displeasure. Sorrow over the deaths of loved ones could be assuaged by the belief that spirits simply passed into another realm to join departed family and friends.

Religion is so much a part of culture that societies are slow to let go of their faith. A poll done in 2018 showed that 90% of Americans said *yes* when asked whether they believed in some kind of higher power. (pewresearch.org/fact-tank/2018/04/25/key-findings-about-americans-belief-in-god, DL 29 Apr 2020)

An ancient manuscript tells how Abraham chopped up all the gods his father had made of wood and stone, save the largest. Then he put the axe in the hands of the largest god and, when his father came upon the scene, blamed the god who was still holding the axe. His father called him a liar saying, "Thou speakest lies to me. Is there in these gods spirit, soul or power to do all thou hast told me? Are they not wood and stone, and *have I not myself made them*." Thus, the young Abram got his father to confess that his gods were soulless and replied, "How canst thou then serve these idols in whom there is no power to do anything?" (Jasher 11:13-61)

The innate human desire to worship something is evident from discoveries of images in the caves and the graves of early hominids. For example, ivory figurines, such as the Lion-man, and Venus figurines, found in German caves, were part of human mythology 35 to 40 thousand years ago (TYA). Ivory musical instruments found in the same layers with those figurines suggest they were used ritually, possibly in the worship of the figurine deities. Funerary caches in early human burial sites from 30 TYA indicate some sort of belief in an afterlife and probably in one or more deities. The adornment of graves was ritualistic in ways that went far beyond mourning for the deceased.

This kind of evidence suggests that humans believed in supernatural beings tens of thousands of years before Adam began preaching the religion of Yahweh. Humans are evidently so strongly programmed for religion that they will invent one if they don't have one. Homo sapiens is the only species whose members want to find meaning in life and to understand their mortal experience. So, this is the only species that creates and/or worships gods.

The so-called "God gene hypothesis" proposes that a particular allele on gene Vesicular Monoamine Transporter Type 2 contributes to the human predisposition to seek spiritual meaning. (*The God Gene: How Faith is Hardwired into our Genes*, Dean Hamer, 2004, p77) Perhaps it was wisdom for God to somehow predispose humans to hunger for spirituality and seek divine help.

Scientists have theorized that natural selection might have promoted faith by endowing people with extra optimism and thereby improving their odds of survival. Multiple studies show that religious people are happier and more likely to survive tough times. Happy folks also tend to have the evolutionary advantage of begetting more children than sad people. (Jinhyung Kim & Joshua Hicks, The Journal of Positive Psychology, *Happiness begets children? Evidence for a bi-directional link between well-being and number of children,* bk11, 2016, pp62-69)

So, while science leans toward evolution to explain the "God-shaped hole in the human heart" that televangelists talk about, believers see God's fingerprints on mankind's inborn hunger for spirituality.

Did God create Himself?

The Egyptians believed that the god of creation, Atum, pulled himself into existence from the energy and matter of the dark waters of Chaos. Never mind how those dark waters came to be. Then Atum created other gods. Why didn't *all* the gods do like Atum and come into existence on their own? The Egyptians probably felt slightly uncomfortable with the self-creation trick and refrained from overusing it.

Likewise, the Greeks kept the number of self-generated gods down to just the first one, Chaos, the personification of the primeval unordered state. Then Chaos created Gaea, the Mother Earth goddess. Then Gaea created other quirky but powerful gods with human faults, feelings, and features. From there, the gods reproduced at will, much like humans.

In Babylonian mythology, the gods Apsu and Tiamat personified the primordial watery abyss and chaos, respectively. At some point they spontaneously became real gods and begat numerous children together. Tiamat begat a grandson named Marduk by her own grandson Ea. Marduk stole the throne from his mother's appointee, Kingu, then killed Tiamat and created all the stars from pieces of her butchered body. Marduk then killed Kingu and created humans from the drops of his blood. (*Enuma Elish*, the Babylonian creation myth recovered at Nineveh by archaeologist Austen Layard in 1849) Yuck!

It's obvious from the stark differences between this myth and Genesis that the ancient Hebrews did *not* borrow from Babylon for their creation story, as sundry scholars say.

As discussed in the previous chapter, nobody, not even a god, can travel from the future to a point prior to the Big Bang and create himself before calling everything else into existence, including the time through which He travelled. That kind of causal loop only works in movies, like *Planet of the Apes*, in which a smart ape goes back in time to become the ancestor of all smart apes including himself.

Rightfully so, prominent atheist activist Richard Dawkins deems it impossible that God "jumped into existence spontaneously". (imdb.com/title/tt1091617/quotes, DL 9 Jan 2018) Wrongfully so, he doesn't have the same problem with a complex universe that jumped into existence spontaneously.

Giving birth to oneself from nothing is just as impossible as completely eating oneself without leaving any scraps. Joseph Smith said that even God "could not create himself". (*Wilford Woodruff Diary*, from Words of Joseph Smith (WJS), 7 Apr 1844, pp343–347)

There's a rule of symmetry to existence; "If it had a beginning, it will have an end." (*King Follett Sermon*, Joseph Smith, 7 Apr 1844) If God will always exist in future, He must always have existed in the past. And, just as He will likely progress in glory and majesty forever in the future, he has likely increased in glory over an eternal past as well.

Was God always God?

Since Jesus changed from incorporeal spirit to a mortal to a corporeal immortal, then it follows that the Father had once done similar because, as Jesus said, "The Son can do nothing of himself, *but what he seeth the Father do*: for what things soever he doeth, these also doeth the Son likewise." (Jn 5:19)

Joseph Smith said, "God himself was once as we are now, and is an exalted man.... Here, then, is eternal life – to know the only wise and true God; and you have got to learn how to be Gods yourselves... the same as all Gods have done before you —namely, by going from one small degree to another, and from a small capacity to a great one; from grace to grace." (*TPJS*, pp345-346)

Is creation proof of God?

Everything in creation is evidence of a creator – for people who already believe, that is. But what is there about creation that would persuade a nonbeliever?

Perhaps the most compelling evidence of God is the fact that the creation seems perfectly tailored for humans, as if a loving Father designed it just for us. We live on a goldilocks planet. Not too hot. Not too cold. Not only is Earth extremely habitable, it is also beautiful in ways that suggest a Creator.

As an ancient American prophet said, "all things denote there is a God; yea, even the earth, and all things that are upon the face of it, yea, and its motion, yea, and also all the planets which move in their regular form do witness that there is a Supreme Creator." (*BOM*, Alma 30:44)

Unlike Earth, our nearest neighbor, Venus is far too close to the sun and has an atmosphere 93 times denser than Earth's. Unlike Earth, it has no water to absorb the carbon dioxide from its volcanoes. This causes a greenhouse effect that elevates the temperature to an average 864°F, too hot for probes and rovers. Venus tilts only about 1/6 as much as earth, so there is no winter. Also, since a Venus day is 5,832 hours long, the planet gets spit-roasted with each rotation.

Mars, our next-door neighbor on the other side of us, is far too cold with an average temperature of -80°F. Even though the planet's atmosphere is mostly heavy greenhouse CO_2, its gravity is too weak to hold enough to trap heat. Jupiter, Saturn, and Uranus are gas giants and have no solid surfaces. Neptune has a surface temperature of -392°F. Thousands of exoplanets (outside our solar system) have been discovered but none seem anywhere near as suitable for humans as Earth. Less than a dozen of them appear likely able to support even the simplest forms of life.

Even the positioning of our solar system is perfect as if planned. If it were too close to the galactic center, life would have long since been stifled by the deadly radiation from the cloud of stars swirling there. But the finely-tuned aspects of nature are inadequate proof of the existence or non-existence of a Creator for the skeptics. Science explains the fine-tuning of creation by pointing out that only in a universe suited to human life would there be humans to marvel at how well-suited it is to human life!

Deep-sea angler fish might marvel at how wonderful it is to have all the conditions suitable for life: near-total darkness, plenty of water, a pH of slightly more than 7, pressures of about 2,600 pounds per square inch, temperatures of about 36°F, food that is attracted to its little glowing lure, et cetera. But, without the ability to leave the water and investigate, the fish will never know whether they are living in the ocean or in a tank where all the variables were set by a scientist in a lab as part of an experiment.

Figuratively speaking, humans can't leave the universe "tank" either. Humans can't even discover the shape of it. So, while the perfect conditions for life may not convince anyone that a "tank" and a "Tank Maker" exist, neither can a secular explanation of the perfect conditions for life eliminate all other explanations.

God said, "The earth rolls upon her wings, and the sun giveth his light by day, and the moon giveth her light by night, and the stars also give their light, as they roll upon their wings in their glory, in the midst of the power of God... and *any man who hath seen any or the least of these hath seen God moving in his majesty and power."* (DC 88:45-57) And yet He leaves it up to each person to see Him in the creation.

Why so many nonbelievers?

Arguments against the existence of God are many and varied and include the following:

♦ Religions all contradict each other so there's no point in believing any of them.

♦ Thousands of years of mythology prove that humans are predisposed to creating gods. The Bible's God is also a myth.

♦ Parts of the Bible have been disproved by science, such as Adam being the first human and Noah's flood covering the whole planet, so the Word of God is lying. God would not lie. So He must not exist.

♦ If there were a God, he would stop the horrible tragedies and suffering. He would have eradicated leprosy instead of just healing a handful of lepers.

♦ God created all mankind as sinners and now expects them to stop sinning. If God existed, He would take the blame for creating them with the tendency to sin.

♦ The Bible asserts that homosexuality is wrong and I can't believe in a God that would condemn my gay friends.

♦ The Bible stipulates death for adulterers. I can't believe in a God who wants to kill me and my lover.

♦ The Old Testament condoned slavery among God's chosen people. I can't believe in that kind of God.

♦ God said working on the Sabbath is a capital crime. (Ex 35:2) I can't believe in a God who would have people killed for working at stores or hospitals on Sunday.

♦ It hasn't been proven that God exists and there's no evidence that He does, therefore He does not.

♦ Now that we have modern science to explain everything, we don't need God.

♦ God said, "Ask and ye shall receive." I asked and did not receive. God is no more real than Santa Claus!

♦ God asks humans to consecrate time, money, and talents to do His work. If an all-powerful God existed, He would provide all the necessary resources Himself.

♦ If God were real, his religions would not have caused wars and all the other problems.

♦ The percentage of prisoners who are atheists is only 0.2%, much smaller than the 10% in the general population. If faith in God makes people more likely to commit crimes, then it must be wrong.

♦ Thousands of verses in the Bible are nonsense, so the Bible is false and God does not exist.

♦ If God existed, there would be miracles today.

♦ If God existed, he would not treat women as second-class humans in the Bible.

♦ The abuse committed by representatives of God prove He's a farce.

♦ The sadness in the world is sufficient proof that God does not exist.

On All Saints Day in 1755, while God-fearing people were in church, an earthquake rattled the city of Lisbon for about five minutes. Many survivors fled the city center where cracks as wide as 16 feet had opened, to seek safety on the seaside docks. For 40 minutes they pondered why God would allow such a tragedy on a holy day and prayed He would help them survive to rebuild. Then, in came a tsunami that wiped out the entire town, followed by two more giant waves.

The earthquakes were felt throughout Europe and parts of Africa. Many other cities were also destroyed. The widespread destruction was followed by a swell of atheism as people pondered how an omniscient and beneficent God could send or allow such a deadly catastrophe – on All Saints Day, no less!

Atheists create strawmen gods and then destroy them with logic. For example, "God is good by definition. But, since His creation is bad, He cannot be real. A good God could not create something bad!" The premise that the creation is bad is wrong. It's good that people have the freedom to choose badly. Many of the laws instituted by governments likewise protect people's rights to do wrong, so long as they aren't stepping on the rights of others. And the sufferer's definition of bad is often wrong. Death, for example, is inevitable but not necessarily tragic from God's perspective.

Some atheists demand evidence of God. But that is fallacious. Absence of evidence is not evidence of absence (unless 100% of everything related to the issue is observable).

Most atheists profess post-conventional morality which posits that situations arise that cannot be decided based on simple rules like those given to Moses on tablets of stone. As a result, atheists can drift with any wind that challenges the moral code they've cobbled together for themselves. Atheists prefer to search for their own purpose in life and reserve the right to pick or devise the principles that suit their fancy. For them there are no moral absolutes.

Studies show that people of faith find more happiness, but many atheists claim that its only true because "ignorance is bliss". (TheHumanist.com, Maggie Ardiente, 20 April 2016, *Ignorance is Bliss*, DL 21 Oct 2016)

Some flavors of atheism base their ethics on Christian values but reject the supernatural aspects of religion. Agnostics are more honest, acknowledging that they don't know whether He exists while patiently waiting until death to find out. But they often drift in the same boat as atheists without a suitable moral anchor.

Lack of faith cannot be blamed on ignorance. Studies show that atheists and agnostics score higher than any religious group on questions about world religions, beating both Jews and members of the Church of JESUS CHRIST of Latter-day Saints. (*U.S. Religious Knowledge Survey*, nationwide poll conducted 19 May - 6 Jun 2010) Perhaps they have searched for the truth about God more earnestly than others people but failed to make sense it. Or, perhaps they study the hardest in order to defend their position.

Why does anyone believe?

Research presented at the Royal Economic Society's annual conference showed that religious people are better than non-believers at coping with adverse life events such as losing a job, losing a loved one, or experiencing divorce, and that they are happier the more they pray and go to church. (telegraph.co.uk/news/uknews/1581994/Believers-are-happier-than-atheists.html, DL 24 Jul 2018)

Many people view the beauty of physical laws and the fine tuning of constants and variables in nature as evidence of God. In the words of Isaac Newton, "This most beautiful system of the sun, planets, and comets, could only proceed from the counsel and dominion of an intelligent and powerful Being." (*Principia*, General Scholium)

More common is that people take the wonderful feelings of peace and happiness engendered by the splendors of nature as evidence of a Creator. Encounters in nature often amount to spiritual experiences for many.

The most important reasons for faith in God have little to do with logic, evidence, scriptures, or explanations for the physical world. God fills emotional needs. He can be a source of solace during sorrowful times, lifting hopes, lending purpose to life, and serving as a Listener for people who share with Him their troubles and triumphs.

Some people believe in God because they were raised that way and it's part of their family culture. As they say, "You don't change religion like you change your shirt."

During times of prosperity, security and ease, atheists are generally comfortable with their position, imputing no purpose to existence, living life as it unfolds, experiencing family, friends, fun, pizza, chocolate, or whatever comes along and deciding what's important on the fly. But, as the old aphorism goes, "There are no atheists in foxholes."

Faith provides a comforting glimmer of hope in the depths of extreme fear or hardship. In such situations, most atheists turn agnostic enough to pray words similar to Robert Ingersoll's, "Oh God, if there is a God, save my soul if I have a soul, from hell, if there is a hell!"

Some people believe in God just to be on the safe side. As physicist and philosopher Blaise Pascal put it in his famous wager: *If God exists, it is best to pray to Him or you'll burn in hell. If God does not exist, you lost nothing by praying. Hence, it is wise to pray just in case God exists.*

For true believers, no evidence is necessary. For true atheists, no evidence is sufficient.

Is God dead?

For the first time ever, on April 8, 1966, Time magazine featured no picture on its cover, just a black background with the magazine title "Is God Dead?" in bold red letters. Just as the publicity-hungry magazine intended, the issue attracted a lot of attention, positive from the elitist critics and outrage from the clergy and general public. Despite all the furor, nowhere in that issue was the question on the cover examined! Instead, it explored the question of whether the societal trend to make God an ever-smaller part of their lives would eventually result in His irrelevance. And, the magazine's point has been proven by the fact that such a cover would be much less shocking nowadays.

Since that issue of Time, increasing numbers of people have become bold denouncers of God. People from all walks of life profane His name, deny His existence, or make a career of openly mocking him. Time's cover paled in comparison to the derisive and mocking tones of Richard Dawkins book, *The God Delusion*, which sold over 3 million copies in 2006, and to the ridicule of religion in Bill Maher's movie *Religulous*, which made $11 million at the box office in 2008. Apparently, such people truly believe God isn't watching. These authors go beyond atheism to anti-theism.

Although it's quite irrational, many stay angry at God long after they stop believing in Him, especially if their transition to atheism involved a tragedy or some kind of painful disappointment. Some express their rage by proselytizing others to abandon their beliefs. These types of emotions are explored in the book *The Rage Against God*, written by Peter Hitchens whose atheist brother is one of the angry ones.

God creates and calibrates conditions in our lives to encourage learning, wisdom, growth, and faith. If He solved all of our problems, prevented all difficulties, and granted every selfish wish, He would deny us the progress that comes from experience. Part of the life equation is that humans must be allowed to make mistakes and bad choices, some of which cause gross injustices, sorrow, and pain to others. But strong faith is the lens through which a person in even the most unpleasant circumstances can see their plight in the context of a loving God's plan, like Abbe Faria who, in the 2002 movie *The Count of Monte Cristo*, taught this perspective to his fellow prisoner and friend while both were languishing in a filthy prison cell:

Edmond: I don't believe in God.

Abbe Faria: It doesn't matter. He believes in you.

The *Time Magazine* cover question was not really about whether God had died – gods don't die. It was about whether humanity's faith in God was dying.

Could science explain God?

Atheist evolution champion Richard Dawkins was once asked whether aliens could be responsible for life on earth; "It could be that, at some earlier time, somewhere in the universe, a civilization evolved by probably some kind of Darwinian means to a very high level of technology and designed a form of life that they seeded onto perhaps this planet… And I suppose it's possible that you might find evidence for that if you look at the details of molecular biochemistry, molecular biology, you might find a *signature of some sort of designer*. And that designer could well be a higher intelligence from elsewhere in the universe. But that higher intelligence would itself had to have come about by some ultimately explicable process." (*Expelled*, 2008 movie)

Yes, Dawkins is comfortable with the idea of a superior being as the creator of life on earth, but only if that alien had come into being through some kind of evolution. Would he stop attacking theists if God's origins could be explained in evolutionary terms? The $10 million he has accumulated from hocking antitheism suggests not.

How evolved does an alien have to be to qualify as God? The Kardashev scale rates civilizations in the universe based on their level of advancement:

♦ **Type I** is capable of exploiting all energy on its planet.

♦ **Type II** is capable of exploiting all energy from its star.

♦ **Type III** is can exploit all energy from its galaxy.

♦ **Type IV** is capable of exploiting intergalactic energy, including the dark variety.

Etc. to clusters, superclusters, and larger scale structures…

♦ **Type Ω**, let's say, is a civilization that exploits energy at the multiverse level and has conquered entropy. Even Dawkins might worship a being from such a civilization.

As the science fiction writer and author of *2001: A Space Odyssey*, Sir Arthur C. Clarke, said, "Any sufficiently advanced technology is indistinguishable from magic." An alien from a Type II civilization would have godlike powers. Think what power the Type Ω God of Genesis has!

The universe extends almost a 100 billion light-years in all directions and is chock full of wonders. If there are no gods out there, it's an awful waste of space!

But let's follow that thread further. Theorists say that the reason our universe is so conducive to human life because the odds were raised to a reasonable level by the vast numbers of universes spawned by the multiverse, each with randomly unique sets of physical laws. And, why is the multiverse so conducive to spawning countless different types of universes? Because the odds were raised to a reasonable level by the countless numbers of multiverses spawned by parent multiverses, each with randomly unique sets of physical laws.

With so many different types of multiverses producing so many universes with so many varieties of physical laws over countless 'verse generations, why shouldn't there be a few universes conducive to Type Ω civilizations?

Jesus walked on water, made wine from water, and calmed the waters. He made the lame walk, the blind see, the sick well, and the dead live. Then He rose from the tomb as witnessed by the soldiers that fled their post and by hundreds of others that saw him afterward. These feats are miracles to us but to Him they are merely advanced science!

Just because Type Ω aliens haven't yet been spotted doesn't mean they don't exist. Humility would suggest being agnostic at worst rather than atheist.

4. God's Nature

Why know God's character?

In His final prayer before being arrested, Jesus said, "This is life eternal that they might know thee, the only true God, and Jesus Christ whom thou hast sent." (Jn 17:3) And, Joseph Smith asserted, "It is the first principle of the Gospel to know for a certainty the Character of God." (*TPJS*, p345)

The better one knows God, the better one can love and become like Him. Jesus said, "Be ye therefore perfect, even as your Father which is in heaven is perfect." (Mt 5:48)

How can God be omniscient?

God doesn't store all knowledge in His brain using networks of neurons connected by synapses. Rather, His whole being is omni-aware via the universal field of intelligence that permeates all normal and paranormal reality.

Is Jesus God?

"The Word was with God, and *the Word was God*. The same was in the beginning with God. All things were made by him… And *the Word was made flesh, and dwelt among us*." (Jn 1:1-14) The 1935 Moffat Bible says, "The Logos existed in the very beginning, the Logos was with God, *the Logos was Divine*." The word "Word" is English for the Greek "Logos", denoting Christ's communicates the Father's word to mankind and they called "his name Emmanuel… God with us". (Mt 1:23)

Jesus was the brightest star in premortality. The night before His crucifixion, He asked the Father, "Glorify thou me with thine own self with the glory *which I had with thee before the world was*." (Jn 17:5) Jesus had been deified before the creation and, by appointment from the Father, became the "God of the spirits of all flesh". (Nm 16:22)

The fact that Jesus was God of the Old Testament (OT) is obvious from its text; "For thy Maker... the LORD of hosts is his name; and *thy Redeemer the Holy One of Israel*; The God of the whole earth." (Is 54:5) "LORD" is translated from the Hebrew "Jehovah," meaning "I Am," the premortal name for Jesus. (Str#3068) It means "Existing One."

When Moses asked God for His name, the answer was, "I AM THAT I AM... Thus shalt thou say unto the children of Israel, I AM hath sent me unto you." (Ex 3:14) And when Jewish leaders once asked Jesus, "Thou art not yet fifty years old, and *hast thou seen Abraham?*" Jesus answered, "Verily, verily, I say unto you, Before Abraham was I AM." The meaning was not lost on the educated Jewish leaders, so "then took they up stones to cast at him". (Jn 8:57-59) They knew Jesus was claiming to be the God who spoke to Moses.

From Genesis 2:4 on, the Hebrew Old Testament uses YHWH (יְהֹוָה) for God. It was the premortal Jesus who talked to all the prophets in the Old Testament. It is He who acts in His Father's stead by divine investiture of authority.

Is the Holy Spirit God?

The "Holy Spirit" or "Holy Ghost" or "Paraclete" has been around since Old Testament times. "For David himself said *by the Holy Ghost*, The LORD (El) said to my Lord (YHWH), Sit thou on my right hand, till I make thine enemies thy footstool." (Mk 12:36) And the psalmist prayed, "Take not thy *holy spirit* from me." (Ps 51:11) The OT also speaks of a "Holy Spirit" or "Spirit of God" in several other places. (Is 63:11; Ex 31:3; Nm 34:2; etc.)

The earliest instance of the Holy Ghost in action was soon after the fall from paradise: "And in that day the Holy Ghost fell upon Adam, which beareth record of the Father and the Son." (*PGP*, Mo 5:9) So, from Adam to Jesus and from Jesus to modern times, the Holy Ghost has been inspiring people and testifying of Christ.

But rabbis do not speak of an individual Holy Spirit. To them there is no trinity, just God. The doctrine of a godhead is not explicit in the OT. So, the Jewish view of the Holy Spirit is somewhat like the Christian concept of the Light of Christ except it radiates from the Father. (Jn 1:9)

Is the Holy Spirit a late member of the Godhead? No. In fact, the Holy Spirit participated in the creation; "The *Spirit of God* moved upon the face of the waters." (Gn 1:2) And, speaking to God about mankind, the psalmist wrote, "Thou sendest forth thy spirit." (Ps 104:30)

The Prophet Joseph Smith said that an "everlasting covenant was made between three person[s] ages before the organization of this earth… God the first, the Creator; God the second, the Redeemer; and *God the third, the Witness or Testator*". (*TPJS*, p190) So, immediately after Jehovah was called to be the Savior, the Holy Ghost was called to be the whispering voice of God to the spirit ears of men through the medium of spirit intelligence.

As His disciples sorrowed that Jesus was about to leave them, He explained, "If I go not away, the Comforter will not come unto you; but if I depart, I will send him unto you." (Jn 16:7) Did the Holy Spirit of the OT leave while Jesus walked among them? Not entirely. He was present at the conception of Jesus and manifested Himself at Christ's baptism.

So, how did the Holy Spirit become a companion for the disciples in a way he wasn't before Christ's ascension into heaven? We must surmise from Christ's comforting comments to His disciples that, although the Holy Spirit was nothing new, he was to be given *as a gift* after Jesus departed. This was most obviously on display when people were reborn in Jesus on the Day of Pentecost. (Jn 3:3)

Was being a *personal companion* a new role for the Holy Spirit? No. The gift of the Holy Ghost or Spirit was available as far back as Adam; "The Gospel began to be preached, from the beginning, being declared by holy angels sent forth from the presence of God, and by his own voice, and by the *gift of the Holy Ghost*." (*PGP*, Mo 5:58)

During the time that Jesus was with them, He served as the guiding light and comforter for the disciples while he was among them. This is explained by John, "Jesus stood and said in a loud voice, 'Let anyone who is thirsty come to me and drink. Whoever believes in me, as Scripture has said, rivers of living water will flow from within them.' *By this he meant the Spirit, whom those who believed in him were later to receive. Up to that time the Spirit had not been given, since Jesus had not yet been glorified.*" (*NIV*, Jn 7:39)

"The Holy Ghost has not a body of flesh and bones, but is a personage of Spirit. Were it not so, the Holy Ghost could not dwell in us." (DC 130:22) The lack of a mortal body has somehow given the Holy Ghost special access to the spirit medium that fills creation, allowing him to communicate and facilitate communications with billions of earthlings simultaneously everywhere even though his actual spirit body is not omnipresent. His ability to connect mankind with God spiritually solves the problem of mortals not being able to communicate directly with celestial beings.

In 1843, Franklin D. Richards, who later served as an apostle, wrote, "Joseph also said that the Holy Ghost is now in a state of probation which, if he should perform in righteousness, he may pass through the same or a similar course of things that the Son has." (*WJS (The Words of Joseph Smith),* p382)

According to speculation, Joseph Smith meant that a certain spirit who was with Jesus before creation volunteered to serve in our godhead as Holy Spirit now, with the understanding that, if he proved sufficiently capable, he would be qualified to serve as a savior to future souls. If true, this would explain why Jesus was the "first-begotten spirit" of all who came to Earth – He had volunteered to serve as a Holy Spirit in a godhead for a previous generation of God's spirit children and therefore was the most qualified to serve as Savior for our world.

So, the Holy Spirit is *a* god, the 3rd member of the godhead. He will not be relieved of his duties until Earth's saga comes to a conclusion. And, if he passes muster, he will be *the* god to a future world of souls.

Is Christianity polytheistic?

Does Christianity worship three gods or one? For most Christians, the answer is, yes – three gods in one.

The vast majority of the 41,000 or so Christian sects believe that God is triune, one Divine Person who has revealed Himself as three co-equal, co-eternal Persons; The Father, The Son, and The Holy Spirit. This is what Christians usually mean when they use the term "The Trinity" – though it is not mentioned in the Bible. Trinitarianism, as the doctrine is called, is a construct so baffling that those who attempt to explain it must resort to obscure terminologies that the average person finds foreign and obfuscational, such as "God is three consubstantial persons or hypostases."

Trinitarianism crept into the Latin Church in the 5th or 6th century and belief in it was made a condition of salvation by the Catholic church in a document known as the Athanasian Creed, which repeats the same theme multiple times, as if verbosity stabilizes nonsense. "We worship one God in Trinity, and Trinity in Unity; Neither confounding the Persons; nor dividing the Essence. For there is one Person of the Father; another of the Son; and another of the Holy Ghost. But the Godhead of the Father, of the Son, and of the Holy Ghost, is all one; the Glory equal, the Majesty coeternal. Such as the Father is; such is the Son; and such is the Holy Ghost. The Father uncreated; the Son uncreated; and the Holy Ghost uncreated. The Father unlimited; the Son unlimited; and the Holy Ghost unlimited. The Father eternal; the Son eternal; and the Holy Ghost eternal. And yet they are not three eternals; but one eternal. As also there are not three uncreated; nor three infinites, but one uncreated; and one infinite. So likewise the Father is Almighty; the Son Almighty; and the Holy Ghost Almighty. And yet they are not three Almighties; but one Almighty. So the Father is God; the Son is God; and the Holy Ghost is God. And yet they are not three Gods; but one God… [etc.]" (*Catholic Athanasian Creed*, translated from Latin into English)

In an attempt to clarify, adherents use a graphic that depicts God as an equilateral triangle. In the center is the word

"God" from which three lines labeled "=" connect to the corners which are labeled "Father", "Son", and "Holy Spirit" respectively while the three sides of the triangle are all marked "≠". So, substituting the first letter for each Godhead member, F=G, S=G, and H=G but G≠S≠H, a blatant violation of the transitive property of equality!

It's a simple picture but so is the Penrose Triangle, one of many impossible objects, easy to draw but still just optical illusions.

When Newton died in 1727, he left behind a substantial estate that included 1,800 or so books and numerous manuscripts by his own hand, many of which contained analysis and opinion on religious topics. In the absence of a will, the Royal Society decided to preserve Newton's genius for posterity and assigned Dr. Thomas Pellet the task of finding material worth publishing in Newton's writings.

Pellet soon discovered that Newton had extensively researched all the available versions of the Bible, every source manuscript he could obtain, and many of the writings of early church fathers, and had deduced that Trinitarianism was not taught prior to Athanasius (298- 373AD). Newton correctly concluded that the godhead of the New Testament consisted of three distinct individuals, one in purpose, and that the Father was the greatest.

Newton also pointed out that somebody added text to the New Testament to support Trinitarianism. Gasp! Up until the 1500s, a section of text said the water, the blood, and the Spirit testify of Jesus;

"This is the one who came by *water* and *blood*--Jesus Christ. He did not come by water only, but by *water* and *blood*. And it is the *Spirit* who testifies, because the *Spirit* is the truth. For there are <u>three that testify</u>: the *Spirit*, the *water* and the *blood*; and the three are in agreement." (*NIV*, 1 Jn 5:6-8)

But a later insertion, marked here with a strike-through, adds another three that testify, Father, Son, and Spirit;

"This is he that came by water and blood, *even* Jesus Christ; not by water only, but by water and blood. And it is the Spirit that beareth witness, because the Spirit is truth. For there are three that bear record ~~in heaven, the Father, the Word, and the Holy Ghost: and these three are one. And there are three that bear witness in earth,~~ the spirit, and the water, and the blood: and these three agree in one." (1 Jn 5:6-8)

Obviously, the extra text about the godhead is out of place in the 3-verse context. The questionable text first appeared in a 5th-century Confession of Faith. It was not in the first two publications of the Greek Textus Receptus of the New Testament but later got included in the Latin Vulgate. Although modern translations exclude the stricken text, it is now the most oft-cited "scriptural evidence" of the triune nature of God. But none of the early church fathers used it when scraping scriptures together to prove their trinitarian doctrine. It's not known who first fabricated this "Comma Johanneum," but it appeared sometime in the 1600s as marginal notes in a few NT manuscripts and found its way into the main text of a small number of later manuscripts which were part of the set that the King James translators used. (bible.org/article/textual-problem-1-john-57-8, DL 3 May 2020)

Newton was right to reject Trinitarianism and wrote extensively about it and other false teachings of the church. So, what did Dr. Pellet do with all of Newton's anti-Trinitarian writings? He labelled each document containing them as "anti-Trinitarian rants" and "not fit to be printed." So, several centuries passed before these portions of Newton's work would be brought to light. (religiouslearning.com/isaac-newton-englands-unknown-heretic, DL 18 Jul 2020)

Another scripture that is used as evidence of God's triune nature is, "I and my Father are one." (Jn 10:30) The Greek word for "one" (*hen*) is not masculine, which it would be if it meant "the same male person", rather it is neuter, meaning one in nature. All such scriptures refer to oneness in purpose and mind, the same kind of unity that Jesus asked His Father to bless His disciples with; "That they all may be one; as thou, Father, art in me, and I in thee, that they also may be one in us… even as we are one: I in them, and thou in me, that they may be made perfect in one." (Jn 17:20-23)

Jesus wasn't asking that the apostles be transformed into twelve different manifestations of one apostle nor for them to be absorbed into the Trinity. Rather that they be united in spirit like highly successful teams usually are. The godhead team and its members are incomprehensible enough without being characterized as a 3-in-1 enigma!

Constantine, the unbaptized Roman emperor, assembled the Council of Nicaea in 325AD, presided over it personally, and threw his weight behind the Trinitarian ideas. Why? To quell the infighting over doctrine and bring more unity to the Roman Empire. "Division in the church," he told the bishops, "is worse than war". (christianhistoryinstitute.org/today/7/4, DL 11 Jul 2020)

The voices of nontrinitarians such as Arian, who had taught the individuality of the Father and the Son, were squelched, but not immediately. The infighting over the nature of God continued for the next couple of years. "Probably more Christians were slaughtered by Christians in these two years (342-343AD) than by all the persecutions of Christians by pagans in the history of Rome." (*The Age of Faith*, Will Durant, The Story of Civilization, bk4, 1950, p8)

Eventually, the squabbling died down and the Trinitarian doctrine was made official in the Council of Constantinople (381AD). The emperor had declared it to be the truth and the church had declared, at the end of the creed, "except a man believe truly and firmly, he cannot be saved." So, all nontrinitarians were supposedly headed for the lake of fire and brimstone.

Thus, God's nature, a most central doctrine, was chosen by a pagan aristocrat who had deferred his baptism to his deathbed so as to have all the horrible sins in his life absolved, and who, in the year following his decision on the Trinity, had his wife and son killed, among others. (roman-emperors.org/fausta.htm, DL 20 Feb 2019)

Nontrinitarian churches include the Jehovah's Witnesses, The Church of JESUS CHRIST of LDS, Iglesia Ni Cristo, Christadelphians, Christian Scientists, Dawn Bible Students, Living Church of God, Oneness Pentecostals, Members Church of God International, Unitarian Universalist Christians, The Way International, The Church of God International, the United Church of God and various branches in Eastern Churches (Syria, Palestine, Anatolia, Russia, et cetera).

With the exception of these denominations, Christianity has ignored the overwhelming evidence in the New Testament that God the Father is a separate personage from Jesus Christ:

♦ *All three godhead members were manifested separately at Christ's baptism.* "And Jesus, when he was baptized, went up straightway out of the water: and, lo, the heavens were opened unto him, and he saw the Spirit of God descending like a dove, and lighting upon him: And lo a voice from heaven, saying, This is my beloved Son, in whom I am well pleased." (Mt 3:16-17) Plugging in "God" for every reference to deity to shows how silly this is, "God, when he was baptized went out of the water and saw God descending and lighting upon God and a voice from God in heaven saying…" God doesn't reveal Himself through avatars!

♦ *The risen Lord had yet to visit the Father.* "Jesus said, 'Do not hold on to me, for I have not yet ascended to the Father. Go instead to my brothers and tell them, *I am ascending to my Father and your Father, to my God and your God.*'" (*NIV*, Jn 20:17) If Jesus were His own Father, why didn't He present Himself to Himself on the spot and spare Mary's feelings?

♦ *Only the Father knew when the end would come.* "But of that day and that hour knoweth no man, no, not the angels which are in heaven, *neither the Son*, but the Father." (Mk 13:32) How could one God simultaneously know and not know the day He planned to return?

♦ *The Son will judge but not the Father.* "For the Father judgeth no man, but hath committed all judgment unto the

Son." (Jn 5:22) How could the same God both judge of all and judge none simultaneously?

♦ *The Father is greater than the Son.* "If ye loved me, ye would rejoice, because I said, I go unto the Father: for my Father is greater than I." (Jn 14:28) This the Father=Son equality espoused by Trinitarianism. The Father can only be greater than Jesus if they are two distinct entities.

♦ *No man saw God.* "No man hath seen God at any time." (Jn 1:18) If the Son and the Father were one Person, then this scripture makes would make sense because multitudes saw Jesus both before and after He resurrected.

♦ *Blasphemy against the Holy Spirit is unforgivable, but not so against the Father and Son.* "Wherefore I say unto you, All manner of sin and blasphemy shall be forgiven unto men: but the blasphemy against the Holy Ghost shall not be forgiven unto men." (Mt 12:31) Logically, if the godhead were all one individual, then blasphemy against any one of them would be unpardonable.

♦ *Jesus thanked His Father.* "I thank thee, O Father, Lord of heaven and earth." (Mt 11:25) It makes no sense for Jesus to pray to Himself and thank Himself. There are dozens of scriptures where Jesus prayed to His Father or referred to His Father as a separate person.

♦ *Jesus taught His disciples to pray to the Father in heaven while Jesus was still on earth.* "When ye pray, say, Our Father which art in heaven…" (Lk 11:2) Jesus was right there with His disciples but never at any time did He teach them, "Pray to me."

♦ *Jesus needed the Father to transfigure Him.* "As he prayed, the fashion of his countenance was altered, and his raiment was white and glistering." (Lk 9:29)

♦ *Jesus asked the Father to forgive others on occasion.* "Father, forgive them; for they know not what they do." (Lk 23:34) Jesus forgave them and then asked the Father to do likewise.

♦ *Jesus reacted to the withdrawal of the Father's presence.* "My God, my God, why hast thou forsaken me?"

(Mt. 27:46) This anguished outburst was not a question to an alternate personality! Why would Jesus ask Himself why He had abandoned Himself?

♦ *Jesus commended His spirit into His Father's care.* "Father, into thy hands I commend my spirit." (Lk 23:46) Jesus was not talking to Himself, handing His own spirit off to Himself.

♦ *Jesus asked His Father to spare Him.* "Father, if thou be willing, remove this cup from me: nevertheless not my will, but thine, be done." (Lk 22:42) This request only makes sense if Jesus and the Father are two distinct individuals, the Son struggling and succeeding to stay on the same page with His father.

♦ *Jesus referred to His premortal existence as separate from that of His Father.* He prayed, "O Father, glorify Thou me with Thine own self with the glory which I had with Thee before the world was." (Jn 17:5) The Father had His own glorious self, separate from Christ's self since before creation. It's nonsensical that God would ask Himself to give Himself more of His own glory.

♦ *The Son is subject to the Father.* "When all things shall be subdued unto him, then shall the Son also himself be subject unto him that put all things under him." (1 Cor 15:28) Contrary to the Trinitarian doctrine of 3-way equality in the godhead, the Son is indeed subject to the Father.

The Bible says a husband and wife become "one flesh". (Mk 10:8) But nobody suspects that marriage transforms a couple into an incomprehensible biunial hypostases. Unity of mind and purpose is a common theme in life for groups with shared objectives. Paul's statement that "ye are all one in Christ Jesus" does not mean His followers are a unary identity of countless co-equal persons. (Gal 3:28)

It is ironic that millions of Christians accuse other Christians of not being Christian because "they believe in a different Jesus," all the while embracing the unbiblical idea of a composite God who is His own Father, conceived by His third self, the Holy Ghost! And which member of the godhead now has the glorified body that rose from the tomb?

Trinitarianism contradicts the scripture that "God created man in his own image" as humans are not composite individuals. Trinitarianism pollutes and hyper-mystifies the true character the godhead, unnecessarily distancing us from the Father of our spirits while depersonalizing our Savior by denying Him His distinct individuality.

But wait! Didn't God say He was the sole deity? "Before me there was no God formed, neither shall there be after me." (Is 43:10) This was the premortal Jesus speaking to a people prone to worship idols in the lands of polytheistic pagans. He was Yahweh, the only Lord God for all earthlings, appointed to that role by His Father. Unlike the pagans that surrounded them, the Hebrews had only one God who spoke to their prophets, Yahweh the Son of the Almighty Father, the mediator between the Father and mankind.

Christians are instructed to "do everything in the name of the Lord Jesus, giving thanks to God the Father through him". (*ESV*, Col 3:17) Jesus explained, "Whatsoever ye shall ask in my name, that will I do, that the Father may be glorified in the Son." (Jn 14:13)

Biblical monotheism means that the Father appointed His one perfect Son to serve as God, proxy for the father, the Son subjugating His will to the Father's in all things. The members of the godhead are united in purpose, never fighting or killing each other as was the habit of the Greek and Roman gods of Christ's day.

Is Christianity monolatrous?

Monolatry is the worship of one god as supreme deity while acknowledging the existence of other gods.

Prominent Catholic Bible scholar, John McKenzie, wrote, "In the ancient Near East the existence of divine beings was universally accepted... The question was not whether there is only one elohim, but whether there is any elohim like

Yahweh." (*Aspects of Old Testament Thought*, as quoted on hyperleap.com/topic/Elohim, DL 7 Jul 2020)

When Moses came down the mountain with the Decalogue on tablets, he said, "Jehovah your Elohim is Elohim of Elohim." The KJV translation; "The LORD your God is God of *gods*." (Dt 10:17)

God did not just say, "Thou shalt have no other gods." (Ex 20:3) He added, "before me," an implication that there are other gods but that He is the primary object of worship. That's monolatry (Gr: monod=one, latreia=worship). OT scholar, John Day, suggested that the other gods were demoted by the ancient Israelites to angels as monotheism became the focus. These angels served God by carrying out His will. For example, God once asked those who were with Him, "Who will go for *us*?" (Is 6:8)

Paul said, "For though there be that are called *gods*, whether in heaven or in earth, (as there be gods many, and lords many,) But to us there is but one God, the Father, of whom are all things, and we in him; and one Lord Jesus Christ, by whom are all things, and we by him." (1 Cor 8: 5-6) Origen Adamantius (184-253AD) commented on Paul's statement; "Admitting other beings besides the true God, who have become gods by having a share of God *[does not lower] the glory of* Him who surpasses all creation. We drew this distinction between Him and them… to us there is one Lord, Jesus Christ." (*Commentary on John,* bk2, sec3)

Early Christianity believed in worshiping one God among multiple gods, irrespective of whether the primary focus was the Father or the Son. Believing that God is the top God of all other good gods elevates Him rather than denigrates Him.

Due to the fact that most Christian churches are trinitarian, believing in a triune godhead in which three Gods are to be worshiped as equals, there are very few monolatrous religions. The Church of Jesus Christ of LDS is nontrinitarian and monolatrous. It teaches people to worship the Father and pray only to Him in the name of His Son as instructed by Jesus. (Jn 15:16; 3 Ne 18:19) When people in ancient America prayed to Jesus during His personal visit to them, He apologetically told His Father, "They pray unto me because I am with them." (*BOM*, 3 Ne 19:22)

Is God everywhere?

Christians have historically been taught that God is an amorphous spirit that exists literally everywhere simultaneously. "God cannot have a body." But laity has asked the question, "What of Christ's body?" The Catholic Church's answer is, "Since the Lord only took on human flesh in these 'last days,' and since God has always existed, without beginning or end, we must still conclude that having a body is not part of God's unchangeable nature: he exists in eternity as pure spirit." (catholic.com/tract/god-has-no-body, DL 3 May 2020)

So, the same Catholic tract contradicts itself. *God can't have a body but Christ, who is God, has one, but that doesn't count because the spirit part of Him is the eternal part, even though His resurrected body is now eternal.* Make sense?

To prove God is not everywhere, one would only need to identify a single spot where He's not. How about hell? By definition, hell is called outer darkness because it lacks the slightest bit of light from God's presence.

Although there are places God's person is not, there is no place of which He is not aware; "The eyes of the Lord are in every place, beholding the evil and the good." (Prv 15:3) God's cognizance and power are omnipresent. His spiritual "light proceedeth forth from the presence of God to fill the immensity of space". (DC 88:11-13)

Could God be in more than one place simultaneously? Well, even lowly scientists on Earth can tease a single subatomic particle to exist in two places at one time. (*If an Electron Can Be in Two Places at Once, Why Can't You?* Discover Magazine, June 2005) If an electron can do it, why couldn't God? His ability to navigate time and space instantaneously would blow the mortal mind.

So, even if God does a literal visit, as He did with Mary and other disciples after his resurrection, He could use His ability to change location instantly to appear in multiple places simultaneously. Vintage flip books ran at 12 frames per second (fps), creating effective illusions for slow-moving objects. Modern animation and video ranges from 24 to 60 fps but fast-moving wagon wheel spokes can still appear to spin the wrong way due to the stroboscopic effect. Videogame rates of 240 fps is perceived as a solid image by humans even for very speedy objects. So, it should be easy for God to appear solid visiting multiple viewers at once.

But people whom the Father has visited have probably all seen Him in *visions* rather with than with *physical* eyes. So, projecting images of Himself to multiple viewers may be His standard modus operandi. The same applies to visits by the Son since His ascension.

Jesus told Mary Magdalene at the tomb, "I am not yet ascended to my Father." (Jn 20:17) If His Father had existed everywhere, then Jesus would have already been in the Father's presence and lingered longer to console the poor Mary. Instead He had to leave for heaven to visit the Father and return to Earth later.

God has been coming and going to and from Earth since the beginning. For example, "The Lord came down to see the city and the tower." (Gn 11:5) If God's person were omnipresent, He would have always been here.

What does God look like?

When Philip asked Jesus to show him the Father, "Jesus saith...*He that hath seen me hath seen the Father.*" (Jn 14:9) So, the Father looks like the man Jesus, with a male body, replete with beard and a kind face. Simple, right? Not quite.

There are only a handful of God sightings in the OT. So, it's somewhat understandable that Jews and Muslims reject the doctrine of an anthropomorphic God. But it's less apparent why the bulk of Christianity says, "God has no body." (catholic.com/tract/god-has-no-body, DL 18 Jan 2018)

One scripture often put in evidence is, "God is a Spirit." (Jn 4:24) But, the original Greek, "Θεὸς πνεῦμα" (theos pneuma), means "God is spirit." Taking out the "a" changes the meaning slightly from being *a* spirit to being *of spirit* or *of spiritual nature*. The Lord's body is not physical, but it is tangible, more than "a" spirit, and with more capabilities than a spirit. In fact, most modern translations say, "God is spirit." (NIV Jn 4:24) The writer, John, had felt the nail prints in His hands and feet and knew that He was tangible. The verse continues, "And they that worship him must worship him *in spirit*," and Paul said, "He that is joined unto the Lord is one spirit." (1 Cor 6:17) That doesn't mean all Christians are one spirit personage!

If Mary could mistake the resurrected Lord for a gardener at the tomb, and two of His disciples could mistake Him for an ordinary traveler on the way to Emmaus, and the Son looks like the Father, then we may conclude that God looks like a normal human and, since He took His body up from the tomb, He is still has it. (Jn 20:15-16; Lk 24:13-31) This agrees with His word in the beginning; "God said, Let us make man in our image, after *our* likeness." (Gn 1:26)

The risen Lord went to great lengths to prove his corporeality. He interacted with people and things as if He were still mortal. He asked the marveling disciples, "Have ye here any meat? And they gave him a piece of a broiled fish, and of an honeycomb. And he took it, and did eat before them." (Lk 24:41-43) Worshippers embraced His feet. (Mt 28:9) More than five hundred people in Palestine were eyewitnesses to His anthropomorphic person. (1 Cor 15:4–6) His disciples felt the wounds in his hands, feet, and side. (Jn 20:27; 3 Ne 11:14-15) He said, "Behold my hands and my feet, that it is I myself: handle me, and see; *for a spirit hath not flesh and bones, as ye see me have.*" (Lk 24:39)

So, God has both body and spirit, like humans. But unlike humans, His body is immortal rather than physical, and was inseparably melded with His spirit through the resurrection.

Of the universal resurrection John wrote, "We are God's children now, and… when he appears *we shall be like him*." (1 Jn 3:1-2)

At one point in the Old Testament, Moses wanted to see God's glory. So, "the LORD said… while my glory passeth by… I will take away mine hand, and thou shalt see my back parts: but my face shall not be seen." (Ex 33:21-23) So, the God of Moses (Yahweh) at least had hands, a face, and back parts like humans.

So, man looks like God and God has the likeness of man: "And above the firmament that was over [the angels'] heads was the likeness of a throne… and upon the likeness of the throne was *the likeness as the appearance of a man above upon it*… This was the appearance of the likeness of the glory of the LORD." (Ezek 1:26-28)

St. Thomas Aquinas (1226-1274AD) taught that the risen Lord only ate food to demonstrate the reality of his resurrection, not because His body needed it, and that the food He ate was not absorbed by natural processes, but rather was dissolved away into energy by divine power. (gloria.tv/post/eL63oW8nFUaw4jzyfyses4cSe, DL 18 Jul 2020) Indeed, this is educated speculation, but still, most people do not imagine the need for latrines in heaven.

God is spirit in the same sense that "God is light." (1 Jn 1:5) He is "a consuming fire". (Hb 12:29) The Father and the Son are alike, their "brightness and glory defy all description". (*Pearl of Great Price,* Joseph Smith, History 1:17) Only one's spirit eyes can see Him, and without pain.

The fact that Jesus kept the body that He took from the tomb in no way diminishes His immortal and splendiferous nature. "The Son can do nothing of himself, but what he seeth the Father do." (Jn 5:19) So, if Jesus obtained a resurrection, the Father must have also.

They can pass through walls. They can change locations in an instant. They don't bleed. But they appear human. "The Father has a body of flesh and bones as tangible as man's; the Son also." (DC 130:22) Joseph Fielding Smith wrote, "Our

Father in heaven and our Savior [have] bodies [that] are quickened by spirit and not by blood, hence they are *spiritual* bodies." (*DOS (Doctrines of Salvation)*, bk1, pp76-77)

God can veil His glory when He appears, as Jesus did when He appeared to people after his resurrection. By contrast, "A spirit cannot come but in glory." (*JD (Journal of Discourses)*, Joseph Smith, bk6, 2 Jun 1839, p240) Why not? Because, unlike God, spirits do not have tangible bodies with which to cloak their spiritual light.

What is God's gender?

Perhaps the most oft-quoted verse in the Bible has been neutered. The *Inclusive Version* of the NT says, "For God so loved the world that *God* gave *God's* only *Child*, so that everyone who believes in *that Child* may not perish but may have eternal life." (Jn 3:16) Granted, the personal pronouns in the Greek manuscripts are without gender. However, translating the Greek Υἱὸν as "Child" rather than "Son" is quite a twist, especially since that same Child was clearly circumcised as a babe. (Lk 2:21)

Are readers of neutered Bibles actually uncomfortable with gender-specific pronouns or are certain devotees of political correctness actually wondering whether God might be female? There's no question if they believe the New Testament they're reading! When Jesus references His Father, the original Greek says, Πατρός, pronounced "Patros" – a common masculine root in the English language.

Since the premortal Jesus was Yahweh of the Old Testament, the New Testament's portrayal of Jesus as a man supports the male gender of Yahweh in the Hebrew OT texts.

If God wasn't male, He wouldn't have had a woman bear His Son. If God wasn't male, He wouldn't have used the bride metaphor for His church. If God wasn't male, there wouldn't be 170 Bible references to God as the Father.

Since most Christian religions don't believe God's glorious body is tangible, they lack standing and doctrinal grounds to argue against His gender neutrality.

There is good reason to believe Jesus did not lose His masculinity during His resurrection. "Every limb and joint shall be restored to its body; yea, even a hair of the head shall not be lost." (*BOM*, Alma 40:23) No exception is made for gender-specific characteristics.

What is the Father's name?

32Learning a name is the customary first step when getting to know someone, even if it's the Supreme Power. Although He has no known personal name (like John or James), He has many titles:

Father. Jesus called His Father "my Father," "our Father," and "the Father." Paul called Him "the Father of Spirits". (Hb 12:9) The term "Heavenly Father" is common nowadays.

God. This is English for deity and a role title used for the Father (as well as His Son). It is sometimes said that "God" was derived from "good" but etymologists disagree. The word was used for gods that were often bad.

HaShem. This means "the Name" in Hebrew, a way to avoid using any of God's actual names. The use of HaShem by Hebrews was a bit like the substitution of "he who shall not be named" for Voldemort by magical folks in the Harry Potter books out of fear or respect.

El. This is Hebrew for "god", pronounced *ale,* El means "mighty one". (Str#410) It is used as a title for the Almighty (as well as for His Son or any type of god, especially in the OT). Not coincidentally, the Jewish creator of Superman included this term in his godlike hero's name, Kal-El, and in his father's, Jor-El, just as ancient Hebrews used it as a suffix in names like Isra-el, Dani-el, Samu-el, Ezeki-el, and Emmanu-el.

El Shaddai. This comes from Semitic languages and means "God Almighty" — used in the OT seven times. For example, God appeared to Moses and said, "I am God Almighty; Walk before me, and be blameless." (Gn 17:1)

Eloah. This is also Hebrew for "God". (Str#433) Eloah occurs 60 times in the Old Testament.

Elyon. This means "the Most High God". (Str#5946) This name is used here and there in the OT but more especially in Genesis and Psalms.

Elohim. This is the plural form of Eloah, sometimes written "eloahim". (Str#430) This term is found 2600 times in the Old Testament. The first chapter of Genesis uses Elohim exclusively to reference God. The Jews and Muslims generally see *Elohim* as a "royal plural" that shows respect. Trinitarian Christians see it as a reference to the three members in the triune godhead. Latter-day Saint (LDS) doctrine sees it as all the hosts of heaven involved in the creation. When Elohim said, "Let *us* make mankind in *our* image, in *our* likeness… So God created mankind… in the image of God he created them; male *and female*," there must have been one or more females in the "us/our" group after whose image Eve and other women were created. Jewish thought is that "Come, let us make man in our image, after our likeness," refers to God and the angelic spirits that were with Him. (sacred-texts.com/chr/bct/bct04.htm, DL 4 Jun 2025)

Eli. This is the Greek transliteration of the Hebrew *El* with the suffix (ī) which means "my". (Str#2241) Used by Matthew to quote Jesus on the cross. (Mt 27:46)

Eloi. This is the Aramaic term for El. (Str#1682) Used by Mark for the same quote by Jesus on the cross. (Mk 15:34)

Elahh. This Chaldean term for El is used in various Old Testament passages such as Ezra 4:24 & 7:26. (Str#1682)

Man of holiness. Refers to the Father. In conversations with Adam and Enoch, God said about himself, "In the language of Adam, Man of Holiness is his name." (*PGP*, Mo 6:57)

Almighty God. This term (Gn 17:1) has the same connotation as "the most high God". (Gn 14:18) It's equivalent to "The Highest". (Ps 18:13)

God of all other Gods. (DC 121:32) This name reflects the Father's status as Almighty God, greater than Jesus or the Holy Ghost. And, He is God of all gods as referenced in Psalm 82:6, "Ye are gods."

Ahman. Adamic language name for God meaning Holy Man. "What is the name of God in the pure language? ... Ahman." (*JD*, Orson Pratt, bk2, 18 Feb 1855, p342) This may have been deduced from the name Adam-ondi-Ahman, the place where God will visit Adam in a grand council at the ushering in of His millennial reign.

Adam. Name title for the Father. "Our spirits and the spirits of all human family were begotten by Adam." (*Brigham Young Papers*, Oct. 8, 1854, call number Ms. d 1234, Church Historian's Office, SLC) Adam is also a name title for those who attend to temple ordinances aspiring to be like God. Male and female aspirants to exaltation are called "Adam" and "Eve" respectively as they are presented for admission into the celestial realm. Adam denotes "First Father". (*PGP*, Abr 1:3)

What are Christ's names?

Of course, Jews don't recognize the Old Testament God to be Jesus. But the following discussion of names will treat Him as such.

Ancient Hebrew was written with consonants only. Without vowels, the exact pronunciation of the tetragrammaton is not known. To complicate things, it was Jewish custom to avoid using God's name in verbal and written communications. Other terms for God were often used instead:

Adonai. This literally means "my Lords" in Hebrew, usually explained as a type of "royal you" plural. (1 Kg 2:26; Strg#136)

Son of Man. This term is used 125 times in the OT, all but one of which are clear references to ordinary men. In the single case where God is referenced, He is described as "one *like* the son of man". (Dan 7:13) In the NT, John twice uses the term "one like unto the Son of Man" to refer to His glorified Lord. The NT uses "Son of Man" as a title for Jesus 83 other times. Christianity generally teaches that "Man" here is a reference to Christ's birth by a *human* mother. But "Man" is also a reference to His Father, *Man of Holiness*; "In the language of Adam, Man of Holiness is his name, and the name of his Only Begotten is the *Son of Man*, even Jesus Christ." (*PGP*, Mo 6:57; 7:35)

Jehovah. This name is an expansion the Hebrew Tetragrammaton YHWH or YHVH, which is based on the name "I AM" that the premortal Jesus told Moses to use when speaking to people (Heb יהוה, *Yahweh,* "to exist"). "Yahweh" is the oldest pronunciation known. "Jehovah" began to be popular in the 16th century AD. יהוה is translated as "Lord" with capital letters in the KJV OT 6828 times. The combination "Yahweh Elohim" (LORD God) appears 891 times with its first occurrence in Genesis 2:4.

Lord of Sabaoth. This is OT Hebrew for "Lord of Hosts."

Messiah. This means "The Anointed". For example, Psalm 2:2.

Christ. This title is Greek for *Messiah* ("the anointed").

Jesus. This proper name is Greek for Yeshua or Joshua, the name given Him by Angel Gabriel.

Son Ahman. Jesus referred to himself as "Son Ahman" on March 1832, at Hiram, Ohio. It means "Son of Man of Holiness". (DC 78:20; 95:17)

Shiloh. Jacob prophesied, "The scepter shall not depart from Judah, nor the ruler's staff from between his feet until *Shiloh* comes," a reference to Christ's earthly advent. (Gn 49:10)

Father. "Because of the covenant which ye have made ye shall be called the children of Christ, his sons, and his daughters; for behold, this day he hath spiritually begotten you; for ye say that your hearts are changed through faith on his name; therefore, ye are born of him and have become his sons and his daughters." (*BOM*, Mosiah 5:7) The Son is also Father by virtue of His role as Creator and because He stands as God in place of the Father by divine investiture of authority.

What are the Spirit's names?

No personal names are known for the third member of the godhead, but the titles used in scripture include "Holy Ghost," "Holy Spirit," "The Spirit," "Spirit of the Lord," "Spirit of God," "Breath of the Almighty," "Comforter," and "Eternal Spirit".

Can God do anything?

It takes an incomprehensible amount of power to create an entire universe. And, justifiably, believers generally call God omnipotent. But, are there things God *can't* do?

♦ Could God create a rock bigger than He can lift?

♦ Could He create a prison from which He cannot escape?

♦ Could He ask a question He can't answer?

♦ Could He make happenings unhappen?

♦ Could He draw a square circle?

♦ Could He erase His existence against His will?

♦ Could He then cause Himself to exist again?

People love paradoxes and the omnipotence paradox is no exception. Some examples are paradoxical because of the semantics of language. For example, "Could God beat Himself at chess?" Although the question can be formulated with a subject, verb, and direct object, it's invalid semantically, like liar paradoxes such as "This sentence is false" or "I am now lying."

Scripture mentions that God cannot "cannot lie." (Titus 1:2) Likewise, He cannot sin, nor abrogate Justice, nor fight Himself, nor destroy Himself. In sum, He cannot even *try* to do anything that is inconsistent with His nature – He won't let Himself!

If He governed in any ungodly way, He would cease to be God. For example, if He allowed "mercy [to] rob justice… God would cease to be God". (*BOM*, Alma 42:25) Similarly, if a positron went negative, it would cease to be a positron, an event more likely than God behaving ungodly. Considering the possibility of God doing anything ungodly is a purely hypothetical exercise because He's perfectly set in His godly ways and would never do anything ungodly; "For I, the LORD, do not change." (Mal 3:6)

In fiction, when powerful fairies, leprechauns, and genies grant wishes, there are usually conditions or restrictions. In Disney's movie *Aladdin*, the genie preempted the boy by stating up front, "ixnay on the wishing for more wishes." But when the evil Jafar acquired the magic lamp, he sidestepped the restrictions by wishing to become an all-powerful Genie, thus granting himself the power to grant himself wishes without limits.

In theology, if God did something ungodly, the hosts of heaven would flee, His light would fade to darkness, and His power would be lost. God is God because He is the Master of eternal laws rather than a sidestepper or rewriter of them.

As Ben Parker told the young Spiderman, "With great power comes great responsibility." God never abuses His power. He takes His responsibility seriously. As it applies to God, omnipotence means, power over all things rather than power to do anything.

Does God know everything?

The scriptures indicate that God knows everything; "Great is our Lord… his understanding is infinite." (Ps 147:5) "God… knoweth all things, and there is not anything save he knows." (*BOM*, 2 Ne 9:20) "The Lord knoweth all things which are to come." (*BOM*, Words of Mormon 1:7) God exists independent of His creation and therefore can observe events inside it like a kid peering into an ant farm. But unlike the kid, God has a panoramic view of time; "Past, present, and future, and are continually before the Lord." (DC 130:7)

Even probabilistic omniscience thousands of years into the future is unfathomable to those of us who often have trouble fully comprehending the present state of things right under our noses. And some things are simply unknowable to mankind, such as what lies outside the observable region of the universe. The expansion rate of the cosmos has pushed light from galaxies beyond a certain distance out of our reach, even if we could chase them at the speed of light!

Human brains learn slowly and forget quickly. So, they try to preserve knowledge in books and other recording media. Even God the Son, while in the flesh, acknowledged not knowing everything, "But of that day [His 2nd coming] and that hour knoweth no man, no, not the angels which are in heaven, *neither the Son.*" (Mk 13:32) But His perfect cognizance was restored when He resurrected. He "comprehendeth all things, and all things are before him, and all things are round about him; and he is above all things, and in all things, and is through all things, and is round about all things, and all things are by him, and of him, even God". (DC 88:41)

If God knows everything, why do Bible texts seem to suggest a lack of perfect knowledge on God's part? For example:

♦ When God heard reports of how wicked Sodom was, He said, "I will go down now and see whether they have done altogether according to the cry of it, which is come unto me; and, if not, I will know." (Gn 18:20-21) Did God really need to descend to Earth in person to discover what the state of Sodom? No. He sent His proxies in the form of angels to be witnesses of its wickedness and warn Lot to flee before he destroyed the city.

♦ God told the Israelites that He'd led them in the wilderness 40 years "to know what was in thine heart, whether thou wouldest keep his commandments, or no". (Dt 8:2) Was God really ignorant of how disobedient His people would be? No. He was testing the Israelites, not so He could know the outcome, but so they would have the opportunity to make good choices in the face of hardships.

God's judgment of people had to be based on actual behavior rather than what He knew would they would do if given the situation.

♦ God said, "They [the Israelites] have set up kings, but not by me: they have made princes, and I knew it not." (Hosea 8:4) Did God totally miss the coronation of Israelite kings? No. The last clause should have been translated, "I approved it not."

Elder Orson Pratt wrote, "The Father and the Son do not progress in knowledge and wisdom, because they already know all things past, present, and to come." (*The Seer*, p117, par96) So, without learning, how does God progress? God progresses by creating worlds, saving their inhabitants to glory, a side-effect of which is that His own dominion and glory increase. It would be a damnation of sorts if He could not progress.

How does knowledge of everything fit into God's head? It doesn't. His head is a processor rather than a storage device. Since He is perfectly aware of all reality everywhere and at all times, He doesn't need to replicate it in His head. He lives on a celestial world and therefore has instant access to knowledge of kingdoms above Him and below Him, as described in the prior section. God is tuned to all reality such that His cognizance, intelligence, and influence are pervasive.

The Lord said, "I… knoweth all things, for *all things are present before mine eyes.*" (DC 38:1-2) He did not say, "All things are in my head." He is aware of all things because, via His light of truth that fills the universe (aka intelligence, glory, light of Christ), He is "in all and through all things". (DC 88:4-13)

Just as happenings in the present that we experience leave memories in our mind, all events, no matter how large or small, are stored as precise echoes in the ubiquitous intelligence that fills the universe and from which all things are created. For example, the blueprint and execution of the creation of all things is preserved in the collective memory of the pervasive light of truth, which is the power "by which they were made" and is the glory of God. (DC 88:7-10)

As for kingdoms not under his purview, there is no reason to believe that God that His ignorant of them. But he is surely more focused on matters amongst His own creations.

Does God's knowledge of the future deprive anyone of freedom to choose? No. Our decisions determine our future, not God's knowledge of them. A parent can know with certainty that a daughter will choose ice cream instead of spinach for dessert. The parent's knowledge does not deprive the child of her freedom to choose.

Why does God create?

God could do without us. "Neither is [God] worshipped with men's hands, as though he needed any thing." (Acts 17:25) God created the universe so that He could populate it with life. God created life to support the existence of humans. So, why did He create humans?

Some Christians speculate that God created humans because He was lonely. But God had already created angels such as Michael, Gabriel, cherubim, and seraphim before He created mortals. Weren't the hosts of heaven enough company for Him? And, if He was lonely for an eternity prior to the creation, why did He wait so long to solve His problem? God is smarter than that! Creating a wife for Himself would have assuaged His loneliness, but mainstream Christianity abhors the idea.

Sending all His creatures out of His presence to a physical earth where they would suffer through mortality's miseries, mistakes, and death is inconsistent with the companionship theory! Unlike the lonely Geppetto, who had no knowledge of Pinocchio's future troubles, God saw our future and created us and sent us to Earth anyway.

Scripture tells us the **why**; "Bring my sons from afar and my daughters from the end of the earth, everyone who is called by my name, *whom I created for my glory*." (Is 43:6–7) His glory is measured by how many souls rejoin Him in heaven. He was not lonely, egotistical, or self-aggrandizing. His purpose is "to bring to pass the immortality and eternal life of man". (*PGP*, Mo 1:39) Earth life for us is the path to His success. He created bodies and a planet for us because He loves us. "God is love." (1 Jn 4:8)

What took God so long?

For centuries people have asked the question: *If God existed unchanging for an eternity past, as Christianity teaches, what was He doing during all that time before He created everything?* Fourth-century Saint Augustine jokingly answered, "Preparing hell for those who ask such questions!"

Nineteenth century atheist Robert Ingersoll quipped that God could not have been thinking before there was something to think about. Of course, most Christians can't imagine God not thinking. But, from the average Christian's perspective, does it make sense that God spent an eternity past just thinking before doing anything? Did He spend an eternity contemplating a future in which He would actually do something?

Some science-savvy Christians say that since there was no time until God created time, the question of what He did *before* creation is meaningless. Such arguments have been used since Saint Augustine first wrote them 1600 years ago despite the fact that the word *before* indicates *sequence* rather than passage of time.

It's hard to fathom but it's true! Mainstream Christianity says that God was the sole existing entity, an unchanging non-spatial and timeless being who knew everything before there was anything to know, was present everywhere before there was anywhere, saw a future before it had a past, was all powerful before there was anything to rule, and did nothing because He hadn't yet invented the time in which to do anything.

Was God sleeping? Immortals don't need sleep! And why, after an eternity of nothing, was God suddenly dissatisfied with the status quo? Sedentariness for that long just seems so ungodly! What triggered the change in God's behavior from doing nothing to creating everything?

The questions seem silly because the assumptions are false. The truth is far better. He was very busy and always has been, as we shall see. It's not in God's nature to be idle or not to be actively doing good.

How is God eternal?

Being eternal is quite different from never changing. Never changing implies no progress. Eternal implies no beginning and no end. It is safe to say that God, in His essentials, is eternal.

"Is it a fact that He never had a beginning? In the elementary particles of His organism, He did not… The light and life of all things, by which our heavenly Father operates, by which He is omnipotent, never had a beginning and never will have an end. It is the light of truth; it is the spirit of intelligence." (*JD*, Charles Penrose, bk26, 16 Nov 1884, pp24-25)

The same is true for every member of the godhead. Each has always existed as a sentient spiritually luminous entity in some stage of progression and is therefore eternal.

Does God count time?

It feels like sweeps even the most reluctant humans along from past to future. It's impossible to go against the current and there's no paddling in place – everyone ages. There are no back eddies. There is no place where events unhappen – all creation is subject to time's forward arrow.

The faster one moves through space, the slower one moves through time. So, people could avoid aging if they could move at light-speed but they would have to be massless to do so, like a photon or a gluon. And there's a catch. At light-speed, people would not experience time and could therefore not luxuriate in their agelessness.

God has zero mass, as physicists would measure it, and does not age. He's made of structured spiritual light and, unlike physical photons and gluons, is aware of time, though He needs no wristwatch. His ability to predict future events with precision is evidence that He has a sense for time.

In fact, whereas humans experience time slightly differently from each other, depending on their individual reference frames, God senses time in every realm and reference frame. It's difficult to comprehend but, "All is as one day with God, and time only is measured unto men." (*BOM*, Alma 40:8)

That doesn't mean God is stuck in an eternal present, rather that He sees all time as a single picture. He is not insensitive to time nor does He ignore it. He laid out the creation into a sequence of time periods. He created the natural cycles by which mortals measure time. "To every thing there is a season, and a time to every purpose under the heaven: A time to be born, and a time to die; a time to plant…" and so on. (Ecl 3:1-3) He told Abraham that the star "Kolob was after the manner of the Lord, according to its times and seasons in the revolutions thereof." (*PGP*, Abr 3:4) "The reckoning of God's time... [is] according to the planet on which [He] resides." (DC 130:4-5) So, even God marks time by cycles in His realm.

There has been much talk of a God/Mankind time ratio; "One day is with the Lord as a thousand years, and a thousand years as one day." (2 Pt 3:8) It's probably best not to make a hard-and-fast mathematical ratio out of this. Paul may have been saying that what seems like a long time to humans is like a mere moment to God. God also said the earth, moon, sun, planets and stars "give light to each other in their times and in their seasons, in their minutes, in their hours, in their days, in their weeks, in their months, in their years—*all these are one year with God*, but not with man". (DC 88:43-44) To take this literally would mean that God's celestial globe orbits a star so slowly as to encompass all the time that has passed since our solar system' components began their cycles.

Without time, either nothing would happen or everything would happen at once. God is aware of the past, present and future simultaneously, but He knows the difference.

Simultaneity or timelessness is not what Einstein meant when he said, "The distinction between past, present and future is only a stubbornly persistent illusion." This was merely Einstein's attempt by letter to cheer the bereaved family of a deceased friend rather than a scientific comment.

Can God do time travel?

In the *Terminator* movie, John Connor sends Kyle Reese back in time where he meets Sarah Connor with whom he has the child who becomes John Connor who sends him back in time, et cetera – a classic causal loop paradox. If God created time, does that mean He can travel forward and back as far as He likes? Does He predict the future by travelling there and back? Can He change the past without causing the kind of temporal paradoxes that made movies like *Back to the Future* and *Bill and Ted's Excellent Adventure* so fascinating?

According to dozens of near-death-experiencers (NDE'ers) who revived with retained memories of the world of spirits, events there happened independent of time as if past, present and future were irrelevant. Some NDE'rs even say time doesn't exist outside mortality. But, since physicists can't agree on an exact definition of time, claims by nonscientists that it doesn't exist in the spirit world must be taken with a huge boulder of salt. Most NDE's include events in a particular causal sequence, the primary feature of time. Without sequential causality, the consequences of crimes could precede the committing of them – not fair and therefore not God-approved!

So, does God know the future because He exists *everywhen*?

The movie *Arrival* portrays benevolent aliens who visit Earth and share their language with a human linguist, thereby rewiring her brain to sense past, present, and future as one panorama in her mind. At the point where the world is about to attack the far superior aliens, she is able to avert a war that the world can't win by using her memory of the future to persuade Earth's leaders to stand down.

Of course, it's likely that God's ability to see the future is independent of His language (if indeed He needs one).

Merely *seeing* the future can create causal loop paradoxes for mortals. So, God is very careful about what kind of future information he gives to humans. His prophecies are usually conditional upon peoples' future choices. If they are wicked, unpleasantness will ensue. If they are righteous, blessings will follow. Thus, there is no possibility of a paradox because God's forecast will be correct no matter what people decide to do.

Conversely, if a prophecy will happen no matter what earthlings do, such as the return of Christ in glory, no causal loop paradoxes can arise because people knowing it provides them no opportunity to change it.

God's knowledge of the future creates no causal loop paradoxes because His every decision is made with full knowledge of how it will play out in the future. So, He doesn't need to change His mind based on what He sees in the future because past, present, and future were part of His decision process at the outset.

Most likely, it is God's cognizance rather than His person that is omnipresent through time, past present, and future. Most likely, He travels forward through time like humans do, and only to the immediate future. But He has many more options when He chooses His reference frames. He may even choose to have none, like a massless particle. He could sit at the center of a black hole and watch the future of the universe play out in but a few moments or He could relocate from one side of a galaxy to the other in zero time.

Even if sci-fi style time travel were possible, why would God need to visit the future when He can already see it? And why would He need to visit the past when He can perfectly recall it? And, since He makes no mistakes, what ex post facto corrections would He have to make? The biblical flood proved that He's a fix-forward God — rather than start over with a new Adam and Eve, He sent Jesus to pay for the world's sins, a gift bestowed on the penitent retroactively back through time and proactively forward through time.

"Nothing which is short of an infinite atonement which will suffice for the sins of the world." (*BOM*, Alma 34:12)

Is God unchanging?

Paul wrote, "Jesus Christ is the same yesterday, and to day, and for ever." (*NIV*, Hb 13:8) This scripture is often quoted to support the mainstream Christian doctrine that God has *never* changed. But "yesterday" ≠ "eternity past".

The Bible tells of many changes. At some point long ago, God changed from not having created anything to being the Creator of everything. And Christ, the sole deity in the eyes of mainstream Christianity, was not always the Firstborn, was not always the Only Begotten Son, and was not always mortal. He changed from baby to toddler to teen to adult. "The child grew, and waxed strong in spirit… And Jesus increased in wisdom and stature, and in favour with God and man." (Lk 2:40, 52) He went from being alive to dead and then back to being alive again. Prior to resurrecting, "Jesus was not yet glorified". (Jn 7:39) But in the future, "the Son of man shall come in his glory". (Mt 25:31)

Did the mortal experience of God the Son change Him? Yes! A prophecy of Christ's first advent stated, "He shall go forth, suffering pains and afflictions and temptations of every kind… and he will take upon him their infirmities, *that his bowels may be filled with mercy*, according to the flesh, *that he may know* according to the flesh how to succor his people according to their infirmities." (*BOM*, Alma 7:11-12)

So, *unchanging* means that God is consistent in His perfect benevolence, justness, love, mercy, and honesty toward mankind, not like the fickle false gods of biblical times whose treatment of humanity could turn malicious on a whim.

What borders on blasphemy is to suggest that the Father *cannot* advance. Jesus added to His Father's glory by completing His earthly mission. That increase in glory was certainly not the first time the Father progressed. Joseph Smith said, "God glorified Himself by saving all that His hands had made, whether beasts, fowls, fishes or men; and He will glorify Himself with them." (*TPJS*, p291)

To suggest that God made good choices that resulted in His progress is not blasphemy.

Could God resign?

When the god Baal failed to light the altars on fire for his prophets, they howled, tore their clothes, and cut themselves in an effort to get his attention. Nothing worked. So Elijah chided them thusly, "Either he is talking, or he is pursuing, or he is in a journey, or peradventure he sleepeth, and must be awaked." (1 Kg 18:27) A modern-day Elijah might have added, "Peradventure he hath resigned!"

Could God could get disgusted enough to resign?

Not to worry! The mess humans make of their lives and the planet were all expected. Things are going just as planned, as horrible as they may seem to blinkered humans. The reasons for Him to stay are many and the benefits of leaving are few. He would never abandon His investment nor quit His job. There would be no severance package. Plus, He doesn't have quitting in His nature. It's ungodly.

And, fortunately for earthlings, it's not God's disposition to love people less when they sin or make a mess of things. Nor it is His disposition to abandon anyone; "I will never leave thee, nor forsake thee." (Hb 13:5)

Why does *He* get to be God?

The prevalent belief that God has always been just as He is now (omniscient, omnipotent, and benevolent) poses a fairness problem. If God was always good, completely predisposed to be good, with no inclination whatsoever to be otherwise – since forever ago – does He really deserve any credit for being good when He had no choice in the matter?

In the same vein, it seems unfair that some people are born princes or princesses or into prosperity and privilege without any apparent effort or merit on their part while others are born with disabilities or into abject poverty in regions torn by violence and disease. Likewise, it seems unfair that some people are born benevolent but others are born psychopathic.

A true interpretation of scripture yields a completely fair scenario. Jesus wasn't always God. The intelligence from which He originated had a perfect affinity for divine light. And, having been born first, He had fulfilled responsibilities for the Father that uniquely qualified Him for the honor of serving as God and Savior of Creation. He was the foremost champion of the Father's plan to create an earth and people it. When the Father asked, "Whom shall I send?" Jesus answered, "Here am I, send me." (*PGP*, Abr 3:27) Then the Father said to Jesus, "Thou art my Son, *this day have I begotten thee.*" (Hb 1:5) "I will be to [thee] a Father and [thou shalt] be to me a Son." (Hb 1:5) Thereupon, the Father appointed Him God of creation and God of salvation!

Because Jesus consistently chose wisely, the Father "hath in these last days spoken unto us by His Son…by whom also he made the worlds… and… bringeth the firstbegotten into the world". (Hb 1:1-6) And, again because He had won the Father's trust, Jesus was "appointed heir of all things". (Hb 1:2)

Jesus emulated His Father who *showed* His Son by example *how* to be God. Jesus said, "The *Father* loveth the Son and sheweth him all things that *himself* doeth… For as the *Father* raiseth up the dead, and quickeneth them, even so the Son quickeneth whom he will." (Jn 5:19-21)

So, the Father's choice of Jesus to be the God of creation and salvation was not a random selection. Jesus had proved Himself perfect for the job! The same principle applies to the Father. He is God because He qualified for the job!

Did Jesus have a grandpa?

Carl Sagan, the famous 20th century astrophysicist who brought science to the populace, asked a series of questions: "If we say that God made the universe, then surely the next question is, 'Who created God?' If we say 'God was always here', why not say, 'The universe was always here?'" (*God and Carl Sagan: Is the Cosmos Big Enough for Them?*, US Catholic, May 1981, p20)

Most Christian religions claim that God the Father has no dad and dismiss sects that claim He does as illegitimate, citing the problem of "infinite regression" as proof it's impossible. (godandscience.org/apologetics/who_created_god.html, DL 3 Aug 2017) Oh! Infinite regression sounds so, well, regressive!

Which is more preposterous, an infinite regression or an effect without a cause? Any complete cosmology will ultimately encounter one kind of infinity or another! The idea that God Almighty always existed is an infinity, but it's not a problem! The popular "Big Bounce" theory suggests that the universe will collapse but rebound to form a new universe as one round in an *eternal* cycle of birth and death. Why should eternal cycles in the past be less logical than eternal cycles in the future?

So, if God the Son has a Father, is it illogical that he might have a Grandfather? The Father groomed a Son to become God of Earth and skies. Wouldn't it make sense that the Father was likewise groomed by *His* father? In turn, Jesus is now being a Father to all mankind, grooming them to inherit thrones of glory. All are God's children, "and if children, then heirs; heirs of God, and joint-heirs with Christ". (Rm 8:17)

Unless He was born talking full sentences, Jesus cried when He was hungry or cold and had to be changed when He soiled His diapers. We think nothing less of Him for this. Neither should we think less of Christ's Father for having similar experiences with *His* father on the path to His celestial throne.

"God himself was once as we are now, and is an exalted man... He may obtain kingdom upon kingdom, and it will exalt Him in glory. He will then take a higher exaltation, and [Christ] will take His place... So that Jesus treads in the tracks of His Father, and inherits what God did before; and God is thus glorified." (*King Follett Sermon*, Joseph Smith, 7 Apr

1844) If the Father was once human, then He must have had a father!

"Jesus said that the Father wrought precisely in the same way as His Father had done before Him… If Jesus Christ was the Son of God, and John discovered that God the Father of Jesus Christ had a Father, you may suppose that He had a Father also. Where was there ever a son without a father? And where was there ever a father without first being a son? … Paul says that which is earthly is in the likeness of that which is heavenly, Hence if Jesus had a Father, can we not believe that *He* [the Father] had a Father also?" (*TPJS*, p373)

Perhaps in the next life, we can become acquainted with our extended family tree as far back as we wish. But for now, we focus on our Father and His Son. "For though there be that are called gods… To us there is but one God, the Father, of whom are all things, and we in him; and one Lord Jesus Christ, by whom are all things, and we by him." (1 Cor 8: 5-6)

Does God create gods?

God has always wanted his children to qualify for an inheritance that includes deification, aka theosis. Paul said children of God could inherit thrones with Christ and be "glorified with Him". (*ASV(American Standard Version)*, Rm 8:17)

Most Christians believe that they should emulate Jesus, thus the WWJD (What Would Jesus Do?) bumper stickers, and that He can save them to heaven. But few people feel that God is able to exalt them to godly glory. The whole reason for creation is that God wants His children to become like Him, not just in life, but also in eternity.

The 1982 movie *The God Makers* is one of many examples of Christians mocking the deification doctrine, yet this is mankind's objective, to qualify for the gift of glorification and receive a throne, and it's Biblical. Paul said, "Henceforth there is laid up for me a crown of righteousness, which the Lord, the righteous judge, shall give me at that day: and not to me only, but unto all them also that love his appearing." (2 Tim 4:8) And Jesus said, "To him that overcometh will I grant to sit with me in my throne, even as I also overcame, and am set down with my Father in his throne." (Rv 3:21) Yes, sitting on God's throne signifies godhood for true saints as well as for Jesus!

At least the Catechism of the Catholic Church still takes deification seriously; "The Son of God became man so that we might become God." (*CCC (Catechism of the Catholic Church)*, 460)

Children following in their parents' footsteps is a pattern for gods as well as for mortals. The doctrine of deification, that God wants his children to "grow up" and become like Him, abounded in Christ's original church as evidenced by statements from early Christian church fathers:

♦ St. Clement (88-99AD), the 3rd bishop of Rome – said, "They who with confidence endured… are now heirs of glory and honour, and have been exalted and made illustrious by God." (*First Epistle of Clement to the Corinthians*, Clement)

♦ St. Justin Martyr (~100-165AD) – "All men are deemed worthy of becoming gods." (*Dialogue with Trypho*, ch124) Martyr quoted several supporting scriptures: "God has taken his place in the divine council; in the midst of the gods he holds judgment... I said, 'You are gods, sons of the Most High, all of you.'" (*ESV (English Standard Version)*, Ps 82) Also, "The Lord said unto Moses, See, I have made thee a god to Pharaoh." (Ex 7:1)

♦ Bishop Irenaeus (~130-200AD) – "We have not been made gods from the beginning, but at first merely men, then at length gods. (*Against Heresies*, bk4, ch38, sec4) How, then, shall he be a God, who has not as yet been made a man?" (ibid. ch39, sec2) Also, "Our Lord Jesus Christ, the Word of God, of His boundless love, became what we are that He might make us what He Himself is." (*The Early Christian Fathers: A Selection from the Writings of the Fathers from St. Clement of Rome to St. Athanasius*, Henry Bettenson, 1956, p106)

♦ Clement of Alexandria (~150-215AD) – "The Word of God became man, that you may learn from man how man may become God." (*Exhortation to the Heathen*, ch1) "But that man with whom the Word dwells… becomes God, since God so wills. Heraclitus [the ancient Greek poet], then, rightly said, Men are gods, and gods are men." (*The Instructor*, bk3, ch1) "Reward and the honours are assigned to those who have become perfect; when they have got done with purification… there awaits them restoration to everlasting contemplation; and they are called by the appellation of gods, being destined to sit on thrones with the other gods." (*Stromata* or *Miscellanies,* bk7, ch10)

♦ Tertullian (~160-230AD) – "We shall be even gods, if we shall deserve to be among those of whom He declared, I have said, You are gods [Psalms 82:6], and, God stands in the congregation of the gods [verse 1]. But this comes of His own grace, not from any property in us, because it is He alone who can make gods." (*Against Hermogenes*, ch5)

♦ Hippolytus (~170-236AD) – "And you shall possess an immortal body… And you shall be a companion of the Deity, and a co-heir with Christ, no longer enslaved by lusts or passions, and never again wasted by disease. For you have become God… You have been deified, and… You shall resemble Him, inasmuch as you shall have honour conferred upon you by Him. For the Deity, (by [this] condescension,) does not diminish anything of the divinity of His divine perfection; having made you even God unto His glory!" (*Refutation of All Heresies*, bk10, ch30)

♦ Origen (~185-255AD) – "The God of gods, the Lord, has spoken… It was by the offices of the first-born that they became gods, for He drew from God in generous measure that they should be made gods… The true God, then, is The God, and those who are formed after Him are gods, images, as it were, of Him the prototype." (*Commentary on the Gospel of John*, bk2, ch2, sec2)

♦ Athanasius (~293-373AD) – "For He [Christ] was made man that we might be made God." (*On the Incarnation of the Word*, ch54, par3)

♦ Jerome (~347-420AD) – "[God] made man for that purpose, that from men they may become gods... [Christ] said that all of you would be exalted as I am exalted." (*The Homilies of Saint Jerome*, pp106–107)

♦ Augustine of Hippo (354-430AD) – "But He that justifies does Himself deify, in that by justifying He does make sons of God. 'For He has given them power to become the sons of God' [John 1:12]. If we have been made sons of God, we have also been made gods." (*Expositions on the Psalms*, Ps 50, sec2)

♦ Bishop Cyril of Alexandria (412-444AD) – "To them who had God indwelling in them, it suffices that they might therefore be truly gods and adored by all, all are gods and to be adored." (*Scholia on the incarnation of the Only-Begotten*, sec18, translated by Pusey)

♦ Bishop Theodore Askidas of Caesarea (ca 540AD) suggested that those who become gods will join in creating other worlds. (*The Hope of the Early Church,* Brian E. Daley, 1991, pp189-90,260, nt65)

And, from Judaism:

♦ Early Midrashic text (ca. 380AD) – "The Holy One... will in the future reveal to all the pious in the World to Come the Ineffable Name with which new heavens and a new earth can be created, so that *all of them should be able to create new worlds.*" (184 Midrash Alpha Beta diR. Akiba, BhM 3:32, quoted in Ralph Patai, *The Messiah Texts* (Wayne State University Press, 1979), p251)

And, from medieval apocrypha:

♦ Armenian Apocrypha – The serpent spoke to Eve and said, "Why do you taste of all the trees, but from this one tree which is beautiful in appearance you do not taste?" Eve said, "Because God said, 'When you eat of that tree, you shall die.'" But the serpent said, "God has deceived you, for formerly God was man like you. When he ate of that fruit, he attained this great glory.'" (*The Armenian Apocryphal Adam Literature*, W. Lowndes Lipscomb, pp120-121, 162-164. Two manuscripts repeat this theosis theme.)

And from modern clerics:

♦ Priest Philip Khairallah of Orthodox Church of Antioch (1990AD), "The one and only aim of human life on earth is union with God and deification." (*The Sanctification of Life*, 1990, Philip A. Khairallah, pp326,395,396-397)

♦ Eastern Orthodox theology group (1987AD), "The cosmos provides the stage upon which humankind moves from creation to deification." (as quoted on fairmormon.org/answers/Primary_sources/Theosis, DL 9 Jul 2020)

♦ Prophet Joseph Smith (1843AD), "God himself, the Father of us all, dwelt on an earth, the same as Jesus Christ." (*TPJS*, p345)

Historically, the biblicality of theosis was recognized through Christianity. Peter promised that the faithful would become "partakers of the Divine nature". (2 Pt 1:4) Jesus had our divine potential in mind when he said, "Be ye therefore perfect, even as your Father which is in heaven is perfect." (Mt 5:48) It is for modern critics to explain why it is now missing from most Bible-based religions!

Does God have a wife?

Who decided that God can't have a wife? Dan Brown's book, *The Da Vinci Code*, sensationalized a chain of events that had been pieced together by the authors of *Holy Blood, Holy Grail*, which theorized that Jesus wedded Mary Magdalene and that their descendants emigrated to what is now southern France where they intermarried with noble families who eventually became the Merovingian dynasty. Fundamental to both books was the idea that the legendary Holy Grail was Mary Magdalene who was the "chalice" from which came God's sacred royal bloodline.

If it is true that Jesus was married, then the God of mainstream Christianity has a wife, unless they got divorced or their marriage was merely until death parted them. If Jesus was baptized solely for the sake of obedience, it should not surprise anyone if He also obeyed the command to "be fruitful and multiply". (Gn 1:28)

All young Jews of Jesus' day were expected to marry, especially those who, like Jesus, were preparing to become full-fledged rabbis. Rabbis typically completed their training and qualifications around thirty years of age. (*Chapters of the Fathers*, 1967, ch5, sec21; "Pirkei Avot" in Hebrew, a compilation of Rabbinical Jewish ethical teachings and maxims) Not coincidentally, thirty is the age when Jesus started His ministry. (Lk 3:23)

The Mishnah, Judaism's first major canonical document following the OT, also gives thirty as the age at which one may speak authoritatively. That Jesus was indeed a genuine rabbi is obvious from the fact that the Pharisees, who hated Him, called him rabbi. (Lk 19:39) Also, Sadducees, lawyers, and rich people addressed Him as rabbi. And, He was allowed to preach the Torah in the synagogues, an honor not granted to John the Baptist, only to bona fide rabbis. And, as a rabbi, Jesus was expected to be married.

Christ began His ministry with a wedding, the one at Cana where He turned water into wine. That wedding is thought by some to be His own because of the apparent responsibilities His mother bore for ensuring the guests had all they could eat and drink. When she ran out of wine, she turned to Jesus for help. It would be odd for Christ to have squandered His first miracle on something as trivial as more wine unless the wedding was His own. "Apostle Orson Pratt wrote, There was a marriage in Cana of Galilee; and . . . it will be discovered that no less a person than Jesus Christ was married on that occasion." (*The Seer*, 1853, p172)

A gnostic Christian text said, "Three women always walked with the master: Mary his mother, [his] sister, and Mary of Magdala, who is called his companion... The companion of the [savior] is Mary of Magdala. The [savior loved] her more than [all] the disciples, [and he] kissed her often on her [mouth]." (*Gospel of Philip*; translator guesses for unreadable manuscript text are in brackets)

It has also been argued that only a wife would have tried to anoint the male body of Jesus in the tomb as Mary Magdalene did. And, the fact that the resurrected Jesus appeared first to Mary Magdalene, even before He reported to His Father and also before He visited His leading apostles or His mother, gives rise to suspicions that a very special relationship existed between the two of them that may have been nuptial in nature.

There are those who argue that Jesus, when He looked down from the cross and asked John to care for His mother Mary, would have also asked someone to watch over His wife, if He had been married. However, any wives He might have had could have been well enough off at the time, such as Martha and her sister Mary who had sufficient means to bury their brother Lazarus in a cave-style tomb.

George Q. Cannon, counselor in the first presidency, said the following in an 1899 leadership meeting, "There are those in this audience who are descendants of the old 12 Apostles and, shall I say it, yes, descendants of the Savior himself." (*A Ministry of Meetings: The Apostolic Diaries of Rudger Clawson*, edited by Stan Larson, Signature Books, 1993, p72)

It should not be shocking that Jesus might practice what He preached, "For this cause shall a man leave father and mother, and shall cleave to his wife." (Gn 2:24) Paul wrote, "Neither is the man without the woman, neither the woman without the man, in the Lord." (1 Cor 11:11)

In particular, Catholics should be receptive to the idea that God has a wife since their priests have been marrying nuns to Him for centuries saying, "Receive this ring, for you are betrothed to the eternal King." (toledosnd.org/faqs, DL 27 Jul 2018) After the ceremony, they wear His wedding ring to their graves.

The belief that God has a wife is found in ancient Judaism. Manasseh "erected the carved image of Asherah that he had made inside the Temple". (*ISV (International Standard Version)*, 2 Kg 21:7) Asherah was the goddess of fertility in Palestine. Although Manasseh himself was not sincere in his worship of Asherah, having simply succumbed to the allure of her hypersexuality and the eroticism of the phallic symbols that marked her places of worship, many of the Jews in his day sincerely revered her. Tangible evidence of widespread adoration of Asherah have been found in excavations of ancient Israeli homes. And, a couple of inscriptions found on ancient artifacts mention "Yahweh and his Asherah". (nbcnews.com/id/42147912/ns/technology_and_science-science/t/gods-wife-edited-out-bible----almost, DL 18 Jul 2020)

And, it may be that the "us" and "our" includes the Heavenly Mother when Elohim says, "Let us make man in our image, male *and female*." (Gn 1:26) This would be consistent with the Bible's use of "man" to mean "man and woman" in almost every case.

Starting around 1839, the Prophet Joseph Smith began teaching the concept of an eternal mother, as reported in several accounts from that period. From his teachings, Eliza R. Snow wrote a beautiful hymn that Latter-day Saints still learn, sing, quote, and cherish, *O My Father*. The 4th prophet of the restored church said, "That hymn is a revelation." (*The Discourses of Wilford Woodruff*, p62)

> "In the heav'ns are parents single?
> "No, the thought makes reason stare!
> "Truth is reason; truth eternal
> "Tells me I've a mother there."

Modern apostle Erastus Snow compared the Heavenly Couple with a pair of scissors to demonstrate that both components must be combined in order to function, and said, "In other words, there can be no God except he is composed of the man and woman united... There never was a God, and there never will be in all eternities, except they are made of these two component parts; a man and a woman; the male and the female." (*JD*, bk19, p270) Women should welcome the idea that, even though every member of the godhead is a male, there are female gods behind the scenes that are equally important in the great plan of happiness for humanity.

The Heavenly Couple makes a sweet celestial family pattern to which mortals can aspire and from which human spirits inherit sparks of divinity. Would the Father command each man and woman to become "one" if He is not "one" with a divine counterpart? (Jn 10:8) Would He be without a wife while having Paul tell the world, "Neither is the man without the woman, neither the woman without the man, in the Lord"? (1 Cor 11:11) Not if He is a good example-setting Father!

So, why doesn't God tell us more about His wife? Perhaps the answer is in the blasphemy that spews daily from the mouths of millions. The name of the mother of Jesus gets dragged through the mud by feeble minds trying to express themselves forcefully, so it's a good bet that the name of God's wife would likewise be blasphemed if it were known. God's refusal to name her would be consistent with his teaching to "give not that which is holy unto the dogs, neither cast ye your pearls before swine, lest they trample them under their feet, and turn again and rend you". (Mt 7:6)

How does God communicate?

It should not surprise anyone familiar with the physics of particle entanglement that God could be in contact with all things simultaneously. Entangled particles are in a relationship with each other such that the quantum state of one cannot be described independently of the other(s) no matter how far apart they are. They be a million light-years apart and one particle will react instantaneously to another's change of state. The 2022 Nobel Prize was awarded for experimental proof of this fact. This is but one example of connectedness that suggests all things throughout the universe are in a relationship with each other.

So, it is not surprising that all things are connected with God in a mutual awareness relationship. "In the presence of God... all things... are manifest... and are continually before the Lord. The place where God resides is a great Urim and Thummim... whereby all things pertaining to... kingdoms of a lower order, [are] manifest." (DC 130:6-9) His planet is the nerve center of a network that is in communication with all the particles of His creations!

Urim and Thummim are biblical Hebrew terms that translate literally as *light(s)* and *perfection(s)*. These were stones kept in a pouch on the breastplate of decision worn over the heart by high priests of Israel. (Ex 28:30) The stones could give or facilitate "an oracular response" to pleas for divine guidance. (*Analytical Concordance to the Bible*, Young) Sometimes only one of the stones was used, as in the case "when Saul inquired of the LORD, the LORD answered him not, neither by dreams, nor by Urim". (1 Sm 28:6)

The Urim and Thummim are just two examples of tangible objects that help bolster the faith needed to open a channel to God. Another example was when God gave Moses the staff that helped Him do miracles, "Take this rod in thine hand, wherewith thou shalt do signs." (Ex 4:17) Though not enchanted in any way, such objects serve the same purpose as training wheels on bicycles for one learning ride. Effectively communicating with God or borrowing His power takes practice. Experienced prophets can be conduits for God's miracles without the help of Urim, Thummim, seer stones, staffs or brass serpents on sticks.

Individuals who inherit a place in the Lord's celestial kingdom will use stones "whereby things pertaining to a higher order of kingdoms will be made known". (DC 130:10) God's promise through John to the churches was, "To him that overcometh will I give... a *white stone*, and in the stone a new name written, which no man knoweth saving he that receiveth it." (Rv 2:17) This white stone will be like a personal COM unit to communicate with parties higher in the hierarchy of worlds. Perhaps God, being highly experienced, no longer uses such a stone. Mortals are currently not using such relics either. Communications from God are usually in response to prayer and answers are transmitted under the auspices of the Holy Spirit and/or from holy angels via the medium of the Light of Christ. (DC 88:6-13)

What's God's native tongue?

People who return from a near-death experience (NDE) say that verbal communication is not needed in the realm of spirits. Pure wordless thoughts can pass from one mind to another. So, God doesn't need a "tongue" per se when He communicates telepathically. But, did His telepathic commands to chaos resemble any kind of language when He created the universe?

Nobelist Richard Feynman said, "To appreciate nature, it is necessary to understand the language she speaks... mathematics." Scientists say that all the workings of the universe could be described by math, numbers, and rules. Indeed, thousands of equations have been discovered that describe aspects of the physical world:

♦ Force = (mass)*(acceleration)

♦ Power = (Watts)/(Time)

♦ (Pressure)*(volume) = constant k.

♦ Voltage = (current)*(resistance)

♦ Etc.

If nature's secrets are all discoverable as equations, then God is definitely a mathematician. MIT physicist Max Tegmark believes the universe is literally made of math. (livescience.com/42839-the-universe-is-math.html, DL 11 Jul 2020) He's wrong – math is simply the best tool we have to model physical laws. And, some of the equations used by scientists to model nature are not very good analogs for the phenomenon they model, even though they may be good at predicting how it will behave.

For example, quantum mechanics uses sum-over-histories equations that produce accurate quantum amplitudes but only after embarrassing infinities are eliminated using a mathematical trick of dubious legitimacy called normalization. In reality, particle paths do not traverse every quantum-mechanically possible trajectory and the infinities are proof that the math doesn't perfectly model reality.

However, the many centuries of successfully using math to predict natural outcomes is strong evidence that the Creator's marching orders for the universe include it. If scientists ever discovered the entire language, they would have their theory of everything, the holy grail of physics. But math is just merely a useful abstract modelling tool.

God can speak anyone's tongue, plus math, music, love, etc. But his primary mode of communication is language-free telepathy and, when He dispenses revelation, He leaves it to the recipient to convert inspired thoughts into the local tongue. So, although the creation may be describable by math, Mandaran, or Moroccan Sign Language, God's thoughts are not bound by word definitions and rules of grammar.

What does God do for fun?

Children are fun and God has lots of them. Keeping in touch with His children on many worlds after they've inherited kingdoms of glory must certainly be an enjoyable pastime for Him. Likewise, His angels were created as "hosts of heaven" to "reside in the presence of God". (DC 130:6-8) He gives them purpose in His plans and provides them with the power to fulfill their roles. He delights in their successes.

God also enjoys the company of animals of various kinds. This is evident from John the Revelator's words as he was "describing heaven, the paradise of God, the happiness of man, and of beasts, and of creeping things, and of the fowls of the air". (Jn 4:6-9; DC 77:2)

God loves music. In his grand vision, John heard "the voice of harpers harping with their harps" and "the voice of many angels round about the throne…having every one of them harps". (Rv 11:5-11; 14:2) So, although the instruments for music in John's Revelation might be merely symbolic, the music is real. The majority of near-death survivors who experienced the "other side" have described musical vibrations as pervading all things.

God also enjoys the beauty of trees and other plants. He adorned Eden with lovely forms of life, patterned after those featured in His heavenly paradise. The prime example, the Tree of Life, will be planted on Earth when it is glorified in the future as described in John's Revelation. (Rv 22:2) All plants go to heaven, even genetically modified organism Genetically Modified (GMO) varieties.

God loves to create. This earth is not His first nor His last. He said, "Worlds without number have I created… There are many worlds that have passed away… as one earth shall pass away, and the heavens thereof even so shall another come; and there is no end to my works." (*PGP*, Mo 1:33-38) God's first love is helping His creations achieve greater felicity, especially those He created in His own image. (*PGP*, Mo 1:39)

Does God have a sense of humor?

Jesus was likely expressing wry humor when He called Herod an "old fox". (Lk 13:32) Aspects of His humor are evident from the irony and paradox in His exchanges with people and in His parables. Lincoln believed God must have had a sense of humor to include people like us in his creation.

Betty Eadie returned from surgery-related temporary death and recalled getting answers from Jesus as fast as she could think of them; "Things were coming back to me from long before my life on Earth, things that had been purposely blocked from me by a 'veil of forgetfulness' at my birth. As more questions bubbled out of me, *I became aware of his sense of humor*. Almost laughing, he suggested that I slow down, that I could know all I desired." (near-death.com/experiences/exceptional/betty-eadie.html, DL 18 Jul 2020)

Of course, God doesn't laugh *at* people in a condescending way. Rather He delights in their feeble thoughts or attempts to do something just as parents delight in funny noises a babbling baby might make.

5: God's Home

Where is Heaven?

Babel's builders thought heaven was low enough for a tower to get them there. (Gn 11:4) At only about 300 feet, the tallest ziggurat in that area hardly seems respectable now that cruising at 30,000 feet is commonplace. But, to the peoples who lived on the vast floodplains of the Tigris-Euphrates River, their tower seemed quite tall.

If the tower had been completed, the person to lay the last brick might have looked up in disappointment, perhaps exclaiming, as did the Soviet astronaut Yuri Gagarin after returning from space in 1961, "I looked and looked but I didn't see God." Yuri was a good Christian but had to follow the script given him by the Soviet leadership for the media.

Billions of people on all sides of the earth imagine their prayers going upward, no matter where they live or what Earth's orientation is. If God's home is in all directions, the geometry of heaven must be extremely exotic. Some Bible pundits suggest that heaven is in the northern sky because God once told Moses to kill animals on the north side of the altar and the OT says salvation doesn't come from any other direction. (Lv 1:11; Ps 75:6) But that's farfetched eisegesis!

Muslims pray facing the Ka'bah, a cubic building in the center of Islam's most important mosque, located in Mecca. In part, because of difficulties determining which direction to pray properly, Islamic leaders issued a fatwa forbidding Muslims from travelling on a 1-way trip to Mars. The smartphone app *Qibla Compass* tells more than 5 million earthbound Muslims which direction to pray. As for Christians, they are free to pray in any direction, even on Mars, because:

♦ Prayers are transmitted instantaneously in all directions so God can pick up the signal no matter which direction a believer faces.

♦ Prayers are transmitted to God through hyperspace.

♦ Prayer messages are routed to Him like mobile phone messages from wherever you are.

♦ God senses our prayers from a realm that intersects the dimensions of our universe.

♦ God's home is a wraparound realm outside our universe.

♦ God's cognizance is everywhere.

Paul writes of a man who was caught up into the third heaven. (2 Cor 12:2) Some pundits say that Paul was counting from the bottom so that the first heaven is earth's atmosphere, the second heaven is outer space, and the third heaven is the beyond the physical realm, the "heaven of heavens" seen only by spiritual eyes. (1 Kg 8:27; Acts 7:55) "Thus saith the high and lofty One that inhabiteth eternity, whose name is Holy; I dwell in the high and holy place." (Is 57:15) It indeed seems like God's holy dwelling place is above and independent of the physical universe He created.

In a grand vision from God, "Abraham… saw the stars… and that one of them was nearest unto the throne of God… the greatest of all the [stars], because it is nearest unto [God]." (*PGP*, Abr 3:2-3, 16) So, if we just knew where that star is, we could know the approximate direction of God's home. But it would still not be necessary to locate that star before praying.

What is God's home like?

Scripture doesn't provide much detail specific to the Father's home. However, since Christ's celestial kingdom will be patterned after His Father's, descriptions of the heaven that the righteous will inherit must suffice.

Joseph Smith saw "the transcendent beauty of the gate through which the heirs of that kingdom will enter, which was like unto circling flames of fire". (DC 137:2) This glory gate is the portal to Planet Heaven through which the unworthy cannot pass.

Jesus said the meek "shall inherit the earth". (Mt 5:5) So, Earth will be Heaven for "joint-heirs with Christ". (Rm 8:17) "This earth, in its sanctified and immortal state, will be made like unto crystal" so that its inhabitants can see all things pertaining to their kingdoms by looking into a planet-sized crystal ball, much like ancient prophets looked into the Urim and Thummim for light and knowledge. (DC 130:6-9)

John said Planet Heaven's surface will be like a "sea of glass mingled with fire". (Rv 15:2) This is the future "earth in its sanctified, immortal, and eternal state". (DC 77:1) Christ and those who conquer the beast will dwell on Planet Heaven, or as John put it, "stand on the sea of glass". (Rv 15:2) "Before the throne there was a sea of glass like unto crystal... glass mingled with fire." (Rv 4:6) The entire planet will be a "globe like a sea of glass and fire... like unto crystal". (DC 130:9)

Heaven's holy capital city will also be "pure gold, like unto clear glass". (Rv 21:18) John was not seeing the precious metal once used as money, symbol AU, atomic number 79. Nor was he describing molten and cooled silicon dioxide. Nor was he referring to a material with molecules aligned in a microscopic lattice. Gold, glass, and crystal are obviously all adjectives to describe the same thing, a glorified world featuring a glorified environment with glorified structures, all of which are translucent to spiritual eyes.

Planet Heaven's capital city will be "gold, like unto clear glass" with walls of "jasper". (Rv 21:18) The city will be square and have twelve gates, three on each side, all of pearl. (Rv 21:13, 21)

Smith saw the "beautiful streets of that kingdom... had the *appearance* of being paved with gold." (DC 137:4) John saw the heavenly streets as "transparent glass," again conveying glorified beauty. (Rv 21:21)

Jesus said that heaven contains "many mansions." (Jn 14:2) Brigham Young spoke of "palaces, walks, and gardens". (JD, bk4, 1857, p268)

In a vision of the celestial kingdom, Joseph Smith saw "the blazing throne, whereon was seated the Father". (DC 137:3) Being a tireless mortal, God doesn't need to rest his feet and legs by sitting. That's not to say there is no actual throne – only that the throne symbolizes kingship.

In his vision of heaven, John saw a fiery altar at the foot of the throne. Since heaven transcends carbon compounds, the burning of incense there does not involve physical combustion ($CH + O_2 \rightarrow CO_2 + H_2O$). Rather, the smoke is a metaphor for the prayers of saints killed for their testimony of Christ. (Rv 6:9-10; 8:3-4)

John saw four horns on the altar through which God directs angels to answer prayers and run other errands. (Rv 9:13-14) These horns are probably also figurative imagery since God can command angels without the assistance of loudspeakers.

So, the structures and features of heaven don't just float around in some kind of celestial space! They exist on the surface of God's home world, whether that's Christ's future celestial planet or the Father's current one. "I was very aware that heaven is a planet," said Erica McKenzie whose spirit went up to meet God while medical professionals attempted to resuscitate her.

Televangelist Kenneth Copeland reasoned that earth was created in the image of planet heaven, "It's a copy of the Mother Planet. Where God lives, He made a little one just like His and put us on it." (*Following the Faith of Abraham I*, 1989 audiotape, #01-3001, side 1)

Does heaven have time?

Two thousand years ago Jesus said, "I come *quickly*; and my reward is with me, to give every man according as his work shall be." (Rv 22:12) It's been 2,000 years since then. So, what seems like a very long time to humans is but a moment in God's perception. Time perception is strange that way. The older humans get, the more quickly time seems to pass for them. Perhaps that's why God, who has been around forever, has a very short sense of time.

After Christ's return in glory, "Satan is bound and time is no longer". (DC 84:100) Also, "There shall be time no longer; and Satan shall be bound... for the space of a *thousand years*." (DC 88:110) It's an oxymoron to say time will cease for a thousand years. So, it must mean that there is no more time to prepare for His coming. Or, as most modern NT translations put it, "There will be no more delay." (Rv 10:6 — NIV, NLT, ESV, BSB, BLB, NASB, etc.)

When Satan is loosed at the end of the millennium of peace, there will be a "little season" of wickedness. (DC 88:111) So, time will continue at least until the earth passes away after the little season.

Will time continue on Earth after God upgrades it to Planet Heaven? John's vision of Heaven's capitol says, "The city had no need of the sun, neither of the moon, to shine in it: for the glory of God did lighten it, and the Lamb is the light thereof. And the nations... shall walk in the light of it." (Rv 21:22) Without the sun and moon, Planet Heaven will have no seasons. How will the inhabitants of heaven schedule important events?

Abraham wrote that the governing star around which other stars move, "was after the manner of the Lord, according to its times and seasons in the revolutions thereof" and "is after the reckoning of the Lord's time". (*PGP*, Abr 3:4, 9) So, if Christ's new celestial world is like His Father's, Planet Heaven will rotate at some speed. Joseph Smith explained, "The reckoning of God's time... [is] according to the planet on which [He] resides." (DC 130:4-5)

6: Intelligence

With what did God start?

"For not without means was your almighty hand, that had fashioned the universe <u>from formless matter</u>..." (Wis 11:17) This statement from *The Wisdom of Solomon* is one of the reasons Protestants left the books of the Apocrypha out of their Bibles. They saw the "formless matter" bit as contradicting the fundamental idea of creation that God started with absolutely nothing.

If God created everything from nothing, what kind of nothing was it? If it was the absence of everything, as most theologians claim, then God, who is 100% good, was the sole existing thing, and all of His creations should have been 100% good. So, how do theologians explain why Lucifer, whom God created, turned 100% evil? Why would God create him knowing beforehand that he would rebel? A logic problem like this suggests some assumptions are wrong.

Let's take another tack and assume God created everything from pre-existing primordial raw material, pure supernatural essence, fundamental intelligence, some of which already had the seeds of imperfect choice-making in it, and that He created all His angels as good as possible from this semi-sentient urstoff. Then, being a champion of free will, He allowed this urstoff, now organized as Lucifer and other angels, individual agency to make their own choices. Voilà – God is off the hook for all the evil choices made by devils.

Can we identify the elemental unit, the minimum possible building block, having no constituent parts, from which God created everything? This urstoff would have been around forever. And, it would have had one very important attribute – sufficient sentience to make very basic decisions. Why is this important? Aside from explaining how imperfections crept into a creation by our perfect God, it also aligns with the principle of agency. God would never create someone against their will, whether it be angels or some other kind of spirit creature or creation. Only stuff sufficiently sentient to decide to join up with God should rightly be part of His creative work. The fundamental unit of all reality must be able to "act for itself… otherwise there is no existence". (DC 93:30)

Since each primeval bit is an entity that acts for itself, the fundamental constituent of all creation is called an *intelligence.* (merriam-webster.com/dictionary/intelligence, DL 17 May 2020, definition 5b) After experiencing temporary death and receiving a tour of the cosmos, guided by an angel, Beverly Brodsky said, "The universe is all one grand object woven from the same fabric." (near-death.com/religion/judaism/beverly-brodsky.html, DL 18 Jul 2020) And there is an infinite supply of this "fabric" in its unorganized primordial form.

God had omnificence and "God had materials to organize the world out of chaos—chaotic matter, which is element, and *in which dwells all the glory*. Element had an existence from the time he had. The *pure principles of element* are principles which *can never be destroyed*; they may be organized and re-organized, but not destroyed." (*HC*, Joseph Smith, bk6, pp308-309) The idea that "all the glory" dwells in the pure and uncreated fundamental stuff from which all else is "organized and reorganized" hints at its paranormal nature. *Element* as used by Joseph Smith is not a reference to the periodic chart, which had not yet been created!

The law of conservation of intelligence is that it cannot be created nor destroyed. It's eternal. Joseph Smith said, "That which has a beginning will surely have an end; take a ring, it is without beginning or end—cut it for a beginning place and at the same time you have an ending place." (*TPJS*, p181) He also taught that spirits were created by God but "the intelligence of spirits had no beginning, neither will it have an end". (*HC (History of the Church)*, bk6, p14)

Combining this law with the laws of conservation of energy, of mass, and of information gives the law of *conservation of anything*; "Stuff, whether it be mass, energy, spirit, intelligence, or information, can neither be created nor be destroyed but can be restructured to manifest in different forms."

At the very beginning of creation, God observed that certain of the primeval wisps of intelligence acted with affinity for Him. So He, out of charity and compassion, determined to pull them from chaos, where there was no order nor rule of law nor protection from evil, and put them on a path of advancement. He drew them forth from their primordial haze and put them in a training program where they, level by level and step by step, could advance towards immunity from the powers of evil at work in the chaos.

How can particles be so smart?

Some physicists say it's actually the quantum fields that are smart rather than the particles. They posit that fundamental quantum fields are pervasive throughout the universe and therefore "know" the conditions around all possible paths so all a particle in the field has to do is behave according to what the field "knows". But, what is a quantum field? What is a particle?

The most well-known example is the omnipresent electro-magnetic field in which photons travel. For example, in forensic science, when luminol is sprayed on surfaces where blood is suspected, the iron in hemoglobin acts as a catalyst for a chemiluminescent reaction and the luminol emits a blue glow of photons. This event creates energy blips in the electro-magnetic field that radiate outward as waves. When the wave hits a retinal chromophore in the eye of an investigator, it collapses into a particle, triggering a chain of events that culminate in a signal to his or her visual cortex that is interpreted as blue and blood.

A computer requires massive memory and millions of lines of code to simulate the complexities of particle behaviors and still cannot predict the outcome of a future random event in a particle experiment. Waves can be described best by probability mathematics. In other words, scientists can calculate with great accuracy the probability of a particle being found at any given location in space but cannot know for certain until it is detected.

Another indication of quantum intelligence is the perfect record of nature's obedience to the rules of physics, defying the common saying, "Rules are made to be broken." From mankind's earliest experience, a dropped stone always falls down rather than up, hands rubbed together always sense increased warmth, and launched arrows always head in the direction of the force applied by the bow.

There are 17 fundamental particles known to science, hundreds of composite particles, and an infinite number of molecular permutations. Protons and neutrons, for example, are composite particles made up of fundamental quarks. When accompanied by electrons, they become atoms. Atoms combine into molecules, all of which obey the rules of quantum mechanics without fail. And never has any particle been known to misbehave!

Even neutrinos, which are so small that almost all of them pass right through the entire earth without any hesitation, follow the rules assigned to them. How do such tiny inanimate objects carry any knowledge whatsoever, much less all the applicable laws of physics? Or is that information embedded throughout the universal neutrino field? What compels total obedience to those laws?

The laws of nature are precisely obeyed no matter how many different kinds of particles are bound together. Galileo said, "Nature never violates the terms of the laws imposed upon her." (lingq.com/lesson/part-2-20105, DL 18 May 2020) But how does a stone know to head straight down and in no other direction? Even when a stone is crushed, each tiny fragment, though no smarter than the original rock, also knows to fall directly to earth. Without this universal intelligence permeating the universe, or without total obedience to all natural laws, chaos would reign.

What waves to make particles?

If particles like photons and electrons are waves, what is waving? Sound waves travel through air and other tangible media. Ocean waves travel through water. But what medium carries radio or light waves?

In answer to this question, theories of a medium called aether were concocted, which was thought to fill the immensity of the universe. But such theories were eclipsed by Einstein's theories of relativity. Now, it is believed that particles of light are actually waves that propagate through quantum fields as do all other fundamental particles. That seems like a type of aether, but it's not proper to call it that.

The "field view of reality" says that the universe is entirely made of fields. Take electrons as an example. They are wavelike concentrations of energy in the electron field, a field that permeates the entire universe. Quantum entanglement experiments show that every perturbation (wavicle) in a field influences every other point in that field instantaneously. Fields are so pregnant with intelligence that they seem able to calculate future events within them!

How intelligent is light?

In some ways photons are smarter than humans. For the human example, let's say a lifeguard sees a child floundering in the water 300 feet directly east from a perfectly north-south shore and 400 feet directly north of the lifeguard along the shore. The shortest path is 500 feet across the water as the crow flies. But, because running is faster than swimming, the quickest route to the child is some combination of running along the beach and swimming toward the child. The typical lifeguard has a gut feeling when to stop running and start swimming to maximize the chances of saving the child. There's no time to do the calculus of Fermat's Principle - the distances and speeds would all be rough estimates anyway.

Now, imagine the same distances and arrangement as before, but with a high glass wall holding the lake off the beach. Instead of a child, an underwater monster lurks at the lifeguard's eye level 300 feet east and 400 feet north. The monster is 500 feet away from the photon's starting point but light travels faster in the air than in the water. So, the photons that hit the monster are the ones that follow the fastest route, not a straight line to the monster, rather a straight path that aims beyond the monster before hitting the water and refracting to a straight line to the monster. But the math that photon uses is precisely perfect every time!

When Albert Einstein was sick in bed as a 5-year-old, his father gave him a compass. The boy was mesmerized by the way the needle always "knew" which direction was north, no matter how he turned the device, an unseen force at work.

Yet adult Einstein balked at what he called quantum field theory's "spooky actions at a distance." Send a photon through a beam splitter and two photons pop out, each with half the energy of the original in a superposition of states such that the act of measuring one particle instantaneously affects the state of its entangled partner regardless of separation distance. The results of the two measurements are always such that the total angular momentum is preserved and thus the total angular momentum of the universe. The two photons could be a million miles apart and yet one twin seems to "know instantly" that the other twin has been measured and, when subsequently measured, yields a state measurement result that preserves the total angular momentum of the pair.

There are many signs of intelligence in nature that scientists cannot explain. But there are a couple of conclusions on which most theorists agree: 1) matter and energy are not the sole and fundamental components of reality, and 2) causality is not a local phenomenon. As Einstein said, "The divine reveals itself in the physical world." (*Einstein and Religion: Physics and Theology*, Max Jammer, 1999, p151)

How intelligent is the universe?

General relativity says there are no absolute positions or locations in space. If two astronauts awoke to find themselves floating past each other in some deep-space void, they would not know which one of them is moving or whether both were.

Einsteinian relativity says it doesn't matter. The laws of physics behave the same from both astronauts' perspectives.

But if one of the astronauts were spinning, it would be easy to know which.

Isaac Newton imagined two rocks connected by a rope in deep space, far from any gravitational object. Correctly assuming that the rocks would pull the rope taut when spun around each other, he incorrectly assumed that this was because they were moving with respect to absolute space. He knew of no other way to define what spinning was. But Einstein totally debunked the idea of absolute space in 1905. So, is all motion really relative? What exactly is Nature's definition of spinning?

Physicist Ernst Mach proposed that spinning occurs relative to all the mass in the universe. Einstein wrote Mach a letter in 1913 assuring him that his relativistic views were correct and referred to the concept in his General Relativity theory published two years later as "Mach's Principle".

Let's pause a moment to ponder what this means. A spinning inanimate bucket of water "knows" the position of all mass everywhere in the universe and the water shifts away from the center of spin accordingly? A gyroscope "knows" to hold itself in line with the net orientation of the entire universe? Not even the smartest and most observant humans can sense the orientation of the entire cosmos! *The universe must be a computer, calculating the stationary orientation for every point in the universe based on the positions at every instant of every other point in the universe!*

Hold on! Except for stuff inside the spinning object's Cosmic Event Horizon, all things in the universe are flying away from each other faster than 186,000 miles per second, meaning that light leaving those things now would never be able to arrive at the object in question. And Mach's Principle might be bounded by the Cosmic Event Horizon, meaning the influence of mass and energy from a nearby object can only affect the spacetime that is not yet out of reach due to the cosmic expansion. But still, determining what's stationary for every possible point in the universe, each with its own

Cosmic Event Horizon radius of 45 billion lightyears, is still an impressive amount of calculations.

The fact that God watched the creation "until it obeyed" suggests that its every part and particle chooses to act in accordance with specific laws. (*PGP*, Abr 4:18) He created the universe from extremely fine urstoff, as yet undiscovered by science, and commissioned it to manifest reality across all frames of reference. A boundless spiritual connection exists between all things via a universal field of intelligence that is as invisible as the universal fields that scientists have already identified, but much less likely to be detected by experiment.

Is intelligence like ectoplasm?

Psychic mediums in the early 1900s popularized the idea that ghosts were made of ectoplasm. Mediums often used cheesecloth dipped in something slimy for the effect. The stuff would often emanate from the medium during a séance. But audiences got wise when debunkers exposed the hokum.

The *Ghostbusters* movies in the late 1900s made ectoplasm famous again, this time with special visual effects. But, obviously, if ectoplasm were sufficiently visible and tangible to cover people in slime, it could not be the ethereal paranormal stuff that fills the immensity of space. It could not be the urstoff of which all things spiritual and physical are composed. Light is a better analog for the primordial fundamentals from which all things are structured!

Joseph Smith said, "There is no such thing as immaterial matter. All spirit is matter, but it is more fine or pure, and can only be discerned by purer eyes." (DC 131:7) But what is this "more fine or pure" matter? Some have speculated that spirit things are made of neutrinos, particles that travel right through an entire planet without slowing down. Others propose that they are made of dark matter, the mysterious invisible stuff that makes up most of the universe. But all these types of mass can be measured by scientists – very uncharacteristic of spirit stuff. The "more fine or pure" matter discernable only to "purer eyes" is the stuff of which all spirit things and physical things are made. It is called intelligence because its elemental units are sapient.

In the *Star Wars* movies, Yoda, Darth Vader and Obi-Wan Kenobi, reappeared as force spirits after dying by sublimating their organic cells into a state of *pure energy*, thus projecting their *life essence* beyond death into spirit forms. This life essence or *pure energy* is analogous to the intelligence (from which all things are composed, including physical bodies) and their *projected life essence* would be analogous to the spirit bodies that animate physical beings.

There is, of course, no Jedi trick involved in becoming a ghost. All freshly dead people find their spirits outside their bodies and, unless they've mired their lives in wickedness, the essence of which their spirits are made glows in humanlike form. (*BOM*, Eth 3:6-9) Spirits of dead men made perfect through Jesus Christ glow with celestial glory. (DC 129:3-6)

What exactly is intelligence?

The cosmos is intelligent because it is composed of intelligence, "which light proceedeth forth from the presence of God to fill the immensity of space". (DC 88:5-13) One NDE survivor described this light as ubiquitous "pure love and intelligence". (youtube.com/watch?v=Pdn4-u2hTg8, DL 14 Mar 2025)

There are several will-known dictionary definitions of intelligence, but perhaps the least used is, "an intelligent entity; especially: angel." (Merriam-Webster.com, 2017) We could illustrate three definitions in a single sentence: Angels are highly intelligent intelligences who deliver intelligence.

Per scripture, there is a fourth definition of intelligence, stated in various ways:

- ◆ the light of Christ (DC 88:6-7)

- ◆ the light of truth (DC 93:29)

- ◆ the power of creation (DC 88:7-9)

♦ the power governing the universe (DC 88:12-13)

♦ the source of revelation (DC 88:11; *TPJS,* p151)

♦ the glory of God (DC 93:36)

These apparently disparate scriptural definitions roll up into one; Intelligence is the power of God that manifests as His glory and conveys truth to all who are tune into its light.

In its most basic form, this type of intelligence is not only analogous to physical light but actually manifests as *spiritual* light. It is the divine beacon that guides all things. "Whatsoever is light is Spirit… And the Spirit giveth light to every man that cometh into the world." (DC 84:45-46)

"Intelligence, or the light of truth, was not created or made, neither indeed can be." (DC 93:29) "[Intelligence] is independent in that sphere in which God has placed it… to act for itself… otherwise there is no existence." (DC 93:30)

Joseph Smith taught, "Intelligence is eternal and exists upon a self-existent principle. It is spirit from age to age, and there is no creation about it." (*Pearl of Great (PGP) Commentary,* M. Hunter, p100) In his book, *Answers to Gospel Questions,* Joseph Fielding Smith explained, "Why [God] cannot create intelligence is simply because intelligence, like time and space, always existed, and therefore did not have to be created." (*AGQ (Answers to Gospel Questions),* bk3, p125)

Being justified (made innocent) and sanctified (made holy) tunes a person through the Holy Spirit to the omniscient, omnipotent, glory, truth, and intelligence that is the spirit field that fills all reality. *(PGP,* Mo 6:60-61)

Being uncreated, this kind of intelligence is the fundamental spirit element on which all existence is founded. God's creation of order from chaos consisted of choosing individual bits from the eternal sea of intelligence and employing them as constituents in the universe. And each bit also chose Him. Only intelligences inclined toward God were drafted. Intelligences not so inclined remain behind.

Where does raw intelligence lurk?

In the search for raw intelligence, it is helpful to acquire a basic understanding of the constituent particles of the universe. Stuff we can see is made of smaller bits we can't see. The smaller bits are made of even smaller bits until we're down to the quarks, leptons, and bosons where the search for fundamental bits ends as far as physicists know.

These "fundamental particles" are excitations of their respective fundamental fields. But, which is more fundamental, fields or particles? We know that there can be fields without real particles but not vice versa. So fields must be more fundamental!

Photons are most often thought of as particles because of the way they behave when they strike something, like a solar panel, or the retina of a human eye. But they behave like waves right up until the event that absorbs and destroys them. So, as physicist Art Hobson at the University of Arkansas put it, "There are no particles, there are only fields." (arxiv.org/pdf/1204.4616, DL 30 Jan 2017)

This was proved by 1995 experiments in which rubidium atoms were cooled to absolute zero (-459.67°F). At that point, their particle structure disappeared leaving a single quantum wave entity dubbed a "super particle" or 5th state of matter. (sciencealert.com/bose-einstein-condensate?form=mg0av3, DL 12 Oct 2024) The 2001 Nobel Prize in Physics went to Eric Cornell, Wolfgang Ketterle, and Carl Wieman for their work.

Just as James Maxwell discovered that the magnetic field and the electric field were actually a single field (the electromagnetic field), physicists hope to craft a working theory for the mother of *all* fields – the Unified Field Theory. Flux lines indicating the presence of magnetic fields were first mapped in 1269AD by Petrus Peregrinus de Maricourt using iron needles. In all the time since, scientists still haven't discovered what makes up a field. Since fields are as invisible as ghosts, a metaphysical explanation is in order — The fields that fill our universe are made of primeval intelligence.

Is there a place in the universe where not even primeval intelligence exists? What about the dark voids between galactic clusters?

Even if you swept away all the random hydrogen and helium atoms, and all the dark matter and other types of mass, the following universal fundamental fields would still fill even the darkest regions of the universe:

♦ 3 lepton fields (electron, muon, tauon)

♦ 3 neutrino fields

♦ 1 Higgs scalar field

♦ 3 weak gauge boson fields: W, Z+ and Z−

♦ 1 electromagnetic field

♦ 1 gluon field

♦ 6 quark fields

♦ 1 gravitational field

So, empty space is quite full! Even if the emptiest pocket of space could be protected from all incoming particles, the energy levels in these ubiquitous fields would still quiver up and down, only obeying the law of conservation of energy by averaging to zero. Truly empty space is impossible! And, having been given a role to play in our universe, none of these fields represent or contain truly primeval intelligences. Once they were in chaos but now they are part of God's creation!

The universality of fields explains how a wavicle instantaneously "knows" what's in its future path or the state of an entangled twin, triplet, quadruplet, etc. Particle entanglements maintain intelligent relationships no matter how far apart the participants may be. A field "knows" all the conditions in it and as well as the appropriate responses to those conditions.

Since fields permeate the universe, intelligence is everywhere! The victim of a mugging found himself on the other side of death, sensed the presence of a metaphysical light, and got "the understanding that *that* light was proof of the existence of a higher intelligence, something nonmaterial that is presence throughout the universe". (Alfred's NDE, youtube.com/watch?v=lK2WI7sfuuw, DL 7 Dec 2022)

Are intelligences stringy?

String theorists say that the single fundamental thing underpinning everything is a stretchy one-dimensional string that vibrates and contorts in different ways through ten or eleven dimensions to manifest as the various subatomic particles. It supposedly vibrates and contorts itself one way to masquerade as an electron, another way to play a quark, and so on, for all the particles in the universe. If string theorists are correct, quantum physicists are wrong that there are 17 fundamental particles – there is really only one fundamental thing, one string to fool them all!

This theoretical wisp can break into pieces, each piece manifesting as an individual particle. For example, a photon is split in two, it's actually a string breaking, with each bit carrying away a percentage of the original energy. Or multiple strings can recombine into one.

Now, it may seem contradictory that a fundamental string could split. However, as in the case of a photon, splitting results, not in constituent parts, but rather in clones of the parent, each carrying a portion of the parent's energy.

If string theory turns out to be true, it would be the Holy Grail of physics. Unfortunately, string theory has proved to be untestable. Over decades of research, string theory has been stretched to accommodate everything but has never predicted anything.

Are preons intelligences?

Preons are theoretical *point* particles, meaning they have no size. If they exist, they would be the minutest building blocks of matter, unless they, in turn, are made of strings. Preon theorists say that repeating patterns in the Standard Particle Model suggest the presence of smaller bits that behave or combine in various ways to portray all the different particles in the zoo.

If preons exist, their quantum field would be the mother of all other fields. And, since preons are thought to be perturbations in this preon field, the field rather than the particle would be the most fundamental thing in physics. But so far, it appears that preon theory clashes in certain cases with experimental data, which is one reason it remains far less promising than string theory.

Is an intelligence fundamental?

If there ever was a mother of all quantum fields, a unitary field, some kind of specialization must have happened to produce the different kinds of fields now in play. The fact that particles can do a switcheroo to one or more other fields, even when only "fundamental" fields are involved, hints at some kind of quantum kinship. For example, a muon can change into an electron and two neutrinos. This suggests that these particles and their fields are not truly fundamental, merely the most fundamental known to science.

Einstein spent the latter part of his life trying to craft a Unified Field Theory (UFT) that would explain all the fundamental physical forces between elementary particles in a single theoretical framework. So far, such a theory has eluded the best and brightest brains. The unitary fundamentals hide behind the physical-metaphysical boundary below which intelligence surreptitiously serves as the basis of all existence.

"The matter composing our bodies and spirits has been organized from the eternity of matter that fills immensity," said Brigham Young. (*DBY (Discourses of Brigham Young)*, p49) So, the ubiquitous fundamental stuff called intelligence underpins all existence, whether spirit and or material. It is the true pure essence. It is analogous to physical light and "proceedeth forth from the presence of God to fill the immensity of space". (*DC*, 88:12) It is living light. That's the glory of God from which all things are made, including man.

This universal, all-pervasive field is sensed by people who die! As one victim of a car crash said after recovering from death, "Everything is connected by an immaterial field." (youtube.com/watch?v=lMQ7Q5NNWCw, DL 13 Feb 2023)

Are intelligences genderless?

Raw, unorganized intelligence, the stuff of which spirit things are created, is fundamental because it "is eternal and exists upon a self-existent principle." (*TPJS*, pp353-54) The definition of "fundamental" denotes minimal attributes or traits. But, since the definition of "intelligence" denotes the ability to make basic choices, we must grant that capability to the fundamental bits of existence.

And, based on revelation to Joseph Smith, we must also grant that bits of intelligence can choose to cleave to other intelligences and to respond to divine light. (DC 88:40) But, since there are many *asexual* things that are created from raw intelligence, reproduction must be a capability that only higher-order creatures can possess.

Cultures have assigned gender to inanimate objects such as the major objects in our solar system. El Sol is masculine, for example. But such gender designations are merely cultural relics of yesteryear and have no basis in science or religion. However, fundamental particles and other genderless bits can be part of a creature that *does* have gender.

For example, human bodies have gender but the individual atoms of which they are composed do not. Likewise, human spirits have gender. (*The Family Proclamation*, LDS) But the individual bits of intelligence from which they are composed do not. The physical is in the likeness of the spiritual. (DC 77:2)

How does intelligence work?

Dianne Morrissey "entered the Light of God" after being electrocuted and dying; "Within the Light was the cure for all diseases… all the knowledge of every planet, every galaxy, every universe... Being one with the Light was like suddenly knowing every grain of sand on every planet, in every galaxy." This light is God's intelligence. He is omniscient because His intelligence permeates and comprises all things.

Intelligence or "spirit of truth... is the light, life, and spirit of all things". (*Millennial Star*, Parley Pratt, 2 Apr 1842) "The light of Christ... is in all things... giveth life to all things... [and] is the law by which all things are governed, even the power of God." (DC 88:4-13) "You may call it the Spirit of God, you may call it the influence of God's intelligence, you may call it the substance of his power; no matter what it is called, it is the spirit of intelligence that permeates the universe." (*MD (Mormon Doctrine)*, "Spirit of the Lord")

The universal field of intelligence provides all-way communication between God and every particle or thread of His creation. It is through this that each knows His will and through this that all things are one in the universe. Through it, His cognizance is everywhere. It is comparable to "the Force" in the *Star Wars* movies except that it obeys God instead of midi-chlorian-endowed characters and is the basis of all existence.

Because its main functions are the reception of truth, obedience to truth, and transmission of truth, intelligence is also referred to as the truth or light of truth. "All truth is independent in that sphere in which God has placed it, to act for itself, as all *intelligence* also; *otherwise there is no existence.*" (DC 93:30) If intelligence decided to quit obeying God, the universe would disintegrate!

Intelligence obeys the original instructions of the Creator of its own will and need not be micromanaged from moment to moment. This is evident from the fact that every jot and tittle of the universe continued to function perfectly even while He lay in a manger.

Intelligence also guides human behavior, not by compulsion, but as the spiritual beacon of truth. It is the conscience or "true light that lighteth every man that cometh into the world". (DC 93:2) It is the medium for communication between God and humans. Joseph Smith said that the "Holy Ghost has no other effect than pure intelligence…. When you feel pure intelligence flowing into you, it may give you sudden strokes of ideas." (*TPJS*, pp149-151) "The light of Christ... enlighteneth your eyes, which is the same light that quickeneth your understandings." (DC 88:4-13) "By noticing it, you may find it fulfilled the same day or soon; i.e. those things that were presented unto your minds by the Spirit of God, will come to pass." (*TPJS*, p151)

Several people who have died and been resuscitated describe a ubiquitous light — an omnipresent consciousness that is alive, making past, present, and future "available". (see youtube.com/watch?v=smNGOyKDVdU, DL 8 Dec 2025)

Studies show that intelligence in an adult, as measured on standard IQ tests, cannot change significantly, no matter how hard the person might train. But *any* person can increase his or her spiritual intelligence by following the Light of Christ. "Whatever principle of intelligence we attain unto in this life, it will rise with us in the resurrection. And if a person gains more knowledge and intelligence in this life through his diligence and obedience than another, he will have so much the advantage in the world to come." (DC 130:18-19) *Spiritual* intelligence is the only kind that matters forever.

A great example of a person being rejuvenated by this light is when "King Lamoni was under the power of God… and the light which did light up his mind, which was the light of the glory of God… yea, this light had infused such joy into his soul, the cloud of darkness having been dispelled… the light of everlasting life was lit up in his soul." (*BOM*, Alma 19:6) He then realized that murdering his servants whenever thieves scattered his flocks was offensive to God.

In the beginning, the Father put Christ in charge of the intelligence that constitutes all existence. Thus, it is called the Spirit of Christ or the Light of Christ. As the most fundamental quantum of reality, each bit of intelligence is connected to God and is therefore part of the spirit light that knows all things; "That which is spirit, even the spirit of truth; And truth is knowledge of things as they are, and as they were, and as they are to come." (DC 93:24) Spiritual beings who tap into this light can have boundless knowledge. "He that keepeth [God's] commandments receiveth truth and *light,* until he is *glorified* in truth and *knoweth all things.*" (DC 93:28)

Is intelligence energy?

Similar to the law of conservation of energy, each tiny bit of the omnipresent sea of intelligence in the universe has always existed and can be organized and reorganized but not created or destroyed. (*EOM (Encyclopedia of Mormonism)*, bk3, "Spirit") Of the "light" of intelligence, Parley Pratt wrote, "Like the other elements, its whole is composed of individual particles." (*Key to the Science of Theology*, ch5, p39) Intelligence, spirit, and matter are on an existential continuum, transformable stage-by-stage in a forward progression. Greater complexity of organization means more potential for greater glory.

Quantum fields of every kind cannot lay dormant, even in the remotest and darkest regions of outer space. Their energy levels fluctuate constantly above and below zero, often reaching thresholds that result in the creation of a particle pair, misnamed "virtual" particles. They are real, despite their typically short lifespan. Real photons can twinkle briefly into virtual electron-positron pairs. The myriad ways in which spacetime's quantum soup bubbles and boils at the smallest known scales is evidence of the perpetually energetic nature of the fundamentals that underlie all reality.

All tangible things consist of structured energy and tend to decay. In the process, energy is released. Any closed system that releases energy is sacrificing some degree of order to do so. Among the 17 fundamental particles identified by science, those with more mass tend to decay into other particles with less mass, as if all creation "wants" to return to pure energy.

Near-death survivors often describe both spirit matter *and* physical matter as vibrations of light, a clue that fundamental intelligence has mobility, energy, the ability to act, the ability to interact, and the ability to combine and express itself in various ways to create physical reality.

Will science ever see the light?

Responding to an atheist, Einstein said, "That which is impenetrable to us really exists. Behind the secrets of nature remains something subtle, intangible, and inexplicable. Veneration for *this force* beyond anything that we can comprehend *is my religion*." (*The Diary of a Cosmopolitan*, H. G. Kessler, London: Weidenfeld and Nicolson, 1971, p157) Einstein's underlying "force" was at the center of his long and futile search for the mother of all fields.

Not only is the amount of knowledge gained by discovery outpaced by the amount of questions engendered in the process, but the difficulty scientists have discovering the next smidgeon of truth increases over time. If we picture human knowledge as an expanding sphere, though the surface area expands exponentially, the radius increases at an ever-slower rate. The further science pushes into new territory, the more discovery-resistant Nature is.

Physicist Brian Green's book "Our Elegant Universe" describes how, when scientists focus on ever smaller bits of vacuum space, the quantum field undulations increase in magnitude as if spacetime gets agitated at being scrutinized too closely. This all-pervasive "vacuum energy" just might be the boundary of what scientists can physically probe.

Quantum physicists have difficulty pinning down what is actually real versus what is merely math. Nobody understands the tangle between entangled particles or the physical nature of a quantum wave. Nobody knows whether physicists can ever quantize gravity as they've done for all the other known forces. Nobody has been able to directly detect dark matter or dark energy. And, black holes pose such indecipherable mysteries that physicist Roger Penrose posited a "cosmic censorship hypothesis" that prevents scientists from making certain discoveries. Yet there is math for all of these things.

If secular seekers of scientific truth stumbled upon irrefutable proof that the universe is made of ordered spirit essence, they would be obliged to believe in spiritual things by nonspiritual means. Since that's not God's formula for acquiring faith, it will probably never happen.

Can God's intelligence increase?

God is already as knowledgeable and smart as He can get. So how could He possibly increase in intelligence? Well, if we're talking about the kind of primordial intelligence that fills the universe and has been around forever, then the answer is that God increases in glory as He draws more intelligence and intelligences to Him; "The glory of God *is* intelligence." (DC 93:36)

The twenty-four elders in heaven praised Jesus, "Thou art worthy, O Lord, to receive glory." (Rv 4:11) "This is [God's] work and [God's] glory — to bring to pass the immortality and eternal life of man." (*PGP*, Mo 1:39) Jesus created the cosmos for the purpose of "bringing many sons unto glory". (Hb 2:10) As He volunteered, the premortal Jesus said, "Father… the glory be thine forever." (*PGP*, Mo 4:2) Glory accrues to Him as more intelligences join with Him in His plans and purposes! The more intelligences honor and glorify God, the more powerful He becomes. To glorify means to "honor, which is my power," said God. (DC 29:36)

It's an upward spiral. God's light and love attract intelligences who desire to progress. Their progression adds to His glory which attracts more intelligence. "Intelligence cleaveth unto intelligence... light cleaveth unto light." (DC 88:40) In unity there is a synergy whereby God and his subjects achieve greater glory together. "That which is of God is light; and he that receiveth light, and continueth in God, receiveth more light; and that light groweth brighter and brighter until the perfect day." (DC 50:24)

Why is intelligence called light?

God told Moses, "Thou canst not see my face: for there shall no man see me, and live." (Ex 33:20) God's glory is deadly to mortal observers. But verse 11 said that "the LORD spake unto Moses face to face, as a man speaketh unto his friend." Yet Moses lived to write about it. How? Moses himself explained away the contradiction when he said, "Mine own eyes have beheld God; but *not my natural, but my spiritual eyes*, for my natural eyes could not have beheld; for I should have withered and died in his presence; but his glory was upon me; and I beheld his face, for *I was transfigured before him.*" (*PGP*, Mo 1:11)

Are spiritual light and physical light related? Both radiate, are energy, and are good. Since intelligence underlies all quantum fields, including the electromagnetic, physical light is based on spiritual light.

As a spirit, the premortal Jesus did not emit or reflect photons. Physical eyes cannot see spiritual beings. Jesus had to appear to Moses' spiritual eyes. And, since the premortal Jesus had no tangible body with which to veil His glory, Jesus could only appear in glory to Moses. (DC 129:6) And, to see God in His glory, Moses had to be transfigured to see in spirit, meaning "quickened by the Spirit of God". (DC 76:11)

The flesh tends to mask a spirit's aura, its most noticeable characteristic. But spirit light can sometimes be perceived by humans despite being in the flesh. When Moses returned from meeting with God on the mountain, "The children of Israel could not stedfastly behold the face of Moses for the glory of his countenance." (2 Cor 3:7) So, in some way discernable to the children of Israel, the intelligence absorbed by Moses from talking with God was visible to others, even though "Moses wist not that the skin of his face shone." (Ex 34:29) On spirit light infusing the physical, Brigham Young said, "Matter is capacitated to receive intelligence." (*DBY*, p48)

7: Spirits

Who are the ghosts?

Whether it's poltergeists, ghouls, spooks, demons, specters, phantoms, elementals, banshees, apparitions, succubae, incubi, or just ordinary spirits, people get thrills and chills from contemplating the possible presence of unseen beings. The tradition of telling ghost stories at sleepovers and around campfires has provided many delightful and frightful hours for children over the years. Authors, entertainers, and movie makers have cashed in on the human desire to be spooked.

For thousands of years, mediums, psychics, and spiritualists have been plundering the pockets of the vulnerable and gullible, claiming to communicate with loved ones on the other side of death's veil. The ghost-related scams will continue as long as people believe in ghosts and recent polls indicated that about 43 percent of US residents do.

Even the great Harry Houdini tried to contact his mother right after she died but failed. He eventually became so outraged by the obvious fraud by mediums that he spent many of his later years debunking them. But before he died, he made a pact with his friends that he would try to contact them from the other side and every year, his fans gather in séances on Halloween, as that was the day he died, and try to get him to say something. But, since his death in 1926, the great escapologist has remained trapped in the silence of death.

Around 1900, "talking boards" began to be commercially produced, renamed Ouija boards by the patent holder (French "yes" + German "yes"). Plenty of scientific studies have been done showing that nothing comes from the Ouija boards other than what is already in the heads of the players. Nevertheless, many Bible believers avoid the boards because God discouraged talking with the dead, the punishment for which was death by stoning under the Law of Moses. (Lv 20:27)

The belief that demons can masquerade as spirits of the departed is another reason many people shun Ouija boards. Nevertheless, Hasbro continues selling the boards and there are myriad online websites that feature interactive electronic versions of the game.

The Bible speaks of evil spirits, unclean spirits, dumb spirits, and other type of spirits that are malicious toward humans. (Mt 4:24) Fortunately, because God is smart, He has set bounds on such ghosts to prevent them from messing with everyone's lives beyond that which serves His purposes. God also sets rules for communication with humans by righteous spirits. Haphazard scaring or sharing by spirits with humans could make a mess of their lives.

For example, a legion of demons drove a fellow crazy. When Jesus arrived, the demons wanted to be sent into a herd of wild swine rather than be kicked out of the country. When Jesus allowed it, the swine ran pell-mell into the sea and drowned. (Mt 8:30-32) Poor pigs! Either the pigs were smart enough to commit suicide rather than be possessed by demons or else possessed pigs are even crazier than possessed people. Why did the demons want to possess the bodies of animals deemed vile by Mosaic Law? "The devil has no body, and herein is his punishment. He is pleased when he can obtain the tabernacle of man, and when cast out by the Savior he asked to go into the herd of swine, showing that he would prefer a swine's body to having none." (*TPJS*, ch17)

Not to be overlooked are the spirits of those who have yet to don human bodies. Alex Haley wrote, "Three groups of people lived in every village. First were those you could see walking around, eating, sleeping, and working. Second were the ancestors. The third group of people...*are those waiting to be born.*" (*Roots*, 1974, p24) It is a beautifully symmetrical categorization of all spirits: the unembodied, the embodied, and the disembodied.

Paul wrote of "the spirit of man, which is in him". (1 Cor 11) And when Jesus died, He said, "Father, into thy hands I commend my spirit." (Lk 23:46) When Jesus raised a little girl from the dead, "her spirit came again, and she arose." (Lk 8:55) So, every person has a ghost inside. It's Biblical!

Right after the earth was created, it was unbeautified and lifeless. The Lord said, "I, the Lord God, had created all the children of men and not yet a man to till the ground; for *in heaven* created I them." (*PGP*, Mo 3:5) Like the spirit of Jesus, the spirits of all humans were created before life was placed on earth.

Paul wrote that Jesus was "the firstborn among many brethren" who were foreordained "to be conformed to [His] image". (Rm 8:29) Who were the many brethren? They were spirits who were called to work in official capacities in Christ's work. They are the spirits who took important roles in God's work. Many have served as angels to earthlings.

A sacred text of Judaism says, "Before the world was created, there was none to praise God and know Him. Therefore He created the angels and the holy Hayyot, the heavens and their host, and Adam as well." (sacred-texts.com/jud/loj/loj104.htm, DL 19 Jul 2020)

Right after saying, "I was in the beginning with the Father, and am the Firstborn", Jesus said, "Ye were also in the beginning with the Father, that which is Spirit, even the Spirit of Truth... Man was also in the beginning with God. Intelligence, or the Light of Truth, was not created." (DC 93:21-23, 29) So, Yes — All humanoid spirits are siblings of Jesus and share a common Spirit of Truth heritage!

Who believes in premortality?

Five hundred years of religious sculptures and paintings depict infant spirits in the form of baby angels or chubby cherubs, aka putti. Indeed, the belief that all spirits were once babies was present in early Christianity. For example, Justin Martyr (100-165AD) said, "If the world is begotten, souls also are necessarily begotten." (*Dialogue with Trypho*, ch5)

But belief in premortality fell into the dustbin of post-apostolic apostasy. Religions whose doctrines expressly exclude it include Adventists, Baptists, Catholics, Episcopalians, Jehovah's witnesses, Lutherans, Methodists, Presbyterians, and Unitarians. These churches teach that spirits are created at the same time the body is created in one of the following ways:

Emanationism

Per Emanationism, human spirits are part of God's omnipresent divine spirit and all things are derived from the original reality flowing from Him. Some emanationists believe that all realities, except the original, emanate from prior realities and, like copies of copies on photocopy machines, fall victim to replication flaws with each iteration. So, mistakes are the fault of the photocopy process but it is not clear whether the Manufacturer of the copy process is to blame. It's also not clear how much of a person's sins can be blamed on a faulty copy process. As copy errors continue to accumulate, later generations should be able to blame virtually all of their imperfections on bad copying.

Emanationism does not explain why Adam felt inclined to partake of the forbidden fruit despite having a spirit that had never been copied.

Traducianism

Per Traducianism, an infant's spirit derives from the spirits of its parents at conception, even in cases of in vitro fertilization. None of the parents suffer any loss of spirit in the process, no matter how many kids they have. Only the soul of Adam was created directly by God. When God created Eve, some of Adam's spirit came with the rib to become Eve's spirit. When Adam and Eve had children, each soul inherited some of the stain of original sin from the father. Per Traducianism, humanity's spiritual weaknesses can ultimately be traced back to Adam but it's not clear why God created a faulty spirit for him. It's also not clear why some people seem to inherit more than their fair share of weaknesses.

The question has been asked whether a clone would even have a spirit since as there would be no moment of conception and no parents in the normal sense. Per Traducianism, it was the absence of a human father that allowed Christ to be conceived without original sin.

Traducianism also engenders questions about the origins of the souls of identical twins, since only one zygote is created at the moment of conception.

Creationism

Per Creationism, God creates every spirit from absolutely nothing and implants it into the fetus sometime between conception and birth. Exactly when that happens would be good for ethical abortionists to know so they can limit their killing to only those babies who have not yet been ensouled.

If God creates the spirit, it must be perfect because He is perfect. Right? So, all the blame for imperfect behavior falls on the body! And, all the mistakes a person makes must be the fault of parents who created the body. The chain of blame goes all the way back to Adam whose body and spirit were both created by God, and there the trail of blame ends! Creationism doesn't explain why everyone's mistakes are not God's fault, especially if He started from perfect nothingness!

Premortality

Per truth, God created all spirits from primordial intelligence, giving them magnificent opportunities to improve their knowledge and character before entering a mortal fetus. Premortality has been called *preexistence*. But that's an oxymoron that has been abandoned as existing prior to existence makes no sense. However, it's scriptural that we existed as spirits prior to the creation of Earth.

How are we God's offspring?

The NT indicates that God fathered our spirits. Paul taught that people are "*offspring* of God". (Acts 17:29) He also wrote, "We have had fathers of our flesh which corrected us, and we gave them reverence, should we not much rather be in subjection to the *Father of spirits*, and live?" (Hb 12:9)

And to the Romans he said, "The Spirit itself beareth witness with *our spirit, that we are the children of God."* (Rm 8:16)

Scripture also indicates that our spirits were created before the creation of Earth. Solomon declared that he "was set up from everlasting, from the beginning, or ever the earth was" and was "brought forth while as yet [God] had not made the earth... when he prepared the heavens". (Prv 8:24-27) The Preacher wrote, "The spirit *comes* to the bones in the womb of a woman with child." (*RSV (Revised Standard Version)*, Ec 11:5) The point is that spirits were created by God *before* being sent to baby bodies in wombs. And at death, "The spirit shall *return* unto God who gave it." (Ec 12:17) Spirits only return to God if they came from God!

God alluded to Jeremiah's premortality, telling him, "Before I formed thee in the belly I knew thee; and before thou camest forth out of the womb I sanctified thee, and I *ordained* thee a prophet unto the nations." (Jer 1:4-5)

God spoke of premortal spirit sons and daughters in his conversation with Job when He asked, "Where wast thou when I laid the foundations of the earth... Or who laid the corner stone thereof; when the *morning stars sang* together, and all the *sons of God* shouted for joy?" (Jb 38:4-7) Here, a Hebrew parallelism equated the singing of *morning stars* with the rejoicing of *sons of God* who were yet spirits when the creation of the earth was undertaken. Consistent with this Hebraism, the term *astral* projection, which means *star* projection, is a reference to a person's spirit leaving its body.

The apostle Paul wrote, "In hope of eternal life, which God, that cannot lie, promised before the world began." (Titus 1:2) To whom would He have made promises before creating the world? His spirit children who aspired to the eternal life He promised them!

"The *spirit* (Heb *niš-ma_t_*) of man is the candle of the Lord." (Prv 20:27) The word *niš-ma_t_* means *spirit* or *breath*. (נְשָׁמָה) "God formed man... and breathed into his nostrils the *breath* (*niš-ma_t_*) of life; and man became a living soul." (Gn 2:7)

Just hours before Jesus was betrayed, He prayed to the Father for His apostles, *"They are not of the world, even as I am not of the world...* As thou hast sent me into the world, even so have I also sent them into the world." (Jn 17:13-18) Like Jesus, who "In the beginning... was with God... [and then] was in the world... made by him," the apostles could not have been sent into the world unless they first existed outside the world. (Jn 1:1-10) Christ called His apostles before the physical creation, as Paul said, "He chose us in Him before the foundation of the world, that we should be holy and blameless before Him." (Eph 1:4)

The aforementioned references to prebirth spirithood suffice as proof that premortality is indeed biblical!

Abraham put it bluntly; God *"took his spirit* (man's), *and put it into him"*. (*PGP*, Abr 5:7) Whether spirits are called the breath of God or the spirits of men, they existed prior to ensoulment! Christ told Abraham that He had selected noble and great spirits in premortality to be prophets on earth, saying, *"These will I make my rulers...* Abraham, thou art one of them; thou wast chosen before thou wast born." (*PGP*, Abr 3:22-26)

The ancient Hebrew Babylonian Talmud says of man's spirit, "With the soul of Adam, the souls of all the generations of man were created." (ms. *B. Chagiga 12B*) Islam also teaches that the spirits of all humans were created at the same time as Adam's spirit. Since Adam's spirit was created prior to his body, so did everybody's.

Another reference to spirit creation comes from a gnostic text, "Jesus said, 'If they say to you, From where have you originated? Say to them, 'We have come from the Light.'" (*The Book of Thomas*, Nag Hammadi Texts, saying50)

Jewish philosopher and biblical commentator Philo of Alexandria (1st century AD) wrote in his *On the Creation of the World* that "Heavenly Man" was created in Genesis 1 (in God's image), and "Earthly Man" was formed from the dust in Genesis 2. This is in harmony with the view that Chapter 1 was a spiritual creation and Chapter 2 was the physical creation.

The Church of JESUS CHRIST of LDS is the only Bible-believing church that has chiseled premortality into the marble of its doctrine; "All human beings… are created in the image of God. Each is a beloved spirit son or daughter of heavenly parents… In the premortal realm, spirit sons and daughters knew and worshipped God as their Eternal Father." (*The Family: A Proclamation to the World*, 1995)

Joseph Smith said, "We were first begotten as spirit babies in heaven and then born naturally on earth." (*JD*, bk4, p218) Brigham Young said, "Our Father in Heaven begat all the spirits that ever were, or ever will be, upon this earth; and they were born spirits in the eternal world." (*DBY*, p24) He reaffirmed it later, "The Father actually begat the spirits." (*DBY*, p50)

Young puzzled many when he said, "The spirits of all the human family were begotten by Adam, and born of Eve." (*The Essential Brigham Young*, Eugene E. Campbell, 15 Apr 1992, pp96-97) But Young was not talking about the Adam who was cast out of Eden. Young was using the same name titles for the Heavenly Parents of our spirits that are used for candidates of celestial glory in LDS temple rituals. Every time Young referred to God as "Adam" (the source of the so-called "Adam-God doctrine") he was using the name title for exalted beings. This was made clear when he told Orson Pratt, who was struggling with Young's statements, that he, Young, "did not believe that Orson would ever be Adam". (*Samuel W. Richards Journal*, 11 Mar 1856, p113) Thus Young coyly gave Pratt the key to understanding his Adam-God statements. Since Adam Jr. and Adam Sr. were both frequently in Eden together, some people conflated the two from Young's statements. And some confusion may have arisen from Young's ambiguous use of pronouns.

That Young had a proper understanding of Adam and God is evident from other statements by him; "Adam was as conversant with his Father who placed him upon this earth as we are conversant with our earthly parents. The Father frequently came to visit his son Adam, and talked and walked with him." (*DBY*, p104) Also, "Some say, 'We are the children of Adam and Eve.' So we are, and they are the children of our Heavenly Father. We are all the children of Adam and Eve, and they are the offspring of him who dwells in the heavens, the highest Intelligence that dwells anywhere." (*DBY*, p222)

Like Young, Paul applied the name "Adam" to Jesus; "The first man Adam was made a living soul; the *last Adam* was made a quickening spirit." (1 Cor 15:45) But nobody misconstrued Paul's statement as a weird "Adam-Christ doctrine". Furthermore, spirit fatherhood was taught in the School of the Prophets run by Smith; "All the world, also, as far as the idea of His existence extends, may have to exercise faith in Him, *the Father of all living*." (*LOF*, p33)

"Adam called his wife's name Eve; because she was *the mother of all living*." (Gn 3:20) To whom does the pronoun *she* refer? Eve of Eden had not yet given birth to anyone. The *Literal Translation* says it this way, "For *she hath been* mother of all living." (*YLT*, Gn 3:20) It must be that Adam gave his wife his Heavenly Mother's name title, just as he had been given the name title of his Heavenly Father's. (eldenwatson.net/7AdamGod.htm, DL 29 Aug 2017)

The handful of modern Christian groups who believe in a heavenly mother include the World Mission Society Church of God and The Church of JESUS CHRIST of LDS. Joseph Smith reasoned, "How could a Father claim His title unless there were also a Mother to share that parenthood?" (As quoted in *Saints*, 2018, bk1, pt4, ch34, pp403-404)

The Encyclopedia of Mormonism says, "God… organized pre-existing intelligence to beget spirits." (*EOM*, bk1, "Doctrine") "Spirit matter is identified with intelligence." (*EOM*, bk3, "Spirit") Apostle Spencer Kimball said, "Our spirit matter was eternal and co-existent with God, but it was organized into spirit bodies by our Heavenly Father." (*The Miracle of Forgiveness*, 1969, p5) And Apostle Marion G. Romney said, "Through that birth process, self-existing intelligence was organized into individual spirit beings." (*Church News*, 7 Oct 1978, p8) President David McKay taught, "We have the glorious truth that uncreated, ever-existent intelligence animated spiritual bodies." (*Gospel*

Ideals, p445) Apostle Parley Pratt taught that during spirit creation the essence of the pervasive spirit is "organized in individual form and [then] clothed upon with flesh and bones." (*Key to the Science of Theology*, F. D. Richards, 1855, pp43-45) And Apostle Bruce McConkie wrote, "The bodies of Deity's spirit children were created from the existing spirit element." (*MD*, "Birth")

So, since intelligence is spirit light, God is indeed "the Father of lights". (Js 1:17) This is "the light which is in all things". (DC 88:12) Just as physical things are made of finer bits of matter and energy, spirit things are made of finer spirit matter and energy.

Apostle Orson Pratt wrote, "The intelligent particles of a man's spirit are by their peculiar union, but one human spirit." (*Absurdities of Immaterialism*, 1848, pp21,26) Orson's brother, apostle Parley Pratt, said, "They [spirits] are organized intelligences… made of the element which we call spirit… organized in the size and form of man… a spiritual body." (*JD*, bk1, 7 Apr 1853, p8) And apostle and prophet Spencer Kimball said, "God has taken these intelligences and given to them spirit bodies and given them instructions and training." (*TSWK (Teachings of Spencer W. Kimball)*, p32)

Orson Pratt wrote how each individual piece of intelligence turned its agency to the benefit of the whole organized spirit; "Particles organized in an infant spirit… must consent, in the first place, to be organized with other similar particles, and after the union has taken place, they must learn, by experience, the necessity of being agreed in all their thoughts, affections, desires, feelings, and acts, that the union may be preserved from all contrary or contending forces, and that harmony may pervade every department of the organized system… from the time of their birth as infant spirits, until the time that they are sent into this world to take fleshly tabernacles, the organized particles are instructed and educated in all the laws pertaining to their union, until they are made perfectly ONE in all their attributes and qualities." (*The Seer*, p104)

A Brigham Young University professor said, "Many early Saints recognized a divinity in the intelligence or spirit in man. They made no essential distinction between the intelligence that constitutes a man's spirit and the intelligence that constitutes God's glory." (*BYU Studies*, Charles Harrell, bk28, p83)

Demons are the strongest evidence that God created spirits from pre-existing essence of some kind, referred to as intelligence by the restored church of Jesus Christ. Why are they proof of intelligent urstoff (*primary/primordial stuff*)? Because if God had created everything from nothing, there would have been no defects in the ingredients and no devils among the finished goods — no demons nor imperfect spirits! Satan's bad choices cannot be blamed on the Creator of his spirit, rather they were a consequence of character-shaping choices made by his constituent bits of intelligence before they were organized into his spirit body. A spirit's mind is the collective consciousness of the intelligence from which the spirit was made.

What did the early church teach?

In one of the earliest post-NT Christian writings, the great St. Clement of Alexandria (d. 215AD), pondered the symmetry of a premortal spirit existence; "If I live after, I must have lived before." (*Clementine Recognitions*, I, 1&2 as quoted by Richard Eyre in *Life Before Life*, p29)

Origen Adamantius, another early Christian teacher (~185-254 A.D.), said that the "primeval spirits" of Abraham, Isaac, and Jacob were created "before any other works of God" and then quoted the *Prayer of Joseph*, a Jewish apocryphal work that identifies OT patriarchs as spirit angels before their debut in the flesh; "I, Jacob, who speak to you, arid Israel, I am an angel of God, a ruling spirit, and Abraham and Isaac were created before every work of God." (*Commentary on John*, bk2, sec25) Origin's writings are the earliest surviving extra-biblical Christian writings on premortality.

The doctrine of premortality was also taught by other early Christian leaders such as Augustine, Cyril of Jerusalem, Clement of Rome, Pierius, John of Jerusalem, Rufinus, Nemesius and others.

In 553AD, the pope acquiesced to pressure from Roman emperor Justinian and officially disavowed the doctrine of a premortal existence at the Council of Constantinople. (*Nicene and Post-Nicene Fathers*, series 2, bk14, 1956, p320, as referenced by Richard Eyre in *Life Before Life p*30)

So, it's apparent from ancient texts that the doctrine was pronounced heretical and teachers of it were considered heretics, a crime for which some were punished by death in the 6th century.

Who remembers premortality?

For the vast majority of Christians, the reason they don't remember a life before Earth life is because it never happened. Some sects, such as Jehovah's Witnesses, go even farther and claim that no human ever had a spirit and no human ever will. (*You Can Live Forever in Paradise on Earth*, Watchtower Bible and Tract Society, 1982, p77)

But how do they explain the following?

1) Some children remember waiting to be born.

2) Thousands of people remember their premortal existence during a near-death experience.

3) Multiple Bible passages support the premortal existence of human spirits, as quoted in a prior section.

The reason most Christians talk of going *home* to God at death even though their churches don't teach they were ever there is because it is a truth that resonates with their spirits. Only The Church of JESUS CHRIST of LDS (and its break-off groups) teach that human spirits had an existence before being born into this world. If we are to return to the Light, we must have come from the Light.

As William Wordsworth wrote in his *Ode: Intimations of Immortality*,

> *Our birth is but a sleep and a forgetting:*
> *The soul that rises with us, our life's star,*
> *Hath had elsewhere its setting,*
> *And cometh from afar;*
> *Not in entire forgetfulness,*
> *And not in utter nakedness,*
> *But trailing clouds of glory do we come*
> *From God, who is our home.*

Most people have mere shadows of pre-birth memories without any conscious recollection but a few have reported lingering or restored memories. In some cases, a parent has been visited by a child yet unborn. In other cases, a person who dies is shown the pre-birth spirit realm and then is resuscitated. Memories of premortality are called Prebirth Experiences (PBEs) and hundreds of cases can be found on the Internet. Lending credibility to PBE accounts is the fact that the vast majority of PBE'ers do not belong to churches that teach premortality.

Many a case of Deja vu may actually be vestigial recollections from observing someone else's Earth life from the vantage point of the pre-birth world, perhaps as a guardian spirit to that person. Such ethereal memories are sometimes mistaken for evidence of reincarnation. Also, some memories could be glimpses of a person's future on Earth given to them in premortality.

The earthly body, by nature, shrouds the spiritual mind, which, until entering mortality, could tap into knowledge of present, past, and future from the all-knowing spiritual light that pervades all physical and metaphysical realms. Only a person whose physical mind has some degree of connectivity to their spirit mind can have genuine residual memories of being without a body.

What was the spirit birth order?

Jesus was "the firstborn of every creature: for by him were all things created". (Col 1:15) Why call Jesus, who was born 4,000 years after Adam's birth, the firstborn of every creature? Paul was not referring to physical birth, rather to the birth of the first Spirit Child of the Father! Furthermore, if Jesus had been the *only* spirit born prior to creation, Paul would not have used the term "firstborn". Paul went on to say that Jesus was also "the firstborn from the dead; *that in all things he might have the preeminence*". (Col 1:18)

An angel told John that Christ was "the beginning of the creation of God". (Rv 3:14) Christ would have to be the firstborn spirit if He was to be the Creator of everything and everyone else!

Proverbs 8:26-30 describes how wisdom, personified in Jesus, was "brought up" with the Father before the creation:
While as yet he had not made the earth...
When he prepared the heavens, I was there: when he set a compass upon the face of the depth:
When he established the clouds above: when he strengthened the fountains of the deep:
When he gave to the sea his decree, that the waters should not pass his commandment:
When he appointed the foundations of the earth:
Then I was by him, as one brought up with him: and I was daily his delight, rejoicing always before him.

Several early Christian church fathers rightly took the term "Firstborn" to mean Jesus was the first spirit born.

For example, Justin Martyr (110-165AD) said, Jesus was the spirit that "God begat before all creatures a *Beginning...* the Son... Lord and Logos." (*Dialogue with Trypho*, ch61)

Also, Origen Adamantius (184-253AD) taught that "Jesus Christ Himself, who came, was born of the Father before all creatures". (*The Fundamental Doctrines*, bk1, Preface 4, 225AD)

And, around 300AD, the Christian advisor to Constantine wrote, "Since God was possessed of the greatest foresight…in order that goodness might spring as a stream from Him, and might flow forth afar, *He produced a Spirit like to Himself*, who might be endowed with the perfections of God the Father… God, therefore, when He began the fabric of the world, set over the whole work that first and greatest Son." (*Divine Institutes*, Lactantius)

In 1833, the Lord told the prophet Joseph Smith, "I was *in the beginning* with the Father, and *am the Firstborn*." (DC 93:21) In 1909, the presidency of The Church of JESUS CHRIST of LDS under Joseph F. Smith confirmed this doctrine, "Jesus, however, is the firstborn among all the sons of God—*the first begotten in the spirit, and the only begotten in the flesh*." (*MD*, "Birth," p84)

Jesus may not have been the first spirit ever born to the Father. But He was the eldest spirit in the batch of souls assigned to Earth. He stepped forward when the Father needed someone to head up Creation, serve as God, and be the Savior.

If the most righteous was given spirit birth first, it was much like the resurrection where the most righteous rose from the dead first. Michael, Abraham, Moses, and other great spirits were also among the elder spirits and became great leaders under the premortal Jesus.

So, why God didn't create all spirits as strong and pure as Jesus and spare the entire His family all the heartache caused by evil ones? If the Father had created all His spirit children out of *nothing* then they all would have necessarily been equally perfect, coming from the same null source. Since God is not careless, any defects or inferiorities must have been intentionally introduced, an act totally inconsistent with His perfectly fair and benevolent character.

The fact is that *God did NOT start with equal or null ingredients*. He created them out of ingredients that had existed eternally, out of intelligences that had achieved varying degrees of wisdom and goodness. He who was best was given spirit organization first.

God explained, "If there be two spirits… these two spirits, notwithstanding one is more intelligent than the other, have no beginning, *they existed before*." (*PGP*, Abr 3:18-19) Spirits previously existed as multifarious intelligences in the vast chaos pool from which the Father created His family. He selected only intelligences that qualified for further progression. Then He did the best that could be done to create good spirits, given their various levels of worthiness and wisdom, without destroying agency.

Also, if the premortal Jesus had volunteered to fill a probationary assignment as a holy ghost in a godhead for a previous batch of spirit children, He would have accumulated comprehensive knowledge the creative process, of the hearts and minds of mortals, and of the process of leading souls to eternal happiness.

We had a Heavenly Mother?

One could just as well ask, "We had a Heavenly Father?" Many cultures believe in a god or gods that are not considered parents of humanity. But the Biblical God is not impotent, genderless, nor hermaphroditic. Holy Writ calls Him *Father of all*. (Eph 4:6) So, reason suggests there exists a *Mother of all*. It's unthinkable that God would be the sole parent, manufacturing spirits in his workshop like Geppetto cobbling together Pinocchio, leaving it to the Blue Fairy to play the mother role.

The psalm book of one group of gnostic Christians, called the Manichaeans (216–276AD), tells of an angel coming to take Adam at his death whereupon Adam asks, "How are my heavenly parents, the father of light and the *mother of all living*?" (*Manichaean Psalm-Book II*, edited by Allberry, pp189-199)

Another early gnostic Christian group included a reference to the mother in heaven at the start of their prayers, "From thee father and through thee mother, two names immortal, progenitors of Aeons." (*Refutation of all Heresies*, Hippolytus, 170-235AD, bk5, ch1)

According to the *Secret Book of John*, the godhead included "three: the Father, the Mother, and the Son, the perfect power". (*The Apocryphon of John*, The Nag Hammadi Library in English, editor Robinson, p103)

If it were true that God created all our spirit bodies with the snap of His fingers, why wouldn't He be snapping all humanity's physical bodies into existence as well? The first step in creating His only-begotten Son was to select a mother.

In 1909 the First Presidency, under Joseph F. Smith, wrote that: "Man, as a spirit, was begotten and born of heavenly parents, and reared to maturity in the eternal mansions of the Father [as] offspring of celestial parentage...all men *and women* are in the similitude of the universal Father *and Mother*, and are literally the sons and daughters of Deity." (*EOM*, bk2, "Motherhood")

"Our heavenly parents provided us with a celestial home more glorious and beautiful than any place on earth. We were happy there. Yet they knew we could not progress beyond a certain point unless we left them for a time." (*Gospel Principles*, CJCLDS, 1978) "There came a day, then, when Mother and Father said, 'Now, my son, my daughter, it is now your time to go.'" (*Teachings of Harold B. Lee*, CJCLDS, 1996)

When Zina Huntington Young's mother died in 1839, the prophet Joseph Smith told her she would not only meet her mother again in heaven but would also become reacquainted with her eternal Mother. (*History of the Young Ladies' Mutual Improvement Association of the Church of Jesus Christ of Latter-Day Saints*, [Deseret News, 1911], Susa Young Gates, 15-16)

Each earthly parent contributes in his or her unique ways to successfully raising children. The same is true of our Heavenly Parents. There is no reason to think that they will ever stop creating spirit bodies from the eternal well of spirit intelligence. It is natural to wonder about the details of spirit birth but, just as God has veiled the details of Christ's virgin birth, so will He likely keep the sacred details of spirit motherhood from being trampled by swine.

Aren't spirits intelligences?

It can be a bit confusing. Much like *fish* can mean *fishes* and *deer* can mean *deers*, *intelligence* can refer to the light or knowledge conveyed by the endless sea of urstoff that pervades all creation. Other times the plural is used to refer to the fundamental or elemental bits of spirit light that make up that sea and from which all things physical and spiritual are created, each scintilla capable of "independent" thought to "act for itself". (DC 93:30)

But sometimes the term "intelligences" refers to spirits. For example, Abraham wrote, "The Lord had shown unto me, Abraham, the intelligences that were *organized* before the world was; and among all these there were many of the noble and great ones… For he stood among *those that were spirits*." (*PGP*, Abr 3:22) The fact that God preordained prophets from among intelligences *that were organized as spirits* suggests that intelligences are not necessarily *all* organized as spirits.

Apostle Bruce R. McConkie wrote, "Intelligence or spirit element became intelligences [plural] after the spirits were born as individual entities… Use of this name [intelligence] designates both the primal element from which the spirit offspring were created and also their inherited capacity to grow in grace, knowledge, power, and intelligence." (*MD*, "Intelligence") One near-death survivor described her spirit as consisting of "pure consciousness or awareness". (youtube.com/watch?v=EvAbeQnkGlA, DL 14 Mar 2025)

What is a spirit's anatomy?

Cartoon characterizations of the bulb-headed, round-figured, snowman-like Casper may be fun for children but fall dismally short of anatomical correctness. The American Heritage® Dictionary defines a spirit as:

1. The vital principle or animating force within living beings.

2. Incorporeal consciousness.

3. The soul, considered as departing from the body of a person at death.

Theologically, these definitions are all correct, but definition number three suffers a shortcoming stemming from a lack of belief in premortal spirts among mainstream religions. It should say, "The *spirit* that enters the body of a person sometime between conception and birth and departs from the body at death."

While still a premortal spirit, Jesus "created man in his own image." (Gn 1:27) So, it follows that His spirit was anthropomorphic. Christ told Mahonri Moriancumer, "Man have I created after the body of my spirit." (*BOM*, Eth 3:16)

When Christ appeared to His apostles on resurrection day, replete with hands, and feet, they mistook Him for a spirit. (Lk 24:36-49) If the apostles believed spirits look like humans after being taught by Jesus for three years, they were probably right. And, when Nephi was given a tour of future messianic events by the Spirit of the Lord, he beheld that He "was in the form of a man". (*BOM*, 1 Ne 11:11)

When king Nebuchadnezzar saw the human form of the Spirit of the Lord in the fiery furnace, he "was astonied… and said unto his counsellors, 'Did not we cast three men bound into the midst of the fire?' They answered and said unto the king, 'True, O king.' He answered and said, 'Lo, *I see four men* loose, walking in the midst of the fire, and they have no hurt; and the form of the fourth is like the Son of God'". (Dn 3:24-25)

So, premortal spirits are like humans. "That which is spiritual in the likeness of that which is temporal… the spirit of man in the likeness of his person." (DC 77:2)

Collecting and processing input is not relegated solely to spirit sensory organs, nerves, and brain. Every elemental unit of their bodies is aware and sapient, a unified network of intelligence. For example, spirit vision does not rely on photon receptors in the eyes. The fact that spirits have eyes but can see things behind them without turning their heads suggests that God designed spirit bodies in His image as a blueprint for the final version of His children, beings glorified and immortalized after the manner of Christ's resurrection.

Spirits are about the same size as their adult human counterparts according to people who have retained or recovered memories from their premortal life. Only mature, full-stature spirits are ready to enter mortal baby bodies.

How can a spirit body match its mortal body in every tiny detail? It can't. Spirit creation is not prone to the same kind of errors that plague earthly conception, gestation, birth, and growth. Unlike spirits, human bodies are created by a very messy gene mixing process involving parents with biological defects. Unlike spirits, mortal development is influenced by environment, nutrition, lifestyle, sickness, and injury. And, spirits don't age whereas mortal bodies do, changing from babyish to decrepit over time.

Using Jesus as an example, it seems spirit-mortal body mismatches are resolved during the resurrection, the eternal body conforming to the spirit body. This explains why many of those closest to Him failed to recognize Jesus after He rose from the tomb. Mary Magdalene mistook Him for the gardener. (Jn 20:1-16) Two of His disciples failed to recognize Him when He joined them in their walk to Emmaus. (Lk 24:13-35) He had to show the crucifixion scars to prove His identity when He appeared to a group of disciples in a closed room. (Lk 24:39) And, when His disciples were out fishing from their boat, they failed to recognize Him when He called to them from the shore. (Lk 5:1-7; Jn 21:1-14)

These scriptural clues strongly suggest that the spirit body of every person is quite anthropomorphic generally but different in the details. Spirit and body will meld in the resurrection, resulting in bodies that are perfect, a pleasant thought for those among us who are dissatisfied with our physical appearance.

Can two spirits overlap?

There are thoracopagus twins that are conjoined in such a way that they share one heart. Having separate heads and identities implies two different spirits in one container. So, yes, two spirits can overlap.

Do spirit beings evolve?

God said that temporal things are in the likeness of their corresponding spirits. (DC 77:2) So, how did the virtually infinite variety of spirit things that animate Earth's diversity of life come to exist?

Could it be that there was once a chaotic cosmic state consisting only of an endless sea of primal intelligence, some of which sorted itself into one or more clusters inclined toward good, eventually progressing to become sapient spirit organisms that amassed sufficient power to create pathways to godhood for others to traverse while fervently striving to prevent the triumph of evil? Could it be that somewhere in this spirit evolution, the form and function that makes the human species so successful in goldilocksy physical biomes was intentionally reverse engineered into the design of spirit bodies to optimize the temporal-spiritual fit?

One can only indulge in wild speculations on this question! If somebody with divine authority knows the answer, they are doing a great job of keeping it from the world.

Do spirits have gender?

Of those people who claim to have temporarily left their bodies, some say that their spirits lacked primary sex characteristics while others tell of intimacy between individuals in the spirit world. Which is true? Neither! "God said, Let us make man in our image, after our likeness... So God created man in his own image... male and female." (Gn 1:26-27) Since the "us" and "our" included spirits, and God is male, it follows that spirits have gender. If roughly half the humans coming to Earth have been female, then about half of God's spirit children were also female.

It has been argued that since spirits do not have male or female sex, that God would not have created them with genitalia. However, if that argument were carried to its ultimate conclusion, spirit anatomy would be reduced to only those features a spirit needs. Spirits can float so they need no legs. They don't chew food so they need no mouths. Carry that logic far enough and spirits would be reduced to nothing more than minds or intelligences, which is what they were before they were organized as spirits!

Biological anomalies exist among earthlings that never happen in premortal spirit bodies. For example, some people have chromosomes that are too ambiguous for gender to be determined. For another example, in Klinefelter's syndrome, Turner's syndrome, and Chapelle syndrome, gender can be undifferentiated or mixed.

Also, there are intersex chimeric people with two sets of DNA. A chimera can have a body that is a mix of male and female, having started life as two fertilized zygotes, one of each sex, that fused together to form a single embryo in a process roughly the reverse of that which results in identical twins. (healthline.com/health/chimerism, DL 20 Feb 2019) The merging usually happens within a day or two after conception. So, if spirits enter their bodies at conception, as some people believe, then one spirit or the other would have to abandon ship once the two zygotes merge. It might muddle final judgment if both spirits remained in the same body.

Where there is only one mind, there can only be one spirit. Ensoulment is probably delayed until the final configuration of the baby body is set. Zygote splitting for identical twins can happen up to nine days after fertilization, meaning ensoulment probably happens after that. Where there are two minds, there must be two spirits, as in the case of Abby and Brittany Hensel who are one body with two heads. Of necessity, their spirits occupy the same physical dimensions.

Species such as fish, snails, slugs, worms, and various echinoderms that are born or hatched hermaphroditically raise the question as to whether their premortal spirits are duel-gendered or genderless. Clown fish are all born male. Only the male in the colony that is allowed to breed can morph to female and that only happens if the sole breeding female dies. Brahminy blind snakes (*Indotyphlops braminus*) reproduce solely by cloning, a process called obligate parthenogenesis. Their spirits would presumably all be female.

Sex switching was made famous in the first *Jurassic Park* movie for reptiles. It also occurs in myriad taxonomic groups such as fish, echinoderms, crustaceans, mollusks, polychaeta, and amphibians. Asian sheepshead wrasses fish are born female but can change gender if there are not enough males around whereas clownfish are born male and can change to female. Did their spirits have gender before coming to earth? Do their spirits change gender with physical gender changes?

If spirit gender is not permanent or does not express itself in the anatomy of spirit bodies, then humans with gender dysphoria might have an opportunity to be resurrected with the gender of their choice. This would be nice for people who married for eternity only to discover later that their chromosomes were different than the gender they were living. Sex reassignment in the resurrection would preserve those marriages into eternity. Otherwise, the marriage will have to be annulled and compensatory steps taken to provide those people with the blessings they thought were already theirs.

"Gender is an essential characteristic of individual premortal, mortal, and eternal identity and purpose." (ChurchofJesusChrist.org/topics/family-proclamation, DL 8 Aug 2018) The closest a church general authority has come to making a statement on the sexual characteristics of spirits is perhaps this; "For divine purposes, male and female spirits are different, distinctive, and complementary." (Ensign, *Marriage is Essential to His Eternal Plan*, David A. Bednar, Sep 1984) Such statements still leave room for us to wonder whether spirits have sexual organs or whether spirit gender is merely a matter of identity and purpose.

It has been speculated that it is life in mortality that determines spirit gender. But what about premortal spirits? Spirit genders are mentioned by PBE people who have memory scraps from their premortal experience and by NDE people who have had glimpses of their premortal spirit existence, but these accounts fail to clarify whether spirits have primary or secondary sex characteristics.

In the case of Howard Storm's temporary death, he looked down to see his translucent spirit body; "Something that disturbed me was that I was naked… I was a human being. I had a body. They [the angels of light] told me this was okay. They were quite familiar with my anatomy. Gradually I relaxed and stopped trying to cover my privates with my hands." (near-death.com/experiences/exceptional/howard-storm.html, DL 27 May 2020)

There is also a growing number of people who claim to leave their bodies and have sex with unembodied spirits on the astral plane. However, such experiences are likely imaginary, delusional, drug-induced, or demonic pleasures that have nothing to do with spirit anatomy.

Can spirits reproduce?

Some preachers like to entertain their audiences by repeating the apocryphal story in which fallen angels, presumably spirits, fathered giants with the daughters of men; "When men began to multiply on the face of the earth, and daughters were born unto them… the sons of God saw the daughters of men that they were fair; and they took them wives of all which they chose... There were giants in the earth in those days; and also after that, when the sons of God came in unto the daughters of men, and they bare children to them, the same became mighty men which were of old, men of renown." (Gn 6:1-4)

An eisegetical reading is required to see this passage as saying spirit angels begat giants with mortal girls. The "sons of God" more likely refer to mortal followers of God rather than angels. "For as many as are led by the Spirit of God, they are the *sons of God.*" (Rm 8:14) The story of angels marrying mortal women comes, not from the Bible, but from a portion of a Book of Enoch which, since it says their offspring were 450 feet tall, must be considered apocryphal. (book-ofenoch.com/chapter-7-sect-ii, DL 24 Feb 2019)

Furthermore, the Bible says nothing about giants being born to anyone, rather merely that there were Nephilim in the land, a proper name for a people at the time. (Str#5303, Strong's Concordance number 5303) The same is a name for a post-flood people as well. (Nm 13:33) The KJV is one of the few versions that, for reasons that escape scholars, translated Nephilim as giants.

Biology alone dictates that spirit bodies are not compatible with physical bodies for purposes of reproduction. Nevertheless, centuries of succubae and incubi traditions in religious cultures indicate that people have long believed spirits to be capable of coitus with mortals. For example, Jewish tradition says Adam's first wife Lillith consorted with the devil and begat spirit demons who roam the earth, sexually violating mortals of all ages. The tradition does not explain why Lillith's offspring are spirits like the devil instead of flesh like her or how they, as intangible spirits, can abuse physical humans.

To prove that He was *not* a spirit, the resurrected Jesus told His disciples, "Handle me, and see; for a spirit hath not flesh and bones, as ye see me have." (Lk 24:39) If He had been only a spirit, they would have felt nothing. Since mortals feel nothing tangible when touched by spirits, coitus with them is quite impossible. Likewise, mortal gametes would not likely react to spirit gametes in a way that constituted conception!

Could genitals be needed for spirit waste elimination? Comprised mostly of water, urine is inconsistent with the nature of spirits and spirit dung exists only in fantasy games. These arguments don't prove that spirits lack genitalia, only that there is no apparent need in that state.

Either way, the presence of anatomically correct sex characteristics on spirits would not guarantee that spirit reproduction is possible. Adam and Eve were fully equipped

172

with reproductive organs in Eden but "would have had no children" until they became regular mortals. (*BOM*, 2 Ne 2:23) The same is likely true for spirits.

Human reproduction involves massive cell division and the death of millions of unused gametes in a single sexual encounter, processes that are quite uncharacteristic of immortal spirit bodies. If spirits could reproduce, God would not need to create them.

Some people who have had near-death experiences say they were unable as spirits to feel temperature, taste flavors, or feel pain or palpable pleasure, senses that are integral parts of sexual reproduction. Some NDE'ers describe how the ghosts of people who were addicted to sex in mortality kept trying to reenact sexual antics to which they were addicted in mortality, but without success as spirits.

Christ said, "In the resurrection they neither marry, nor are given in marriage, but are as the *angels* of God in heaven." So, angels, whether spirits or resurrected beings, do not mate.

In summary, there is no definitive evidence to prove whether spirits of any sort have primary or secondary sex characteristics. One thing is certain, if spirits could reproduce sexually, Satan and his minions would be reproducing with wild abandon with the aim of overwhelming the planet with demons of darkness. The lack of sexual and reproductive capability is one of the reasons that evil spirits are envious of mortal bodies.

Do animals have spirits?

The word *animal* comes from the Latin *anima*, which means *soul* or *spirit*. Animism is a worldview held by many Buddhist, Shinto, Pagan and Neopagan groups. It is the belief that animals and all other living things have spirits. To reanimate something is to bring the spirit back into it.

But, the Catholic Church and its offshoots have historically taught that only humans have souls. It is true that only the spirits of humans are children of God and that other species not created in His image cannot become joint-heirs with Christ in eternity. (Rm 8:17) But God's plan of happiness includes animals because He loves them; "Are not two sparrows sold for a penny? Yet not one of them will fall to the ground outside your Father's care." (Mt 10:39)

Why would God require respect to animals if they have no part in eternity? "Blood shall not be shed, only for meat, to save your lives; and the blood of every beast will I require at your hands." (*JST (Joseph Smith Translation)*, Gn 9:11)

Many cultures have referenced animals as fellow soulful creatures, some worshiping their images, others thanking them for their meat. The Greek philosopher Pythagoras (570-495BC) said, "The animals share with us the privilege of having a soul." (inspiringquotes.us/quotes/gUjT_Tv6oFfMd, DL 17 Jul 2020)

The following points argue that animals indeed have souls:

1. Animals are created physically by the same process as humans. So, from a theological standpoint, whether it's emanationism, traducianism, creationism, or birth, animal spirits should be produced by a process similar to that of human spirits.
2. Scriptures support the idea of animal spirits. The OT affirms the existence of both "the spirit of man" and "the spirit of the beast". (Ec 3:19-21) God created the spirit of everything "before it was in the earth". (Gn 2:5) The beasts John saw in heaven must had spirits. (Rv 4:6) Joseph Smith taught, "John heard the words of the beasts giving glory to God, and understood them. God who made the beasts could understand every language spoken by them. The four beasts were four of the most noble animals that had filled the measure of their creation, and had been saved from other worlds." (HC, bk5, pp343-344)
3. Balaam's donkey expressed itself rationally to its abusive owner, a likely case of spirit-to-spirit communication since donkeys lack the necessary anatomy for physical vocalization. (Nm 22:28)

4. Dogs, dolphins, pigs, and parrots display mental abilities, personalities, and psychological traits once thought to be exclusively human. The line between the hairy and the hairless apes thins as science learns. It seems logical and proper that God would grant such intelligent and charismatic creatures an afterlife.

5. Many people who've had near-death experiences, pre-birth memories, or out-of-body glimpses of spirit realms tell of animals in greater variety there than seen on today's earth. Among the myriad adults and children who have come back from death, some reported that they were greeted by animals in the spirit world, some of which were beloved pets in life.

6. The most heavenly places on earth include beautiful fauna and flora. It's hard to imagine that heaven could be a paradise if it were devoid of all life forms except ours.

7. Religions that have historically denied animal ensoulment are now softening. Christianity has been slowly coming to the viewpoint that animals have spirits, one cleric and sect at a time. Perhaps church leaders and theologians nowadays have more experience with animal companions than in yesteryear. If this trend continues, all Christianity will soon embrace the idea:

   - Catholics: Pope John Paul II said, "Animals possess a soul."
   - Evangelicals: Billy Graham said, "If we need animals around us to make our happiness complete, then you can be sure God will have them there [in heaven]."
   - LDS Christians: "The spirit of man [is] in the likeness of his person, as also the spirit of the beast, and every other creature which God has created." (DC 77:2) This unambiguous doctrine has been embraced by the entire Church of JESUS CHRIST for the last two centuries.

Joseph Smith said, "The Lord intends to save not only the earth and the heavens, not only man who dwells upon the earth, but all things which he has created. The animals, the fishes of the sea, the fowls of the air, as well as man, are to be recreated, or renewed, through the resurrection, for they too are living souls." (*DOS*, bk1, p74) That includes amoebas!

Why aren't you a wombat?

The privilege of becoming a spirit child of God came with royal inheritance possibilities that were not available to other spirit creatures. Why did God make some intelligences into spirit children of God while others were made into spirits of pythons or pitcher plants? Were there qualification standards? Did God do some kind of classification on intelligences before assigning them to a particular spirit species, like the *legilimency* of the Hogwarts sorting hat assigned students to the appropriate house in the Harry Potter book series?

Since individual agency is at the core of God's plan for His children, it's likely that He allowed intelligences some degree of choice in the matter of their species assignment. Perhaps performance and preference were both part of the choice to qualify as higher-order creatures. Perhaps some intelligences shied away from the risks and responsibilities of becoming a spirit child of God as it held the possibility of failing and suffering hell, whereas all other life forms go to heaven.

So, perhaps you're not a wombat because you were overqualified and had demonstrated the strong desire to belong to the same species as God. In any case, it is likely that every non-human spirit was happy with its lot before deciding to don the trappings of mortality on Biosphere One, considering it a blessing to be supporting actors in the drama engineered by premortal humans and their divine Father. After all, there is a happy ending for them all in the end!

What powers do spirits have?

President Lorenzo Snow said, "In our spiritual birth our Father transmitted to us the capabilities, powers and faculties which he himself possessed, as much so as the child on its mother's bosom possesses, although in an undeveloped state, the faculties, powers and susceptibilities of its parent." (*JD*, bk14, 1972, p302)

Okay, let's start with the obvious. Spirits can pass through solid walls. The stuffing of spirits doesn't interact with the physical. If a spirit tries to shake your hand, "you will not feel anything". (DC 129:4-8) Spirits cannot use mechanical means to manipulate physical objects.

A spirit body can sense when its physical body is about to shut down or already has and leaves. No external power yanks the ghost out. Ensoulment involves a special connection between a spirit and its body's mind, often referred to as the "silver cord" in NDE accounts and even once in the OT; "People go to their eternal home and mourners go about the streets. Remember him (your Creator) — *before the silver cord is severed.*" (Ec 12:5-6)

Spirits can float unaffected by gravity and travel to distant places faster than the speed of light just by deciding to be there. Brigham Young said that spirits "move with ease" and are "free to travel with lightning speed to any planet, or fixed star, or to the uttermost part of the earth, or to the depths of the sea, according to the will of Him who dictates". (*JD*, 1868, bk13, pp76-77) Instantaneous spirit travel is confirmed by dozens of near-death experiences. Faster-than-light travel is just as possible for spirits as for spacetime as it expands.

Spirits communicate telepathically with each other without moving their lips. They can use this telepathy to "whisper" thoughts to the minds of humans. They do not rely on physical sound waves to hear. Nor is their vision limited to forward-looking eyes. Rather, they are aware of their surroundings in all directions without turning their heads and can see as far as they desire to look, even through physical barriers. Near-death experiencers say they sense all things through the pervasive spirit medium like radio transmissions.

Evil spirits can assert control over a vulnerable mortal mind. Sometimes more than one spirit enters illegitimately into someone. (Mk 5:1-9) If one evil spirit lacks the strength to possess and control a person alone, "Then goeth he, and taketh with himself seven other spirits more wicked than himself, and they enter in and dwell there: and the last state of that man is worse than the first." (Mt 12:45)

But spirits have limits placed on the by God. Spirits waiting to be born into mortality must participate in training to prepare for life on Earth. The spirits of the living must remain joined with their bodies except in cases of impending death. The spirits of the deceased are constrained to places and conditions commensurate with their characters.

Evil spirits roam Earth trying to find relief through the wickedness and misery of mortals. Although they cannot harm people *directly*, they can mess with people's minds to wreak havoc as God permits, as in the case of Satan afflicting Job and family. (Jb 1:6–12) And, even when there is no danger, an evil spirit can make a person feel that they are doomed. Conversely, righteous spirits with angelic license can make a person feel loved and secure.

Are spirits bound by time?

Spirit angels come and go as if distance and time mean nothing. It's quite a journey from heaven to earth and yet they are always right on time. They are like Peter Jackson's Gandolf in *Lord of the Rings*; "A wizard is never late, nor is he early, he arrives precisely when he means to." Angel punctuality was quite fortunate for Isaac who was saved from his father's knife in the nick of time.

Angels can take the spirit of a mortal on a long journey instantaneously. Paul said, "I was caught up to the third heaven." (*NLT (New Living Translation)*, 2 Cor 12:2) Similarly, the prophet Nephi "was caught away in the Spirit of the Lord, yea, into an exceedingly high mountain". (*BOM*, 1 Ne 11:1) If there was any sense of time during these journeys, neither of these prophets thought it sufficiently significant to warrant mention.

In addition to being ageless, spirits can travel across the cosmos without experiencing relativistic time dilation, i.e.

they don't return to earth from afar to find everyone they know dead from old age.

So, do spirits even sense the passage of time? Well, the Son of God, while still a spirit, numbered the time periods in the Genesis Chapter One creation saga. So, if nothing else, sequentiality and action are part of a spirit's experience.

Can spirits visit the past? Yes, but it's like viewing holograms of the past or replaying 3D recordings rather than travelling back in time. Brigham Young opined, "If we [as spirits] want to behold Jerusalem as it was in the days of the Savior; or if we want to see the Garden of Eden as it was when created, there we are, and we see it *as it existed spiritually*, for it was created first spiritually and then temporally, and *spiritually it still remains.*" (*DBY*, p380) He was referring to the 3D animated plans for Creation or the spiritual designs crafted by the Council of the Gods.

Literal travel to the past would create unresolvable paradoxes even for spirits. For example, a spirit might return to the moment Satan rebelled, tell him about his horrible future, thus persuading him to choose good, so the horrible future from which the spirit came never happened, so the reason for the spirit to travel back in time ceases to exist so, so nobody notifies Lucifer, so he behaves badly after all, and so on. There's the impossible causality loop —a classic time paradox.

A version that avoids free-will elements is Polchinski's Paradox in which a billiard ball falls into a wormhole, backward through time, and out the other end to collide with its younger self such that no ball falls into the wormhole in the first place. A paradox means something is false or impossible. Happenings can't be unhappened, not even in the spirit world. So, backward time travel is impossible, even for spirits!

Can spirits skip forward in time? No. Spirits must wait for certain events, just like mortals, though the sense of time passing is different for spirits. For example, spirits must wait for the resurrection of their physical bodies. When the resurrection of Christ was imminent, the spirits of long dead righteous folks "were filled with joy and gladness, and were rejoicing together because the day of their deliverance was at hand. They were assembled *awaiting* the advent of the Son of God into the spirit world, to declare their redemption from the bands of death… the dead had looked upon the *long absence* of their spirits from their bodies as a bondage." (DC 138:15-50) So, waiting for something important to them can be unpleasant even for spirits.

Who are the spirit angels?

Contrary to what Zuzu's teacher said in the movie *It's a Wonderful Life*, angels do not get wings every time a bell rings. Even the best angels lack wings. They don't need them.

Most Christians think that angels are some kind of peculiar spirit creature that God created before He decided to make humans. But the word *angel* means *messenger*. When Jacob wrestled with an angel and established a hold on him from which there was no escape, if it had been a spirit it would have easily slipped through Jacob's grasp. (Gn 32:24–32) The angel must have been a human messenger, likely sent by God from Melchizedek's country, the high priest to whom Abraham paid tithing. (Gn 14:20; Heb 7:2)

Like the Lord, who descended from the presence of God to become Jesus in mortality, other angels in positions of great authority have or will come to Earth as prophets:

Michael – Hebrew for "Like God." The archangel and commander of angelic forces against Satan is identified by early Christian writers as the angel who, in the premortal realm, was second in command to Jesus: "I placed on the earth, *a second angel*, honorable, great and glorious… I called his name Adam." (*2 Enoch,* ch30) "The Lord appeared unto [those attending Adam's last family meeting before he died], and they rose up and blessed Adam, and called him Michael, the prince, the archangel." (DC 107:54)

Gabriel – "My strength is God." Counted as one of the archangels, Gabriel's earthling identity was revealed by Joseph Smith who said, "Noah, who is Gabriel… stands next in authority to Adam in the Priesthood." (*HC*, bk3, ch26, p385) Since Gabriel's angelic missions were in the spirit of an Elias, such as announcing the birth of John the Baptist and Jesus the Christ, "Elias" is one of his aliases in modern revelation. (DC 27:5-7)

Raphael – "God heals." According to the Catholic Bible's *Book of Tobit*, Archangel Raphael rescued Tobias and a few other people from death and performed other miracles. He is also mentioned as one of those who, as part of the 19th century restoration of all things, declared "their dispensation, their rights, their keys, their honors, their majesty and glory, and the power of their priesthood." (DC 128:21) But Raphael's earthly identity remains a mystery.

Uriel – "My light is God." The *Second Book of Esdras*, still part of the Catholic Bible, says Uriel will be one of the rulers at the end of the world. His earthly identity will likely remain a mystery until then.

Joseph Smith said, "There are no angels who minister to this earth but those who belong to it." (DC 130:5) The premortal service of spirit angels (Raquel, Sariel, Ramiel, et cetera) was part of their preparation to come on earth as mortal prophets, prophetesses, leaders, fathers and mothers.

Was Satan Jesus' brother?

Let's jump to the point. The definition of brother is: "A male having the same parents as another or one parent in common with another." (Wordnik)

Is Lucifer a spirit? Yes, he is a fallen angel. (Is 14:12) Is he a son of God? Yes, God is "the Father of spirits"! (Hb 12:9) God is the Father of angels! (Satan attended the meetings "when the sons of God came to present themselves before the LORD". (Jb 1:6; 2:1) So, God is the Father of Lucifer!

Is God also the Father of Jesus? Yes. (Jn 3:16)

Since God is the Father of both, the conclusion is inescapable — they are brothers! That logic is watertight.

Which two sons of God are called "Morning Star" in the Bible? Jesus and Lucifer! (Isa. 14:12; Rev. 22:16)

Christ is the eldest spirit child of the Father, "the firstborn of every creature". (Col 1:15) Some scholars believe Satan was the second spirit created by God but there is no support for that in sacred texts, although Joseph Smith said that God had considered Lucifer to have had been the "next heir and had allotted to him great power and authority" after Christ. (*George Laub Journal*, WJS, 7 Apr 1844, p362)

Whenever and however He fathered Satan, God is at least as much the Father of Satan as Dr. Lanning was of Sonny in the movie *iRobot*, by creation!

The brotherhood/sisterhood of *all* spirits is as old as the Christian church. One ancient Gnostic Christian text says that the Holy Ghost appeared to Mary and asked, "Where is Jesus, my brother, that I may meet him?" (*Pistis Sophia*, in Fragments of a Faith Forgotten, by G.R.S. Mead, 1900, p475)

Arguments against the brotherhood of Jesus and Satan often veer into a micro-sermon on how different the two are without mentioning how similar they once were. Lucifer was once very righteous and highly respected in heaven. The name *Lucifer* (from the Hebrew for *Shining One, Light Bearer*, or *Morning Star*) was historically a reference to the Venus, which was most noticeable just before sunrise, an allusion to his star-like and light-bearing nature. (Str#1966) But Lucifer was not the only one to be called that in scripture. "I Jesus... am the root and the offspring of David, and the bright and *morning star*." (Rv 22:16) It is instructional and poetic that both Jesus and His nemesis brother would be called after the *morning star*.

The simplest truths can sometimes be made to sound shocking. But why should the idea of Lucifer being Christ's spirit brother be shocking? One brother took the highest road and the other took the lowest. There's a lesson in that!

Does it shock anyone that both Hitler and Jesus are members of the same human family? Is it shocking that Jesus not only welcomed the traitor Judas as a friend but actually was the One who asked Judas to join His Twelve!

It shocked the Jews that Jesus mingled with a variety of vile sinners! If any of these relationships denigrated Him in the least degree, then everyone who has a relationship with Jesus is guilty of denigrating Him, "For all have sinned, and come short of the glory of God." (Rm 3:23) If God's love is unconditional, then He still loves and mourns the loss of every fallen angel, even His rebellious son Satan.

It is bewildering how dogmatic Christians can be, without Biblical support, that Jesus cannot be anybody's spirit brother. He had sinful *mortal* brothers without any stigma. "While he [Jesus] was still speaking to the people, behold, his mother and his brothers stood outside, asking to speak to him." (Mt 12:46) "Then his mother and his brothers came to him, but they could not reach him because of the crowd." (Lk 8:19)

If it's not shocking that Jesus had *mortal* brothers, why should it be shocking that He had *spirit* brothers?

Is the Holy Spirit female?

Since the role of the Holy Ghost includes some stereotypically feminine activities, such as comforting people, some have speculated that he is female. As evidence, they point to the fact that only 9 of the 84 instances of the word "spirit" in the Old Testament are masculine terms. The others are either decidedly feminine or neuter. In Aramaic, the language believed to have been spoken by Jesus and his disciples, the word spirit (*rukha*) is feminine. The word for spirit in Hebrew (*rauch*) is also feminine. The Greek word for spirit (*pneuma*) is neuter. And the Greek pronouns referring to the Holy Spirit in the New Testament are genderless.

But *grammatical gender* of a word is often not the same as the gender of the thing the word represents. In German, for example, Mädchen and Fräulein are both neuter despite meaning maiden and girl. The German terms for ocean – der Ozean, das Meer, and die See – all can be used interchangeably but are three different genders as can be seen by the article that precedes them (der, das, die). So, the linguistic gender of *spirit* cannot be used as proof that the Holy Ghost is female.

Some simply think it fitting that the Holy Ghost be a female so that the godhead is not without the feminine component. Such speculation dates back at least to the 3rd century AD when the writer of a gnostic Christian gospel said, "Some said Mary became pregnant by the Holy Spirit. They are wrong and do not know what they are saying. When did a woman ever get pregnant by a woman?" (*Gospel of Philip*, 55:22-23) Also from 3rd century theologian Origin is this quote, "Just now my mother, the holy spirit, took me by one of my hairs…" (faithfutures.org/JDB/jdb134.html, DL 18 Jul 2020) Another 3rd century text calls the Holy Spirit "mother" in several places and also praises the "Holy Spirit, the mother of all creation." (*The Acts of Thomas, 39*) The Greek version says, "We praise and glorify you [Jesus] and your invisible Father, and your Holy Spirit *and* the mother of all creation." (*The Apocryphal New Testament*, Elliot, 1993, p464)

What would be the implications of the Holy Ghost being female? It would make sense that the Holy Spirit be God's wife, making a nuclear family in the trinity – Father, Mother, and Son. And, if the Holy Spirit is Christ's spirit mother, then she is also the mother of other spirits. And, if our tangible celestial God used a tangible *mortal* mother to sire His tangible *mortal* Son, then He would logically need an intangible *spirit* mother to sire all His *spirit* children. But this line of thinking is beyond the fringe.

The bottom line is that virtually all modern theologians and prophets refer to the Holy Spirit as masculine. The Triune God of mainstream Christianity has no room for a female member. Heber C. Kimball, apostle and close associate of the prophet Joseph Smith gave an emphatic answer, "Well, let me tell you, the Holy Ghost is a man; he is one of the sons of our Father and our God; and he is that man that stood next to Jesus Christ." (*JD*, bk5, p179)

Who impregnated Mary?

When the angel Gabriel told Mary she would bear a son she asked how she could possibly conceive a child without being intimate with a man. "The angel answered and said unto her, The Holy Ghost shall come upon thee, and *the power of the Highest shall overshadow thee*: therefore also that holy thing which shall be born of thee shall be called the *Son of God*." (Lk 1:35)

The Holy Ghost was necessarily with Mary so that she could withstand the presence of the Father. Perhaps it doesn't matter to people who believe in a three-in-one God but, according to Gabriel, the Highest was present at the conception of Jesus. President Brigham Young said, "When the Virgin Mary conceived the child Jesus, the Father had begotten him in his own likeness. *He was not begotten by the Holy Ghost*." (*JD*, bk1, p50)

The strongest statements from LDS Christian leaders say that Christ was literally the Son of God in the same way that a human baby is offspring of mortal parents. Although none of the following statements are canonized doctrine of The Church of JESUS CHRIST of LDS, they have been sensationalized by critics as if they were:

♦ "Now, we are told in scriptures that Jesus Christ is the only begotten Son of God in the flesh. Well, now for the benefit of the older ones, how are children begotten? I answer just as Jesus Christ was begotten of his father ... Jesus is the only person who had our Heavenly Father as the father of his body." (*Family Home Evening Manual*, Joseph F. Smith, 1972, pp125-126)

♦ "Christ was begotten by an Immortal Father in the same way that mortal men are begotten by mortal fathers." (*MD*, "Only Begotten Son")

♦ "We believe absolutely that Jesus Christ is the Son of God, begotten of God, the first-born in the spirit and the only begotten in the flesh; that He is the Son of God just as much as you and I are the sons of our fathers." (*Millennial Star,* Heber J. Grant, p2)

♦ "Elohim is literally the Father of the spirit of Jesus Christ and also of the body in which Jesus Christ performed His mission in the flesh..." (*The Articles of Faith*, James E. Talmage, pp466-467) What is meant by *literal Father*? Elohim created Christ's spirit, though the details are unknown and unimportant. Likewise, Elohim created baby Jesus, though the details are unknown and unimportant.

♦ "The Church of JESUS CHRIST of Latter-day Saints proclaims that Jesus Christ is the Son of God in the most literal sense. The body in which He performed His mission in the flesh was sired by that same Holy Being we worship as God our Eternal Father. Jesus was not the son of Joseph, nor was He begotten by the Holy Ghost. He is the Son of the Eternal Father!" (*Come Unto Christ,* Ezra T. Benson, p4)

♦ "And when the Virgin Mary was… with Child it was by the Father and in no other way only as we were begotten." (*Wilford Woodruff's Journal*, April 9, 1852) Woodruff went on to say that God charged His body with the tangible material necessary to make the process of conception possible by eating food grown on Earth.

These brethren obviously interpreted the scripture phrases "the power of the Highest" and "the Son of God" in a very literal way, perhaps encouraged by a statement from the Book of Mormon that Mary "is the mother of the Son of God, *after the manner of the flesh*". (*BOM*, 1 Ne 11:18)

The risen Jesus had a *tangible* body of "flesh and bone" and ate of "broiled fish, and of an honeycomb" with His disciples. (Lk 24:39-43) So, since God the Father has a tangible body like Christ's, He could have eaten physical food, as Wilford Woodruff suggested, and contributed a physical gamete from His own body to make Jesus His Son in a literal sense. Christ, being male, could not have been a clone of Mary. His conception required a Y chromosome!

Each mortal human has 46 chromosomes. Reduction division (meiosis) in parent cells produces 23-chromosome haploid cells, sperm in males and eggs in females. Every sperm cell that does not fertilize an egg must die. If God's cells are all immortal, then He must have somehow synthesized at least one functional physical 23-chromosome haploid male gamete for the conception of Jesus.

In sum, the doctrine only says that Jesus is God's only begotten son but does not say whether the conception of Jesus involved coitus. If humans have discovered various ways to precipitate a zygote and implant it in a woman without her losing her virginity, as we say in the current vernacular, God could certainly do something similar. The Book of Mormon makes a solid point of calling Mary a "virgin" *after* Jesus was born. Nephi said, "I looked and beheld the virgin again, bearing a child in her arms." (*BOM*, 1 Ne 11:20)

Is there evidence for spirits?

Let's just say it up front; there is no physical evidence for ghosts. Nobody has ever collected ectoplasm and there are no certifiable photos of ghosts. "Virtually all ghost hunters claim to be scientific, and most give that appearance because they use high-tech scientific equipment such as Geiger counters, Electromagnetic Field (EMF) detectors, ion detectors, infrared cameras and sensitive microphones. Yet none of this equipment has ever been shown to actually detect ghosts." (livescience.com/26697-are-ghosts-real.html, DL 28 May 2018) In fact, it seems silly that ghost hunters would use physical equipment that only measures physical phenomena to detect the presence of nonphysical beings.

What about mediums who sense ghosts or read them? Couldn't they be using their spirit senses to do so? These scam artists have been repeatedly debunked as merely making guesses based on probabilities. For example, a reader might choose an older person from the audience and say something like, "I sense your mother's presence near you. She had trouble getting around last you saw her and was on some kind of medication... At times you doubt whether you've made the right decision or done the right thing..." Well, of course an older person's mother is likely dead, took medicine before expiring, and suffered debilitation toward the end. And most people have doubts about one or more decisions regarding a dying parent.

Or, a reader might mutter something that sounds like a horoscope, balancing each sentence to allow the client to selectively identify; "Sometimes you are outgoing and sociable, while at other times you feel more withdrawn. You prefer a certain amount of change and variety but can feel stressed when change comes unexpectedly..." People usually remember only those topics from the reading that resonate with their life experience and end up marveling at how much the medium got right.

Cases where people "see" or sense ghosts can be caused by medical conditions or by environmental factors that trigger the brain's imagination, such as subtle drafts or infrasounds which, though inaudible, can cause a reaction in the brain akin to trepidation. (science.howstuffworks.com/science-vs-myth/afterlife/ghost3.htm, DL 6 Sep 2017)

Also, in evidence are the testimonials of people who retain memories of their premortal existence as spirits. In one case, without knowing about the miscarriage, a young boy remembered the sister who had gone to earth before him but died in the womb. In another case, a three-year-old boy asked his mother for a story about Grandpa Robert, her father who had died many years before. When the mother asked the boy how he knew about her father he explained, "He's the one who brought me to earth." (liveabout.com/life-before-birth-2594548, DL 25 May 2020)

Common memories of the premortal spirit existence include being in the presence of a spirit advisor, being shown future life situations, being apprehensive about leaving the peaceful spirit realm for earth, choosing one's parents, waiting to descend into a fetus until being given the signal to do so, actually descending into a baby body and trying to adjust to it. God does not just reach into a grab bag of spirits and pull out the one that happens into His hand when a baby needs ensoulment. Progress from spirit to mortal is orderly.

In some cases, children remembered things from their time in the womb, such as the music playing on the car radio on the way to the hospital for delivery, a stop for fuel on the way, or a hospital phone call. Awareness of such minor details suggests that spirits who have attached to the unborn baby body can still sense events outside the womb via their spirit awareness. This has implications for expectant mothers who talk about aborting their babies whilst their unborn baby listens. Interested readers can search the web for "PBE NDE" to find and evaluate hundreds of such narratives for themselves.

There are dozens of books full of testimonials. For example, in *Coming From the Light*, 150 narratives describe instances where people, often young children, have seen spirits who were later born into the same family. A 2013 (Public Library of Science) PLOS study compared NDE memories with other intense real-life memories and imaginary thoughts and dreams and found that the NDEs were the richest and most vivid of all. A 2014 Electroencephalogram (EEG) study found that NDEs are stored in the same part of the brain as events in which the person actively participated rather than where learned facts are stored. The conclusion was that NDEs show electrical patterns that are quite different from imagined memories. (*A Brush With Death,* by Amy Paturel, Discover, May 2016, p24) One would expect the similar results from comparable PBE studies.

It's unlikely that such pre-birth and post-mortem spirit memories are a product of previous exposure to religious ideas because the vast majority of children and adults who have such spirit memories are associated with churches whose official doctrinal stances exclude any kind of premortal existence. Most Christian churches have taught that deceased people go straight to heaven or hell, so when members of those churches have a posthumous experience that involves neither, it can't be attributed to religious bias.

Some people whose religion denies a premortal existence died with an I-didn't-ask-to-be-born attitude but then encountered a spirit who gave them a PBE glimpse in which they saw themselves choosing to come to Earth and making other premortal decisions that determined the time and circumstances into which they were born. Dozens of atheists and agnostics returned from temporary deaths to relate their out-of-body experiences as spirits. (For a small sampling, see near-death.com/philosophy/atheism-and-ndes.html, DL 25 May 2020)

So why doesn't everyone remember their premortal lives as spirits? Well, according to PBE/NDE accounts, a veil of forgetfulness or, more precisely, a disconnectedness from the universal intelligence that permeates all reality, occurs at ensoulment and blocks most premortal memories from infancy. This amnesia allows people to choose anew to love others, develop mentally, grow spiritually despite the vicissitudes of mortality and the challenges thrust upon them by the choices of others. Also, without the veil, most people's lives would be derailed by anticipation of, and a desperate longing to return to, the peaceful and painless realm of spirits from whence they came.

Is there only one spirit world?

When you die, will you return to the same wonderful realm from which your premortal spirit came?

One thief on the cross next to Jesus was rebuked by the other who defended the Savior. Jesus told His defender, "Today shalt thou be with me in paradise." The fact that He made no such promise to the first thief suggests that there is more than one destination for the spirits of deceased people and that afterlife spirit realms, tailored as they are for people exiting mortality, are not suitable places for pristine premortal spirits awaiting their turn on Earth.

Some PBE memories of the premortal the spirit realm include glorious campus experiences where spirits prepare themselves for Earth. There is also a less-glorious staging area seemingly closer to Earth where spirits wait who are about to enter their mortal baby bodies.

190

Where is the spirit world?

Jesus said, "I beheld Satan as lightning *fall* from heaven." (Lk 10:18) So, since Satan is *on* the earth now, the premortal spirit "heaven" from which he fell must be *above* the earth.

Was the premortal "heaven" as high as the heaven where God dwells in celestial glory? Satan's claim, in the premortal spirit heaven, "I will ascend into heaven," suggests that the premortal spirit heaven is below God's personal home in the hierarchy. (Is 14:13)

This fact is supported by John's vision of precreation events: "There appeared a great sign in heaven in the likeness of things on the earth: a woman… And the woman, being with child, cried, travailing in birth and pained to be delivered. And she brought forth a man-child, who was to rule all nations with a rod of iron; and her child was caught up unto God and his throne." (*JST*, Rv 12:2-3) So, the woman, who represents the precreation church of God among spirits, brought forth the premortal Jesus, the Christ, who was caught up to the Father's throne in the celestial heaven and made King over all creation. (*JST*, Rv 12:7)

Jesus spoke of His premortal descent from God's celestial heaven down to the spirit heaven, *"I came down* in the beginning in the *midst of all the intelligences…*that were *organized* before the world was." (*PGP*, Abr 3:21-22) After coming down, the Firstborn announced the creation project that would unfold on the lower physical plane, *"We will go down, for there is space there,* and we will take of these materials, and we will make an earth whereon these may dwell…" (*PGP*, Abr 3:24)

So, there are at least three premortal heavens, God's celestial home, a lower premortal spirit realm where spirits prepared the earth and prepared themselves for Earth life, and the even lower realm where spirits are staged for entry into earthly bodies, as per PBE testimonials.

Where do spirits get energy?

All physical forms of life need fuel to function. But what about Spirits? They constantly move and act without even sleeping. Must they eat or recharge to keep going?

Ghosts consist of structured spiritual light or energy that is massless. They move and act effortlessly independent of physical laws. So, a spirit's energy is never expended by exertion. The 2nd law of thermodynamics does not apply to the spirit world.

A spirit's energy level is determined by its nature. Malevolence results in loss of spiritual light. Benevolence attracts and accrues more of the ambient or unstructured spiritual light that fills all reality, aka intelligence. Spirits do not die for lack of spiritual light. They simply darken down to dim as noted by people who've returned from NDEs having seen evil spirits in the realms of the dead.

According to many an NDE, spirit beings and things hum with frequencies of light. Can spirit light vibrate forever? Why not? That's exactly what physical light does! A photon travelling in a vacuum will never die. Photons are massless like spirits. So, spirits should be at least as exempt from decay as photons.

Doctors intentionally flat-lined the brain and heart of Pam Reynolds to perform a brain surgery they nicknamed standstill. After the blood drained from her head, they removed an aneurysm and then restarted her systems. When she awoke, she described seeing all the people and equipment involved in the operation while outside her body. She also told of meeting deceased relatives as a spirit and being strengthened by them; "They were feeding me. They were not doing this through my mouth…but they were nourishing me with something… something sparkly… I definitely recall the sensation of being nurtured and… made strong." (near-death.com/science/evidence/people-have-ndes-while-brain-dead.html, DL 9 Aug 2018)

How do spirits pass through walls?

Solid matter is 99.999% space. So, there is plenty of room for "ghost particles" – that's science's pet name for neutrinos – to pass through solid concrete. Trillions pass through each person every day. Neutrinos could travel through a wall of pure lead as thick as the galaxy is wide without slowing one bit and without bumping into anything. About 65 billion neutrinos from the sun zip through each square centimeter of earth's surface and out the other side of the planet every second. (timeblimp.com/?page_id=1033, DL 28 May 2018)

So, if physical things like neutrinos, X-rays and radio waves can pass through solid barriers, why should it strain credulity to say that spirits can? Spirits could not have a physical interaction even if they wanted to. This should be taught to little children everywhere so they stop fearing ghosts. Oh yes, there are cases where spirits have appeared to people. But such visions are seen through the spirit eyes of the perceivers.

God has safeguards in place that protect the innocent from undue manipulation by evil spirits. Even righteous spirits have to get divine authorization before revealing themselves to a person. God has provided a way to test a spirit's intentions. "When a messenger comes saying he has a message from God, offer him your hand and request him to shake hands with you… If he be the spirit of a just man… he will not move, because it is contrary to the order of heaven for a just man to deceive; but he will still deliver his message. If it be the devil as an angel of light, when you ask him to shake hands he will offer you his hand, and you will not feel anything; you may therefore detect him." (DC 129:4-8)

So, since spirits lack tangibility in the physical world, if you hear something go bump in the night with your physical ears, it may be a sign of danger, but it is not a ghost.

Can spirit bodies have flaws?

The premortal Jesus revealed himself to a prophet of old and said, "Man have I created after the body of my spirit; and even as I appear unto thee to be in the spirit will I appear unto my people in the flesh." (*BOM*, Ether 3:16) Did Christ's spirit have the tiny skin imperfections that His earthly body had? Did His spirit have the slight misalignment of teeth that His mortal mouth might have had, or other imperfections typical in human bodies? Since Jesus was born of a mortal mother who, like everyone else in the human family carried some imperfect genes, it is likely His earthly body had a few imperfections.

So, yes, spirit bodies look like mortal bodies – generally – but do not have any of the flaws that are inherent in mortality. "Spirit beings have the same bodily form as mortals except that the spirit body is in perfect form." (*Gospel Principles*, CJCLDS, 2009, p242)

Can spirits be destroyed?

Paul said that Jesus would "destroy him that had the power of death, that is, the devil". (Hb 2:14) However, the word "destroy" is translated from a Greek word meaning *to render inoperative*. (Str#2673)

And Jesus claimed "all power, even to the destroying of Satan and his works at the end of the world". (DC 19:3) But "destroying" could mean conquering, as is the case when one professional sports team "destroys" another in a game.

At least some evil spirits feared total destruction, as was expressed by one of a number who were about to be exorcised by Jesus; "Art thou come to *destroy* us? I know thee who thou art; the Holy One of God." (Lk 4:34) Here the word "destroy" is translated from a Greek word meaning *to kill* or *to destroy utterly*. (Str#622) Is that proof that evil spirits be completely obliterated, spirit body and mind? Are they deconstructed and recycled back into the chaos from which they came? No. Fear evoked the hyperbolic "destroy" in the case of the evil spirit's words to Jesus.

While Satan might wish he were erased out of existence in the future, God said he must remain to "be tormented day and night for ever and ever". (Rv 20:10) Like all other spirits, Satan is immortal. The destruction of the devil simply means the termination of his career as tempter and the complete unwinding of his legacy. The spirits who commit the unpardonable sin in life, for whom "there is no forgiveness in this world nor in the world to come" are to be "with the devil and his angels in eternity". (DC 88:34-44) That prophecy can't be fulfilled if evil spirits are completely obliterated out of existence before eternity expires.

Spiritual creatures are indestructible. The hottest lake of fire and brimstone cannot destroy a spirit. Joseph Smith said, "The spirits of men are eternal." (*TPJS*, p208)

Why not stay just spirits?

Without a body, your spirit can flit around unaffected by gravity and pass through walls and zoom through outer space. All that seems like fun. Your spirit is immortal and impervious to injury, sickness, or pain. That seems pleasant. It can commune telepathically with other spirits and tap into the intelligence that radiates from God. That seems useful. So, why not stay a spirit forever? Why get born into mortality or resurrected?

Firstly, without a body in the resurrection, you will not achieve God's brand of full potential, "knowledge, power, glory, and intelligence". (*TPJS*, p354) Secondly, God said, "Man is spirit. The elements are eternal, and spirit and element, inseparably connected, receive a fulness of joy; And when separated, man cannot receive a fulness of joy." (DC 93:33-34) Earth life enables us to fully appreciate the gifts God has prepared for His children!

Thirdly, without tangible bodies, "our spirits must become subject to that angel who fell from before the presence of the Eternal God, and became the devil... And our spirits must have become like unto him, and we become devils". (*BOM*, 2 Ne 9:8-9)

God saves, not just from the sins we commit in mortality, but from being dominated by evil and power-mongering spirits! The Father has a glorified body, a key factor in His dominance over evil. Joseph Smith said, "Beings who have bodies have power over those who have not." (*TPJS*, p181) Smith also said, "The spirits of all men were subject to oppression and the express purpose of God in giving it a tabernacle was to arm it against the power of Darkness... All his acts [are] for the benefit of inferior intelligences--God saw that those intelligences had not power to defend themselves... therefore the Lord calls them together in counsel, organizes them, and agrees to form them tabernacles." (*WJS*, pp62-68) Smith further explained how righteous spirits achieve immunity against the forces of evil forever. "This is done by our taking bodies ... and having the Power of the Resurrection pass upon us whereby we are enabled to gain the ascendancy over the disembodied spirits." (*WJS*, p208)

What do spirits wear?

Do spirits have clothes? Do they set them aside when they enter their earthly baby body?

According to Rabbinical writings, every spirit descending to Earth from the treasury of souls "divests itself of its heavenly garment, and is clothed in a garment of flesh and blood". (*Tree of Souls*, Howard Schwartz, 2004, sec200, p166)

Do spirits put the same clothing on again when they leave mortality? Is modesty even an issue with spirits? Is there a spirit clothing factory? Is there a clothing dispensary?

Howard Storm was self-conscious about his nakedness in the spirit world during his temporary death experience but his new spirit angel friends assured him that his nudity was okay and that they were quite familiar with his anatomy. (near-death.com/experiences/exceptional/howard-storm.html, DL 27 May 2020)

Among all the angel sightings in scripture, a few include some clothing details. An angel "clothed in linen, whose loins were girded with fine gold of Uphaz" visited Daniel. (Dn 10:5-6) Did Daniel literally mean that the angel wore linen and gold from Uphaz? No. He was using familiar physical things to describe white and yellow brilliance.

The angel who greeted Mary Magdalene at Christ's tomb wore "clothing white as snow". (Mt 28:3) Mary also "saw two angels in white" inside the tomb. (Jn 20:12) As His disciples gazed into heaven at His ascension, Jesus sent two angels "in white apparel" to comfort and instruct them. (Acts 1:10) And, John saw a "mighty angel come down from heaven, clothed with a cloud". (Rv 10:1) Based on the pattern so far, it was probably a white cloud. He later saw "the armies in heaven, clothed in fine linen, white and clean". (Rv 19:14)

Indeed, most people who return from death describe angel clothing in the spirit world as white robes. But occasionally, a spirit is seen wearing something else. For example, Ranelle Wallace saw a late friend wearing the same kind of clothes he wore when he was killed, while her grandma wore the white dress in which she was buried. Ranelle concluded that spirit clothes reflect personal preferences. (near-death.com/experiences/exceptional/ranelle-wallace.html, DL 28 May 2020)

Since nobody returns from the spirit world to tell of naked angels, clothing must be the norm. Deceased people's spirits that don't return to life must eventually get clothes. And, since the clothing of angels emits its own light, clothes must be made of intelligence which automatically assembles itself to suit the spirit body. Spirit clothing even adjusts its brilliance to match a spirit's progression status.

When Jesus was transfigured, "his raiment became shining, exceeding white as snow; so as no fuller on earth can white them". (Mk 9:3) So, His holiness was marked by the presence of glory, the same intelligence that lights up spirit clothes. But unholy spirits wear drab clothing, or don whatever serves their wicked purpose, such as white clothes when appearing as "an angel of light". (2 Cor 11:14)

How do spirits progress?

Although Jesus and Satan were both spirit sons of the same Father, Jesus had waxed great in intelligence while Satan had waned, giving Jesus far more power and light than Satan. Likewise, spirits who follow Jesus amass intelligence in proportion to their righteousness as they increase their capacity to love others and emulate God.

Righteous spirits are very glorious. Without a body to veil his brilliance, an angel who is "the spirit of a just man made perfect... will come in his glory; for that is the only way he can appear". (DC 129:6)

Increasing in glory is spiritual growth, like gaining weight for a starving person; "Come unto the Holy One of Israel, and feast upon that which perisheth not, neither can be corrupted, and let your soul delight in fatness." (*BOM*, 2 Ne 9:51) Whence cometh the additional "fat"? From the omnipresent pool of primal spirit light – from the pervasive spiritual intelligence.

When does a spirit enter its body?

Elizabeth was six months pregnant with baby John when Mary came to visit. When "Elizabeth heard the greeting of Mary, the baby in her womb leaped," and Elizabeth, "filled with the Holy Ghost," exclaimed, "The babe leaped in my womb for joy." (Lk 1:41-44) The unborn John must have already been ensouled to recognize the presence of his Master's mother.

Jesus was not yet born when He told Nephi, "Be of good cheer; for... on the morrow come I into the world." (*BOM*, 3 Ne 1:13) So, either Jesus sent this message to Nephi from His baby body or He delayed entering His baby body so He could send the message or He left His baby body temporarily to do a few last-minute tasks before being born into Earth life.

In modern times, premature babies have survived birth in the second trimester. For example, Lyla was born at 21 weeks and survived, thereby refuting the American Association of

Pediatrics decision that babies aren't viable until 23 weeks. A handful of other babies have been born at 22 weeks and survived.

During Carter Mills' NDE, a memory was restored of coming to Earth at the moment of fertilization. "He had to choose hair color and eyes out of the genetic material available to him and [other] genes that might give him the body he would need. He bypassed the gene for club footedness." (near-death.com/religion/christianity/berkley-carter-mills.html, DL 10 Sep 2017) Mills said he *came to Earth and watched cell division and fertilization occur*, not that he entered the egg at the moment of conception. And, of course, few would believe ensoulment *precedes* fertilization.

This kind of NDE doesn't necessarily prove that *every* spirit chooses all their genetic traits, or even some. But there are lots of spirits willing to take on the extra physical challenges posed by genetic diseases. It seems God is the type of deity that might at least grant spirits some choice in matters of biology and circumstance on their way to Earth, perhaps the selection of parents and certain challenges, privileges, and potentials within divinely-specified limits such that we are co-creators of our earthly experiences. And it seems logical that God would grant people whatever talents are necessary to fulfill their missions in life.

While a freshly formed zygote might be considered a live human, it would hardly be expedient to cram every spirit into a single cell so they can wait for nine months to be born. The mother will not even get a positive pregnancy test until two weeks later when the zygote has grown into a blastocyst and implanted itself in the uterine wall. Implantation often fails. So, it would be even more senseless to have *every* spirit come to earth in the first two weeks.

Brigham Young said, "When the mother feels life come to her infant it is the spirit entering the body preparatory to the mortal existence." (*JD*, bk17, p144) As sweet as this might sound, the mother is unlikely to know whether fetal movement is neuro-reflexive or initiated by the baby's spirit. But Brigham Young also believed in second chances for spirits whose bodies die in the womb; "When some people have little children born of 6 & 7 months pregnancy & they live but a few hours then die… I think that such a spirit has not a fair chance for I think that such a spirit will have a chance of occupying another Tabernacle." (*Wilford Woodruff's Journal*, 15 Oct 1867)

In a modern case, a loving husband was extremely distraught about his wife, whose pregnancy with twins was threatening her life. "One night, he woke up and looked toward the bedroom door. A light was shining in the hall… The light grew in brilliance as it came down the hall, then turned into their bedroom. Within the light was a young man wearing a white robe. He came and hovered next to the bed… 'Dad,' he said. 'My sister and I have talked it over, and decided that she will come first. It'll be better for Mom this way. I'll come in about two years…'" The next day, the mother miscarried the boy but the girl was born at full term. Twenty-two months later, a boy was born. (paranormal.about.com/od/lifeafterdeathreincarnat/a/Life-Before-Birth_2.htm, DL 12 Feb 2017)

So, it seems that the timing of the ensoulment of unborn babies is situational and that the angels in charge of ushering spirits down to the designated wombs probably make judgment calls depending on a number of factors. For example, if they know in advance that the baby will be killed early in gestation at a Planned Parenthood clinic, they probably have the option to suspend ensoulment of that baby.

The policy of The Church of JESUS CHRIST of LDS states, "Temple ordinances are not performed for stillborn children… However, this does not deny the possibility that a stillborn child may be part of the family in the eternities… It is a fact that a child has life before birth. However, there is no direct revelation on when the spirit enters the body." (*General Handbook*, 2020, 28.3.3)

In view of the increasing probability of ensoulment as gestation progresses, it seems reasonable that the closer a tiny human life is to becoming a viable birth, the greater the tragedy of losing it or killing it. The Church of JESUS

CHRIST of LDS opposes elective abortion for personal or social convenience and counsels members not to submit to, perform, encourage, pay for, or arrange for such abortions. (ChurchofJesusChrist.org/topics/abortion, DL 12 Feb 2017)

Are spirits fearless?

If spirits were fearless, Satan would have been unable to scare a third part of them into following him in premortality. (Rv 12:4) Although all the spirits who followed Jesus come to Earth, many of them experience trepidation as their time to be born approaches, some even expressing second thoughts. But the loving reassurance of strong leaders holds them to their commitments to enter mortality, fulfil their missions there, and move forward in their progression.

Do plants have spirits?

When God created life, He made "every plant of the field *before* it was in the earth, and every herb of the field *before* it grew". (Gn 2:5) With that statement from the Bible, it might surprise some that only Latter-day Saints believe that plants were created spiritually before they were created physically.

Many NDE accounts of visits to the spirit world include descriptions of beautiful gardens where plants radiate the light of intelligence, voicelessly communicate praise to God, and convey happy feelings in response to visitors. If spirit plants exist in the postmortal spirit world, they must have been created in a premortal spirit world. And, like the spirits of people, plant spirits await their turn on earth.

"All living things—mankind, animals, and plants—were spirits before any form of life existed upon the earth." (ChurchofJesusChrist.org/scriptures/gs/spirit?lang=eng&lett er=S, "Spirit", DL 30 May 2018)

Do all objects have spirits?

History is littered with religions that venerated the earth, sun, moon, planets and stars as gods. Were they crazy? Maybe not. Speaking to the earth and expecting it to respond has a precedent in creation where God commanded it to bring forth life and it obeyed. (Gn 1:11, 20, 24)

Enoch actually heard the earth groan, "I am pained, I am weary, because of the wickedness of my children." (*PGP*, Mo 7:48) And, at the death of the Creator, the earth convulsed; "The veil of the temple was rent in twain from the top to the bottom; and the earth did quake, and the rocks rent." (Mt 27:50-51) That the earth is alive is evident from the fact that "it shall die [and] it shall be quickened again". (DC 88:17-26)

All physical creations have eternal spirit. As NDE veteran Beverly Brodsky said, "Everything was alive and had a consciousness... There is love in every molecule." (ashtarcommandcrew.net/profiles/blogs/your-focus-task-in-life-and-the-movement-of-the-heart-field-by, DL 15 Jul 2020) From a spiritual perspective, there is no crisp line separating living things from inanimate things.

Does the cosmos have a spirit?

If the earth, moon, sun, and stars have spirits, then why stop there? The spirit of the universe enables it to know and obey the will of God. It is "the light which is in all things, which giveth life to all things, which is the law by which all things are governed". (DC 88:13) And, if our universe is pervaded with spirit, then any possible superverse or multiverse from which it sprang likely has spirit also.

Nobel laureate Eugene Wigner and other physicists have argued that "there must be a cosmic consciousness pervading the universe" that would explain the inexplicable quantum state phenomena. (*Physics of the Impossible*, Michio Kaku, 2008, p243)

How do spirit & mortal bodies relate?

While in a Missouri prison, Apostle Parley Pratt wrote, "Matter and spirit are the two great principles of all existence." (*HC*, 4:55) Matter and spirit are both structured from the more ethereal and all-pervasive field of intelligence. Apostle Orson Pratt wrote, "It is the actual presence of this Spirit that produces all the phenomena ascribed to the laws of nature, as well as many of the deviations from those laws, commonly called miracles." (*Absurdities of Immaterialism*, Liverpool, 1848, p32)

God's mastery over all physical things stems from the allegiance of the elemental intelligence from which all creation is molded. There is no known chemical or nuclear reaction that can change water to wine. That miracle came about because fundamental essence – the intelligence in the H_2O – obeyed Jesus by reorganizing into grape juice.

Benjamin Johnson, one-time private secretary of Joseph Smith, remarked that Joseph Smith "was the first in this age to teach... that light or spirit, and matter, are the two first great primary principles of the universe, or of Being... and *from these two elements both our spirits and our bodies were formulated*". (*Benjamin Franklin Johnson: Colonizer, Public Servant and Church Leader*, E. Dale LeBaron, as quoted by Charles R. Harrell; BYU Studies v28, #2, p83)

In 1915, Apostle John Widstoe explained how spiritual intelligence dovetails with physics; "The forces of the universe do not act blindly, but are expressions of a universal intelligence... since intelligence is everywhere present, all the operations of nature, from the simplest to the most complex, are the products of intelligence." (*A Rational Theology*, John Widstoe, p13) Without intelligence, "there is no existence". (DC 93:30)

So, this is the spirit creation story in short: God saw numerous pockets of intelligence languishing in chaos but yearning for divinity so He organized them into spirit entities and put them on a path of progression toward godliness.

8: Cosmogony

Why care about cosmogony?

What is cosmogony? You guessed it, the Greek word part, *gony*, denotes coming into being, the same as the root in <u>Genesis</u>. So, cosmogony is a model, theory, or study of the origins of the cosmos. Why not say cosmology? Okay – it *is* easier to pronounce, but it's not specific to cosmic origins.

"I have traced my lineage back to King Edward," says one. "There's no way *I* came from an ape," says another. This innate desire to know our origins has spawned myriad creation mythologies, some of which share common elements, almost all of which begin with one or more divine beings:

♦ Australian aborigines told of a "dream time" time when everything on earth was asleep except for the Father of All Spirits who awakened others to get creation rolling.

♦ American Iroquois told of a Sky People, one of whom created all things earthly after being pushed from the sky by an angry husband.

♦ An African bush myth says people came from a blissful place below the earth where the Grand Master ruled over peaceful creatures.

♦ A Japanese creation myth begins with a germ of life that stirred everything on the face of the earth into a muddy sea from which creator gods spontaneously formed.

♦ The Mayans' story began with various animal-like gods who created humans from different colors of corn.

♦ Greek mythology says that a dark void or chaos existed from which gods were born who created everything else.

♦ The Tao creation myth describes a nothingness that morphed spontaneously into an existence from which sprang the yin and yang that produced everything else.

This tiny sampling from among hundreds suffices to show that myths reflect the culture from which they spring. Some reflect the problem humans historically had imagining an absolute beginning. "If we discover a complete theory... of why it is that we and the universe exist... it would be the ultimate triumph of human reason -- for *then we should know the mind of God." (A Brief History of Time*, Hawking, 1990, p193)

Is the cosmos a figment?

"If a tree falls in a forest, and nobody is there to hear it, did it make a noise?" This question has reverberated through the last three centuries despite its obvious ridiculousness. Whales make sounds too low-pitched for human hearing and bats make sounds too high-pitched for human hearing but subsonic and ultrasonic vibrations are still sound! Hearing is not a requirement for an audible event any more than a radio receiver is necessary for transmitters to broadcast.

This question was first based on a premise that reality depended on human observers. And, amazingly, modern scientific experiments resurrected a similar idea. Let's assume a billion-dollar lab is set up to do the following experiment:

Shoot a single photon at a barrier with two slits in it, a photographic screen behind it, and a detector between one slit and the screen. The photon will pass through one slit or the other as a particle. But, without the detector, the photon passes through both slits as evidenced by a two-wave interference pattern on the screen. So, scientists wonder whether the very act of observing creates the particle reality. More complicated experiments in this vein of research prove the *mere knowability of the path* creates the particle reality *from the instant the photon leaves the gun*!

The mysterious results of these experiments have impelled scientists to say things that sound as ridiculous as the tree in the forest assertion. For example, Nobelist Eugene Wigner said, "The content of the consciousness is the ultimate universal reality." (beliefinstitute.com/quote/wigner-consciousness-ultimate-reality, DL 15 Jul 2020) And even crazier is this question, "How can we be certain that the universe around us actually exists?" (*Through the Wormhole,* Science Channel, 17 Jul 2013) Silly scientists – reality got along just fine before there were humans to observe it!

Nothing goes entirely unobserved, whether by humans or by insentient things. When I enter my shop, a sensor detects my movement and energizes a light bulb. When a tree falls, air molecules "feel" its movement as it whooshes to the ground. Plants on the ground "sense" being crushed and respond by releasing pheromones. The dirt is disturbed where the tree lands. Furthermore, in the unlikely event that the impact of the tree is just enough to trigger an avalanche that sets off a massive earthquake, thousands of people who never heard the tree fall may be convinced that something somewhere made a noise!

So, if a tree falls in a forest, it makes a sound, even if nobody is there with ears to hear. Every event in the universe impacts something else. It's a network of "sensing". And, of course, God observes everything. Not one sparrow falls to the ground without Him noticing. (Mt 10:39)

Did God backfill history?

One of my advanced placement High School students wondered, "How can so many scientists be so stupid?" He couldn't fathom how cosmologists could say the universe is 13.8 billion years old when it seemed obvious to him from Genesis that God just finished creating it only 6,000 years ago. Did an omnipotent God instantly create the entire cosmos to include all in-flight light from distant galaxies such that it now appears to have taken billions of years to arrive at telescopes on Earth?

Backfilling the past seems like a crazy idea but some physicists seem receptive to the idea, at least in regard to some of the cleverer two-slit experiments. Rather than assume a photon or its quantum field senses the path ahead, some theorists have speculated that nature revises its behavior after the fact.

For example, John Wheeler concocted the following thought experiment: Imagine that a photon left a distant star 44 billion years ago on a collision course with the future Earth. Midway in the journey it encountered an object that acted like two slits. At that point, earth was still another 16 billion years in the future and a device for detecting a single photon was 22 billion years in the future. Did the photon know at that time whether a human would ultimately deploy or not deploy the photon detector based on a coin toss? Did the photon split like a wave or choose one of the two paths like a particle? Or, did reality stay undecided until the photon arrived at the detector at which point Nature decided how to backfill 44 billion years of history based on whether the device had been activated?

This "delayed-choice/quantum-eraser experiment" still bewilders scientists. Retroactive causality is no longer a totally wackadoodle idea to them! And from a young-Earth creationist's viewpoint, if a lab worker could force the backfilling of billions of years of a photon's history by simply observing it or not, why couldn't God Almighty create billions of years of cosmic back history by simply willing it?

Around the turn of the century, clever experiments showed that the first twin of an entangled photon pair always matches the particle or wave behavior of the second twin no matter how far into the future the second twin encounters any choice-forcing obstacles. Some say these results show the early short-path photon's decision is backfilled after the later long-path photon's decision is made so that the behavior of the two photons always match. Information on these experiments can be found by searching for "delayed choice quantum eraser" on the Internet.

John Wheeler concluded, "We ourselves have an undeniable part in shaping what we have always called the past." (*The Ghost in the Atom*, Davies & Brown, 1999, p67) Physicist Asher Peres concluded, "Quantum effects mimic not only instantaneous action-at-a-distance but also, as seen here, influence of future actions on past events, even after these events have been irrevocably recorded." (*Delayed Choice for Entanglement Swapping*, 2000, p141)

So, what's the point of all this quantum mechanical mumbo jumbo? It is to highlight the possibility that God created everything only 6,000 years ago replete with the 13.8-billion-year history that scientists see today.

But, wait! Would a truth-obsessed God manufacture faux history on any scale, much less that of the entire universe? Fake events like ice ages and dinosaur deaths with fossil evidence all over the planet would be deception on an unimaginable scale! If we accept that God does that kind of thing, why not accept that the universe was created in the last picosecond, replete with all the relevant history, including the existence of this book and the memories of writing it in the author's head?

If God created all history one second ago, nobody would really responsible for their past choices. Furthermore, each person would be making future decisions based on past experience fabricated by God! The principle of agency would be violated.

It's far more likely that the universe is made of quantum fields of intelligence capable of anticipating chains of causality far into the future non-locally. If God needs a 13.8 billion-year-old past, it's a certainty that He had the patience and foresight to start 13.8 billion years ago rather than wait until the last moment and backload everything.

Did God follow a recipe?

Einstein wondered, "Did God have a choice in creating the universe?" In other words, could there have been another set of physics that would result in a cosmos just as perfectly programmed to bring forth life as is our universe? If God used a recipe, how many workable recipes are there? Many theorists have pondered the question but none can suggest a single improvement to the existing set of physical laws.

Science endeavors to reverse engineer Nature's recipes. Kepler and Newton found that the physical laws lent themselves well to being described in beautiful mathematics. Newton concluded that a Master Mechanic must be their author. Kepler likened geometry with divine patterns for creation. And Einstein said, "Things should be as simple as possible, but not simpler," and "God always takes the simplest way". (azquotes.com/quote/87346, DL 1 Apr 2020)

Our universe is so exquisitely complex that scientists will still be discovering and marveling at it until the end of humanity. They will never be bored. It's a perfect blend of particles, forces, energies, masses, action, and interaction for the emergence of life and its luxuries. When Einstein wondered whether "God could have made the universe in a different way," he was acknowledging its elegance and intimating that "the divine reveals itself in the physical world". (*Albert through the Looking Glass*, Z. Rosenkranz, 1998, p80)

What are the theories?

Philosophers and scientists have speculated about models of cosmic origins for over 3,500 years – plenty of time to boil it down to just one theory. But, even after much debunking of old ideas by modern science, there is still a multiplicity of models. Here are the most serious contenders:

Steady State

Newton described a steady state universe in his famous *Principia Mathematica* in 1687. And Einstein's bias for this steady state model was so strong that when his relativity equations predicted the universe would collapse, he introduced a repulsive constant into his equations just strong enough to steady everything.

It did not occur to anyone that the universe wasn't stable until Edwin Hubble's tenacious stargazing proved it is expanding. So, Einstein regretfully called his cosmological constant the "biggest blunder of his life." But, by the end of the century, he was somewhat vindicated by the discovery that the expansion is accelerating. So, his cosmological repulsive constant got dusted off, renamed dark energy, and assigned an antigravity-like value.

Expanding Steady State

In 1922, Alexander Friedmann saw an expanding universe in Einstein's General Relativity math. A minister named Georges Lemaître discovered the same thing a bit later and called the idea "the hypothesis of a primeval atom or cosmic egg." But when Lemaître ran it past Einstein, the renowned genius called the idea "abominable" and the world of physics ignored it.

Then along came the Hubble with hundreds of hours of telescope data to prove that our galaxy is just one of many. His painstaking observations and measurements also showed that virtually all galaxies are fleeing away from each other. At that point, most cosmologists concluded that the entire universe had burst from a single event eons ago, just as Lemaître had posited.

But there were a few holdouts. The famed English astronomer, Sir Fred Hoyle, could not abide the thought that the expanding universe would eventually slow to a halt and collapse cataclysmically or expand into frigid darkness. Using his BBC radio show, he derided the new theory, calling it the "Big Bang." The nickname was uncharacteristically comical for cosmology, but was much shorter and catchier than Lemaître's and has been used ever since by even the most somber scientists. Hoyle thought the Big Bang theory was a "form of religious fundamentalism". (*Matter and Spirit in the Universe*, Kragh, 2004, p236)

But the evidence was irrefutable. So, Hoyle proposed that, as the expansion creates new voids in space, new matter backfills them spontaneously, eventually coalescing into new galaxies to keep the cosmos evenly populated. He reasoned it should not be difficult for this to happen as it would only require one new atom of hydrogen per cubic meter of space every 10 billion years to keep up with the observed expansion. However, serious consideration of this theory died with Fred Hoyle in 2001 and no version of it has never been resuscitated. Perhaps "expanding steady-state" seems a bit too oxymoronic even for nerdy physicists.

Big Splat

The Big Splat idea is that the universe was born from a cataclysmic collision of preexisting universes in an extra-dimensional superverse or multiverse. Such a splat would explain the tiny variations in the universe's ubiquitous Cosmic Microwave Background radiation (CMB) that the inflation stage of the Big Bang model failed to predict. To produce the evenly distributed matter we now see everywhere, the universes had to broadside each other in an extremely well aligned 3-dimensional way.

A slight variation on this theory is that two big branes (membrane-shaped universes) collided and either scattered or fused to produce our universe.

Big splat theories propose that such colossal collisions are quite common in the "incredible bulk" region (higher dimensional space, hyperspace, superverse, or multiverse).

Evolution

Parent universes produce countless baby universes and only those that grow up to produce their own baby universes bestow the proclivity to reproduce on the next generation. Over the course of many generations of universes, the laws of physics that lent themselves well to reproduction survived. And, as it inexplicably happens, such universes are also predisposed to support the emergence of life, like ours.

How are baby universes born? According to Dr. Lee Smolin's hypothesis of *cosmological natural selection*, mass falling into a black hole can blast right through into the multiverse and emerge as a baby universe. Such a new universe would tend to "inherit" similar physical laws and constants as its parent. Universes whose physical laws deviate too much would not reproduce. After many generations, less productive universes would tend to dwindle in number.

Big Bang

Hubble spent many nights with the Hooker (a 100-inch Telescope in California) tediously logging the motion of galaxies and found that the farther away a galaxy is, the more stretched its light. What is stretched light? Well, just as sound from a car gets stretched if it's travelling away from you, resulting in a lower pitch, light from a galaxy gets stretched if it's travelling away from you, resulting in a redder color.

Only a fraction of a percent of the galaxies observed have blue-shifted light. So, almost all galaxies are fleeing away from us. It's the same in every direction, which means the galaxies are also fleeing away from each other.

Cosmologists imagined the universe running in reverse and concluded that everything sprang from a single small event long ago. Whether that was a singularity or some kind of quantum fuzziness is still debated.

Perhaps the "big bang" happened in a portion of normal space in universe that has always existed. Perhaps two highly massive objects collided with an energy that converted the objects into the hot and dense plasma from which everything known by science sprang. But so far, the Big Bang's pervasive heat signature, the Cosmic Microwave Background, has yielded no clues as to what set it off.

From a scientific perspective, early in the Big Bang where the math of physics still works is best starting point for investigation. Any earlier than that and the math collapses in a flurry of infinities.

Big Bounce

The big bounce idea is that our universe is but the latest instance in a series of big bangs where each is a rebound from the previous universe's implosion. What causes the bounce? Why doesn't the universe stay collapsed?

No universe collapses to a singularity. Rather, as a universe reaches infinite density, the behavior of space's quantum foam and fundamental physical constants change such that it all springs outward again. As the new instantiation of the universe expands, the laws of physics go back to behaving in the usual ways.

Conformal Cyclic Cosmology

This is Roger Penrose's name for an idea that, like the Big Bounce, is cyclical but doesn't involve a collapse per se. After everything reaches thermal equilibrium and the last black hole has evaporated into radiation, the only stuff remaining will be the massless particles like gravitons, gluons, and photons, none of which experience time. And, without time and mass, space and distance become meaningless and the end of the universe is the same as it was at the beginning, pure light poised to expand into a new Big Bang from energy without spatial dimensions.

Per Penrose's theory, the universe did not have a beginning. Its expansion never really stopped —the finish line is simply reset to the starting line at the point that time and distance phase out. It is one continuous Big Bang with a frozen blackout phase separating hot phases.

Ekpyrosis

This theory proposes that the universe was formed from the collision of two 3-dimensional worlds on a hidden fourth dimension.

There are other less popular theories, like the one that proposes this universe was born through a quantum tunnel from previously existing universe or like the one that proposes this universe is a local big ban that was spewed out of a wormhole from elsewhere in a metaverse, a cosmic birth canal. So far, no theories whatsoever can be sufficiently verified by science to be pulled from the "theory" category.

Which theory do churches like?

Commenting on the Big Bang theory in 1951, the pope told the Pontifical Academy of Sciences, "It would seem that present-day science, with one sweep back across the centuries, has succeeded in bearing witness to the august instant of the primordial Fiat Lux [Let there be Light], when along with matter, there burst forth from nothing a sea of light and radiation, and the elements split and churned and formed into millions of galaxies." (*Measuring Eternity*, Martin Gorst, 26 Mar 2002, pp254-55) In other words, the Big Bang was the light of Genesis Day One, per the pope.

However, fundamentalist Christian leaders generally see the Big Bang in conflict with the Bible, citing issues such as time frames for creation and differences in the sequence of creation in Genesis versus the Big Bang. Official statements by a number of Baptist groups in the USA reject the Big Bang. For the most part, however, official position statements by major religious denominations embracing one scientific theory or another are difficult to find.

Which theory is most likely?

Physicists predicted long ago that if the Big Bang really happened there should be an afterglow permeating the entire cosmos. And, because the light would be stretched by the expanding universe, it would be redshifted into the microwave spectrum. Furthermore, this Cosmic Microwave Background (CMB) would be about 5°Kelvin, as Alpher and Herman predicted in 1948.

A mere sixteen years later, a couple of Bell Lab researchers found static in a newfangled antenna they were testing. They cleaned off the bird droppings but the static persisted. So, unable to identify the source, they concluded that it had to be the ubiquitous CMB. They won a Nobel Prize.

The CMB is now hailed as the most robust evidence for the Big Bang theory, although scientists are still tweaking their various versions of it in an effort to align them better with observations. Big Bang deficiencies include the discovery that the expansion is accelerating, requiring the postulation of a repulsive force field dubbed dark energy, the observation that galactic rotation behavior can only be explained by acknowledging the existence of vast amounts of enigmatic mass we can't see, and dozens of other observations that conflict with the Big Bang theory.

More stunning than those challenges is, that with the deployment of the James Webb Space Telescope (JWST), the Big Bang model has been significantly overwritten. It is no longer a viable theory of the beginning of the universe or of its behavior across all space. Rather, it works merely as a theory of a post-creation local expansion of space out to some 30 billion lightyears in all directions.

Firstly, scientists calculated that the images of objects beyond a certain distance would be magnified slightly by the extreme stretching or negative curvature of spacetime over that distance. Negative curvature means the density of mass is too low to ever slow the outward expansion. To virtually everyone's surprise, the data collected by the JWST contradicts what the math predicted. Distant galaxies, though quite large physically, were not magnified by spacetime curvature, contradicting the Big Bang model's prediction.

Secondly, due to the time it takes for light to travel to Earth, the JWST saw galaxies as they were 350 million years after the Big Bang began. Scientists expected them to be dim, chaotic, low-mass and embryonic. Instead, the galaxies are well-formed, nicely spiral, mature, huge, and as smooth as galaxies that are close. Science has no explanation for how such mature-looking galaxies could have formed so quickly and soon after the Big Bang began. It's like finding photos of a baby taken at its first birthday showing a full-grown adult.

Thirdly, some of the stars within these galaxies appear to be more than 1 billion years old at the time the light from them began its journey. How could stars in half-billion-year-old galaxies be more than a billion years old?

Fourthly, the regions of space observed by the JWST contained roughly 100,000 times more galaxies than the Big Bang model predicts!

Fifthly, the JWST observed large-scale structures similar to closer ones photographed by the Hubble Space Telescope (HST), the kind that the Big Bang model says requires billions of years to form, meaning the beginning of their formation must have been long before the Big Bang began!

There are several other surprises from the JWST data but it is sufficient to say that the Big Bang Model has received so many wounds that it will take decades for scientists to craft bandages for it. If, for distances greater than 30 billion lightyears away, spacetime is actually behaving differently from what physicists expected, it may mean that the universe this side of that horizon is a relatively "small" lump in a much larger superverse, multiverse, or omniverse pudding that could contain other universe lumps, making the term *universe*, which comes from the Latin word *universum*, meaning *the whole* or *all things*, somewhat misleading.

Where did the Big Bang begin?

So, the Big Bang is not an explosion *in* the universe. It's an expansion *of* the universe. The CMB is almost the same across the entire sky because it has expanded with the universe just as the color of a perfectly symmetrical balloon fades the same over its entire surface as it expands.

If the Big Bang had come from a single sizeless point, as many versions of the Big Bang theory have stated, then everything in the universe would still be at the center, just as any randomly-chosen point on the surface of a perfect ball would equal status as center with every other point on the surface. Then, you would be at the center of the universe! Yay! But, since everyone else would also be, we would all be equally special — or perhaps unspecial.

But cosmology is drifting away from the idea that such a singularity was the cosmic "seed" of the Big Bang, leaving each of us in our own special place in the universe after all.

How tiny was the early universe?

If we imagine playing a movie of the expanding universe backward, it ends with everything all infinitely compacted into a single point. But there is no infinity-free math to describe it. When infinities pop out of mathematical models, it's a sure sign something is wrong.

There is no consensus on what laws of physics were in play at the exact moment the Big Bang popped. Stephen Hawking said, "All the matter in the universe, would have been on top of itself. The density would have been infinite. It would have been what is called, a singularity." (*Primordial Nucleosynthesis and Evolution of Early Universe*, Sato & Audouze, 2012, p130)

But, as time goes by, more scientists are backing off the idea of a singularity; "The Big Bang singularity is the most serious problem of general relativity because the laws of physics appear to break down there," said Dr. Ahmed Farag Ali of Benha University. He has worked to revise Einstein's relativity equations to allow a singularity *not* to have happened.   (phys.org/news/2015-02-big-quantum-equation-universe.html#jCp, DL 12 Aug 2017)

Why did the Big Bang pop?

Some physicists think that a primordial quantum field hiccupped to kick off all creation. Quantum fields are always fluctuating. But a field needs space to exist, something that the theory says wasn't there until *after* the bang.

If the gravity of black holes is so extreme that even a photon zooming at top speed can't get out, how could the entire mass of the universe have escaped the gravity of a singularity containing all the mass of the universe? Some posit that repulsive particles dubbed *inflatons* formed at the moment of the Big Bang and drove the early expansion of the universe, but even that would have happened after the initial pop.

So, if the universe was indeed once a very tiny thing, it's very tempting for a theistic theorist not to consider the possibility that its sudden expansion was a *miracle*; "An event that appears inexplicable by the laws of nature and so is held to be supernatural in origin or an act of God." (Wordnik)

How old is Creation really?

The universe must be at least as old as the oldest objects in it. The oldest objects discovered in deep space are slightly younger than 13.8 billion years. Why do scientists think that? Galaxies that were formed right after the Big Bang were clumpy. In contrast, galaxies that have been swirling for more than 13 billion years, like our Milky Way, are "smoother" after going through 60 or 70 full rotations. Physicists also look at the size, brightness, phase, redshift, and other information. The redshift of the ubiquitous CMB also confirms the age of the universe to be 13.82 billion years.

When televangelist Pat Robertson conceded that the scientific evidence proved creation to be ancient in 2014, Ken Ham, president of Answers in Genesis, responded, "God could have created everything in six seconds if He wanted!" (freerepublic.com/focus/religion/3156645/posts, DL 7 Mar 2018)

If the universe were only six seconds old or even 6,000 years old, God would have had to violate His own laws of physics with wild abandon to make it appear 13.8 billion years old to scientists. He would have had to stretch light beams almost instantaneously across 46 billion lightyears of space, replete with the appropriate redshifts for each, to fool astrophysicists. And, He would have had to create a counterfeit CMB. Why would the God of truth create such a vast deception?

But the scale of God's "deception" worsens the more science learns. The JWST discoveries have forced scientists to consider the possibility that their 13.8B year estimate of Creation's age based on previous "shortsighted" observations may be super wrong. - To reconcile JWST's findings, physicist Rajendra Gupta proposed a new model that suggests the universe is 26.7 billion years old.

How big is Creation now?

If the universe has enough matter to halt its expansion, scientists say it is denser than the critical density Ω and will eventually implode. But, if the universe is less dense than Ω, it will expand forever in what's called a saddle shape.

What if the universe happens to have *exactly* the critical density Ω? It will expand forever as a 3D version of flatness (unwarped on average). This is the only case where the angles of a triangle large enough to span several superclusters would still add up to 180°.

Unfortunately, measurement limitations prevent science from being 100% sure, but every indication is that the universe just happens to be flat with very few physicists disagreeing. (*How Big is the Universe*, Cathal O'Connell, Cosmos, 17 Jul 2017) So, it will likely expand forever.

Light from objects more than about 46 billion lightyears away will never reach Earth due to the accelerating expansion of the universe. That's the radius of the *observable* universe. How far beyond that observable boundary does the universe extend? Astrophysicist Robert Trotta and his colleagues at Imperial College London calculated that the entire universe must be *at least* 250 times greater than the observable universe by volume, or 7 trillion light-years across. (space.com/24073-how-big-is-the-universe.html, DL 1 Jun 2020)

If nothing can travel faster than light, how did the universe get many times broader than 13.8 billion lightyears across in only 13.8 billion years? The answer is that nothing prevents the universe itself from expanding faster than light. The lack of large-scale curvature in space suggests that the universe inflated much faster than c from the moment it was born.

Is the Big Bang miraculous?

Aside from its ability to spring into action without any identifiable cause, the Big Bang has several amazing features:

Flatness

It appears that the Big Bang produced a universe that is exactly balanced on the razor's edge between *open* (expanding forever) and *closed* (expansion will eventually halt). In 2013, data from the European Space Agency's Planck telescope failed to detect any curvature across the universe with an error margin of $\pm 0.4\%$. To produce this result, the early universe had to have a density of a particular value, correct to 1 part in 10^{60}. Even the slightest deviation from this early value would have caused a humongous deviation from the current perfect balance.

Antimatterlessness

The laws of physics demand that particles popping out of quantum fields come in pairs, one particle being ordinary matter and its evil twin being antimatter. So, the Big Bang should have produced just as much antimatter as matter, yet antimatter is very scarce in Nature, a problem called matter-antimatter asymetry. That is why the fuel needed by Star Trek's *USS Enterprise* for warp travel must be generated by Starfleet fuel facilities (☺).

Dan Brown's blockbuster movie, *Angels and Demons*, was correct that a lot of energy pops out of a combination of tiny bits of antimatter and regular matter! One-fourth of a gram of antimatter would destroy Rome in an instant. But, in reality, it would have taken the CERN collider 100 million years to generate that much antimatter.

Equal amounts of matter and antimatter should have mutually annihilated each other, releasing mindboggling amounts of pure energy as per $E=mc^2$. But the only light left over from the early Big Bang is the CMB, which is far too dim to include radiation from a matter-antimatter annihilation event. Nevertheless, cosmologists are forced to accept that the entire cosmos consists of leftovers from early matter-antimatter annihilation.

Darkness

Essential to the Big Bang theory is the existence of dark matter and dark energy which permeate the universe. These two exotic components of the universe constitute roughly

95% of universe's ingredients, 70% being dark energy and 25% being dark matter, leaving only 5% of matter and energy not made of mysterious invisible stuff. (phys.org/news/2021-03-composition-percent-universe.html, DL 10 Apr 2021)

Without dark matter, galaxies could not form. It was the scaffolding on which the first galaxies and stars formed. Without dark energy, the expansion of the universe would not accelerate. After decades of research, science is no closer to identifying a dark matter particle nor a dark energy wave.

The Hubble Tension

The Hubble tension is a major puzzle in cosmology where two precise methods of measuring the universe's expansion rate — the Hubble constant — give conflicting results. One method uses nearby objects like supernovae and the other uses the CMB. The disagreement, now over 4 standard deviations, suggests either unknown systematic errors or new physics beyond our current model of the universe.

Conclusion

The science is not settled. There are myriad little problems with existing cosmogonal models and scientists are still ironing out the wrinkles in their theories. It seems, however, that the more information they wring from the universe about its origins, the more reticently it divulges additional secrets.

What are the Big Bang proofs?

The Big Bang can be observed right now, as it is still underway, evidenced by the continued outward expansion of the universe as proven by the fact that more distant galaxies are more redshifted. The CMB data prove that the universe is expanding and are consistent with observations of large-scale structures in outer space. Also, the prevalence of hydrogen and helium is just what the Big Bang model predicts.

Due to the inscrutabilities of initial conditions, modern cosmologists now define the Big Bang as the expanding universe, as it has been proven empirically by observations that almost all galaxies are flying away from each other.

However, the Big Bang model, oft-modified in the past, certainly needs tweaked again to fit all recent observations.

When did time begin?

Physicists say that time was created by the birth of the universe. And some religious apologists use this idea to answer the question why God wasted an eternity doing nothing until the creation 6,000 years ago, claiming that He didn't waste any time because He hadn't yet created time.

If it's true that time did not exist before the Big Bang, does it even make sense to ask *when* the Big Bang happened? The answer would be, t=0, a self-referencing universe time.

But there is a chicken-and-egg problem. As far as we know, every event takes time, even the very first event. So, time must have existed for its duration. Stephen Hawking proposed that only "imaginary time" existed before the Big Bang, a type of time perpendicular to "real time" (clock time). Imaginary time then morphed into real time when the Big Bang first popped. Imaginary time makes trying to trace history back to pre-time conditions as silly as asking what's north of the earth's North Pole. Perhaps imaginary time is merely a math trick like *i*, the square root of negative one.

A cleverer solution than imaginary time is stand-still time. Roger Penrose proposed that the universe started the way it will end, with only light, which, since it is massless, doesn't experience time nor distance and has no reference frame. Light doesn't experience space either. Without mass, time and space are undefined. So, time stood still until mass emerged from the flash of the Big Bang. (*Cycles of Time*, Roger Penrose, 3 May 2011)

If the data from the JWST truth is correct that there are stars in distant galaxies older than the Big Bang, then time indeed existed before the Big Bang and has therefore probably always existed wherever there are things with mass.

What is time?

Saint Augustine speculated 1,600 years ago that a razor-edged "now" separated the future from the past. That is the normal human sense of time. But if *now* is infinitely thin, how does an infinite number of them add up to anything? There is no basic unit of time that is indivisible into briefer instants. Separating the past from the future with an infinitely thin "now" is simply a useful mental device.

Grayson Hugh's song says, "Time is Like a River." This simile describes time as laypeople experience it. Future dates get closer and then flow into the past. Time compels cause and effects to happen sequentially. Ray Cummings wrote in his 1921 science fiction novel *The Time Professor*, "Time is what keeps everything from happening at once." (*Time Machines*, Paul Nahin, p98)

Though the topic is much discussed, seldom do any two physicists agree on the definition of time. Physics has no standard.

Many quote Einstein as saying that the sharp difference humans perceive between past, present, and future may only be an illusion. This oft-cited opinion was actually part of a consolation letter from Einstein to a friend who had lost a loved one and therefore cannot be taken as a cosmological definition of time.

Life forms too simple to suffer from illusions, such as tiny bacteria, have internal clocks such as circadian rhythms and experience senescence. So, time is not merely an illusion.

Is time a fourth dimension? That is an arbitrary ordinal designation. Because it is the only nonspatial dimension, time deserves to be the first. It is a dimension in the sense that it can be represented by a series of values, but one value is real – *now*. A single instant is only meaningful in the context of the spatial dimensions of the reference frame defined by the object whose time is the focus. There is no absolute *now*.

Representing time as a dimension is a way to make it useful but says no more about what time actually is than the duration equation, $t = d/s$.

Theorist Brian Greene wrote, "Just as we think of all of space being out there, we should think of all of time being out there too." If he is correct, then past *present* and *future* are pages in the book of time as legitimately as *here* and *there* are pages in the book of space. (*The Illusion of Time*, PBS, 8 Nov 2011)

But, if time is like a book, why can't we "see" future pages in our memories as easily as past pages? And, why can't we navigate time as we do the pages of space? Why do choices made on the current page change future pages but not prior pages? The answer to all these questions is; because, for any point in space, only *now* is real and time is causation in the direction of the effect.

"Past, present, and future, and are continually before the Lord" and all time is "as one day with God." (DC 130:7; Alma 40:8) Does that mean God has no *now*? No. it means God sees the past because He has total recall and He sees the future because reality is deterministic for an omniscient being.

It is morally dangerous to go down the rabbit hole of thinking that the future already exists because, if it did, there'd be no point in striving to make it better with today's choices! Choices followed by consequences is central to God's plan of happiness.

Just as there is a limit to how fast dominoes can fall, there is a limit to how fast any causal chain of physical events can play out as observed from any particular reference frame. It's commonly called "the speed of light" or "the cosmic speed limit" or simply c for *celeritas*, Latin for *speed* — the top speed of causality. And, since c is the same for all observers, measurements of time (and distance) may be different for observers in different reference frames so that the laws of physics can be the same for all observers everywhere.

What's causality? It is the relationship between one state and a subsequent state, wherein the first state plus a process is at least partly responsible for the second, and the second state is at least partly dependent on the first and the process. If nothing at all were happening in the universe, time would not be passing. Time is the unfolding of causality. Time existed as soon as the universe had mass. Since time is causality, it *started* when something *happened*. Time *is* something happening like fire is when something is burning!

If there were no mass, there would be no reference frames. Space and time would be undefined. Although spirit beings are aware of causal chains of events, they do not sense the passage of time because they are not physical.

In our physical reality, time is the changing of the states of things that have mass by causal events. No change $\rightarrow$ no time! By contrast, in nonphysical realms, time perception comes from experiencing the spirit environment, learning, thinking, character change, et cetera, instead of being marked by Earth's rotation or the ticking of clocks.

Why the cosmic speed limit?

Spacetime has a maximum rate at which cause and effect involving mass can propagate. Light happens to travel at that rate because it has no mass — the rate itself is a property of spacetime, not of light. You're always moving through spacetime at a constant rate. But that motion is divided between space and time. The motion of massless objects is 100% *through space*, zero through time. Nothing can go faster than light — zero motion through time is impossible.

Are space and time one thing?

In reference to Cumming's quip that time is what keeps everything from happening at once, physicist John Wheeler joked, "Space is what prevents everything from happening to me!" True! Pure space would simply be simple separation between things and nothing more – no structure, no curvature – just distance. If things were absent, distance would be meaningless. Quantum fields pervade the universe. No pure space exists anywhere.

Although Han Solo could accelerate forward or backward, right or left, and up or down, the Millennium Falcon of Star

Wars fame had no way to brake to an absolute stop except with respect to something he selected, like a landing pad! But every landing pad would have its own motion.

Just as space is no absolute *standstill*, time has no absolute *now* in relativity. Relativistically speaking, observers in different reference frames won't always agree on the duration or sequence of events, merely on causes and effects.

Time and *space* are like *supply* and *demand* in economics. One is meaningless without the other. Space and time dance together in the causality web that is often misleadingly called the "fabric of spacetime."

Once in a while somebody like Petr Hořava gets attention with a theory that splits them apart. She proposed that time and space were separate at the birth of the universe and then went through a phase change that stitched them together into spacetime as the universe expanded. (*Splitting Time from Space*, Scientific American, 1 Dec 2009) But there is no sign of agreement from other physicists.

General Relativity (GR) requires that every unique inertial reference frame be graphed with different values for both space and time.

What is spacetime?

Somebody joked, "Space and time are relative; the more time I spend with my relatives the more space I need." But seriously, as it relates to all physical reality, spacetime is the system of causal relationships among quantum fields and their energetic interactions where space is the distance between the system's things and time is the rate of causality in the interactions. Spacetime is the network of causality that constitutes all reality.

Energy is a feature of spacetime. Energy fields and the objects and events within them define spacetime. Matter and energy structure the causal topography that is spacetime in ways that change how everything from quantum fields, to particles to planets behave. This structuring is termed spacetime curvature.

How does space curve?

Is there a fourth dimension of space into which the first three dimensions curve? No, there is no right angle to reality! When spacetime is warped it simply means the most efficient path between two things is not necessarily a straight line. Renowned physicist John Wheeler said, "Matter tells spacetime how to curve, and curved spacetime tells matter how to move." (*Black Holes, and Quantum Foam,* Geons, p235) Since the gravitational field is universal, it can be interpreted as the geometry of the entire causality network.

Let's use the electromagnetic field to illustrate. The shape of such a field can be revealed by laying a magnet on flat cardstock and sprinkling iron filings on it. As they fall, they respond to the shape and strength of the field to order themselves throughout the magnetic fieldscape. Analogously, objects with mass affect the shape of the gravitational field and everything — not just ferrous flecks — responds to and drives the shape of spacetime.

Particles that interact with the universal Higgs field acquire mass which in turn increases gravity which in turn warps the network of causal relationships called spacetime, the blend of fieldscapes that constitute reality's substrate.

Ripples in spacetime were first observed experimentally in 2016 when the Laser Interferometer Gravitational-wave Observatory (LIGO) detected gravity waves arriving at Earth from the collision of two black holes a couple of billion lightyears away. How much did spacetime quiver? Almost a millionth of the diameter of an atomic nucleus! The gravity waves changed the lengths of everything in their paths, including the spacetime traversed by lasers in the detectors!

Light always travels the most efficient path, the one that takes the least time to traverse. When light from a distant star veers in toward the sun on its way to Earth, it has followed the most efficient path to the telescope, though not necessarily the shortest path. Gravity doesn't change what a straight line is. It just changes what the most efficient path is. That's the General Relativity view.

Now for the Quantum Mechanics view. A photon follows the most efficient path which, in a gravitational field, is changed by interactions with force-carrying gravitons generated by the mass of objects in space. So, the photons in the laser beams at LIGO measured changes in graviton interaction energies.

Why is the universe so repulsive?

A prize-winning discovery in 1998 rocked the world of physics. The science was that gravity was decelerating the expansion of the universe but teams studying deep space found that a mysterious force, behaving like gravity's evil twin, is *accelerating* it instead. That put the final nail in the coffin for the "Big Crunch" scenario. And, what Einstein called his biggest mistake, inserting a repulsive cosmological constant into his equations, turned out to be correct after all, even though its value and his reason for doing so were wrong.

Scientists say that the repulsive ghost force, which is eerily nicknamed *dark energy*, makes up 73% of the universe. Dark matter, discovered about the same time, is an additional 23% of the universe. So, the future of our ever-expanding universe looks cold and dark with the last objects being black holes slowly evaporating via Hawking radiation in the final scenes.

Many physicists think that dark energy might be vacuum energy generated by all the quantum fields that permeate space. Together these fields create tension that results in an outward push. And, as spacetime expands, these fields expand with it, adding more vacuum energy with the increase in volume. Gravity can't win.

Why is the cosmos goldilocksy?

Why is Earth the only planet in the solar system that harbors life? Because it is in our solar system's Goldilocks zone, the distance from the sun that is not too hot and not too cold (the circumstellar habitable zone). And Earth's single sun makes for moderate temperature swings. Doctor Malcolm was right in *Jurassic Park* when he said, "Life finds a way." The reason? Earth is perfect for it.

The entire universe is likewise goldilocksy! If any of the fundamental universal constants, ratios, rates, or characteristics were slightly different, conditions in the cosmos would be too inhospitable for the emergence and survival of life. "The remarkable fact is that the values of these numbers seem to have been very finely adjusted to make possible the development of life." (*A Brief History of Time*, Hawking, 1988, p126)

Examples:

♦ If the rate of expansion one second after the Big Bang had been smaller by one part in one hundred quadrillion, the universe would have collapsed long ago.

♦ If there were even 1/100,000[th] more or less dark energy, the universe would have expanded too fast for galaxies to form and life would not have arisen.

♦ If the strong nuclear force that holds atoms together were 2% stronger, stars would be unable to forge the elements of which Earth is made. (*The Accidental Universe*, Paul Davies, 1993, pp70-71)

♦ If the electron's charge were slightly different, fusion would be impossible in stars like ours.

♦ If the universe had less than three spatial dimensions, humans couldn't exist. If it had more than three spatial dimensions, knots would not stay tied. If it had more time dimensions than one, causality would be chaos due to more than one *now*.

It's as if a Master Engineer tuned all the dials and programmed all the functions just right for life to arrive and thrive. Scientists call it the "fine-tuning problem."

Why is it a problem? Because science doesn't like use coincidence or God as an explanation. Instead, scientists cite the "anthropic principle" which goes something like this; "Since the universe is being observed by intelligent life, it is unremarkable that it has all the properties that would lead to the emergence of such life." Stephen Hawking called it the "Selection Principle" because, he said, of all the countless universes out there, only those with the precise conditions for the rise of intelligent life contain creatures who can ponder how goldilocksy their universes are.

As their appreciation for the fine-tunedness of the universe grows, more and more scientists are concluding that only two possible explanations exist:

1) Almost infinite numbers of universes have popped into existence to raise the odds of having one with life to a reasonable level. Of course, if almost all universes are lifeless, it seems like an awful waste of space.

2) One or more Master Engineers created and fine-tuned the universe. As God said, "There are many kingdoms; for there is no space in the which there is no kingdom." (DC 88:37)

Did scripture help cosmologists?

Here is a brief trail of the significant discoveries that led to science's current view of the universe:

♦ 1754 – Time acts like a fourth dimension mathematically. (Jean le Rond d'Alembert)

♦ 1868 – Light and magnetism are two aspects of the same phenomenon. (James Maxwell)

♦ 1883 – Only relative motion between material bodies is meaningful. (Ernst Mach)

♦ 1887 – Light travels at the same speed relative to the observer no matter how fast the observer moves. (Albert Michelson)

♦ 1885 – Inertial frames must be defined for observers in relative motion. (Ludwig Lange)

♦ 1889 – (4/3)E=mc² as applied a sphere's electric field. (Oliver Heaviside)

♦ 1895 – $E=mc^2$ for electromagnetism in general. (Henri Poincaré)

♦ 1903 – Speeds faster than light require an infinite amount of energy for objects with mass. (Wilhelm Wien)

♦ 1904 – The coordinates for one inertial frame can be transformed into the coordinates for another inertial frame. (Hendrik Lorentz)

♦ 1905 – The laws of physics are the same from the perspective of any inertial reference frame. $E=mc^2$, or more completely, $E^2 = (mc^2)^2 + (pc)^2$ in all cases. (Albert Einstein)

♦ 1915 – Space and Time work together as spacetime, warped by Gravity. (Albert Einstein)

♦ 1923 – There are galaxies other than the Milky Way. (Edwin Hubble)

♦ 1927 – General Relativity says the universe is expanding. (George Lemaître)

♦ 1929 – Redshifted galaxy light proves most other galaxies are receding. (Edwin Hubble)

♦ 2011 – The expansion of the universe is accelerating. (Nobel Prize reception date for Saul Perlmutter, Brian Schmidt, & Adam Reiss)

Although the list includes some brilliant discoveries by scientists who believed in God, there is no evidence that any of them credited their breakthroughs to His written word or to revelation from Him.

What is the "God Particle"?

In 1993, when the US cancelled the project to build an 87-kilometer $12 billion particle accelerator, dismayed physicist

Leon Lederman wrote a book about the lost opportunity to find the "God Particle." As it turned out, that particle was found by a team at the enhanced Large Hadron Collider (LHC) in 2012 and was named after Peter Higgs who predicted it back in the 1960s.

What is so special about the Higgs? Finding new types of particles was easy as shooting fish in a barrel in the early 1900s. So easy that Robert Oppenheimer once quipped that the next Nobel Prize should go the physicist who did *not* find a new particle. But, of the hundreds of particles found, most are made of other particles. The Higgs is special because it is the latest *fundamental* particle to be found. Physicists long ago created the "Standard Model Chart" which is for fundamental particles what the Periodic Chart is for elements. The Higgs is the 17th particle on that chart.

Why talk about Higgs bosons with lay people? Well, without them we would all float away. The particles of which our bodies and the earth are made all interact with the Higgs field to get the mass needed for gravitational attraction.

The Higgs only interacts with elementary particles but not with all of them, massless photons, gluons & theoretical gravitons being the exceptions. Uninhibited by the Higgs field, they travel at c in a vacuum – all of their momentum is through space and none through time (they do not age or decay).

For objects that interact with the Higgs field, the faster they move through it with respect to you, the more mass they acquire. So, if a spaceship takes off, and zooms past you, it will have more mass than when it was on the launch pad. But, if you were riding in the spaceship from its launch, it would gain no mass at all from your perspective. Thus, the Higgs field is responsible for adjusting perceptions of reality so that all the rules of physics are consistent in every reference frame as predicted by Einsteinian relativity.

Michio Kaku wrote, "What put the 'bang' into the Big Bang? In quantum physics, it was a Higgs-like particle… All the masses of the particles were the same, i.e. zero. But the presence of Higgs-like particles shattered this perfect symmetry [allowing] the particles to assume the masses we see today." (*The Spark That Cause the Big Bang*, WSJ online, 5 Jul 2012, US Edition)

So, the God Particle is in a class by itself, the *sole scalar boson* in the Standard Model. It makes the Standard Model work. Being an atheist, Peter Higgs was vociferously against calling it the God Particle. But the nickname stuck until the excitement over its discovery subsided.

Did God concoct the laws?

For most of human history, it never occurred to people that Nature might be a clockwork of rules. Finally, Descartes (1596-1650) went on record as the first to say the universe is governed by immutable laws. Since then, scientists have discovered hundreds of natural laws. For example:

♦ Laws of motion

♦ Law of gravity

♦ Laws of conservation of mass and energy

♦ Law of conservation of momentum

♦ Laws of thermodynamics

♦ Laws of electrostatics

♦ Laws of electricity and magnetism

♦ Laws of quantum mechanics

♦ Law of invariance of c

Descartes also proposed that God chose these laws, not at random or out of whimsy, rather because these laws are the only ones that would accomplish His design. But, whether some other set of laws would have served His purposes equally well is unanswerable.

Descartes also said that the universe operates moment-to-moment without God's intervention. Being a man of faith, he did not rule out miracles, of course.

Science is still discovering and exploring the natural laws. Those that are well understood exhibit a certain beauty and precision that hints at the existence of a Master Engineer. True, the universe seems to runs all on its own. But, God still has the knowledge and ability to hack the natural laws or tap into higher laws when intervention is needed, Christ's water-to-wine miracle being one example.

Scripture says, "God… hath *given* a law unto all things…" (DC 88:41-42) It does not say God created each law given. So, it's likely that God applied laws that He already knew rather than inventing them from scratch.

Does the universe spin?

Does our universe rotate? Well, according to some physicists, the answer is yes. A research team studied over 15,000 galaxies in the northern sky and found that significantly more than half of those galaxies are rotating in one direction. This would be expected if the universe was born rotating. Dr. Michael Longo, head of the research team said, "The excess is small, about 7 percent, but the chance that it could be a cosmic accident is something like one in a million." (phys.org/pdf229329774.pdf, DL 17 Aug 2017)

But, according to other scientists, the answer is no. They found that the Cosmic Microwave Background (CMB) is sufficiently isotropic that the odds of the universe rotating is only 1 in 120,000. (livescience.com/65882-does-the-universe-rotate.html, DL 28 Dec 2020)

All motion is relative. So, if the universe comprises all reality, which is what the word means, then with respect to what is it moving? And, if its movement is rotational, where is the center or axis of its rotation?

What Einstein called Mach's principle is the idea that rotation in our universe is determined by an object's movement relative to large-scale mass distribution, perhaps relative to all mass in the surrounding space whose gravity influences the object in any degree. So, in order for our universe to have linear or rotational movement, it would have to be defined as a local expansion phenomenon in a superverse/superverse or as one of many universes in a multiverse where large-scale mass distribution would provide an extended reference frame for determining rotation.

What's a multiverse like?

What's beyond our universe? Other universes? If so, are they different than ours? Are all of them infinite? Did they all come from big bangs? Are some of them barren? If so, do they serve a purpose?

Stephen Hawking postulated that the universes in our multiverse are interconnected via a network of tunnels, aka wormholes, aka Einstein-Rosen bridges. (*Of Wormholes, Time Machines, and Paradoxes*, Astronomy, Feb 1996) He also proposed that a Black Hole in one universe might open as a White Hole in another universe via such a tunnel.

Do universes interact with each other in a multiverse? One team of scientists has used computer analysis to detect anomalies in the CMB that might be scars and bruises on our universe from primordial collisions with other universes. Another team claims that some kinds of quantum phenomena arise from multiple universes interacting with each other. (*Ghost universes kill Schrödinger's quantum cat*, By Michael Slezak, NewScientist.com, 5 Nov 2014, DL 20 Jul 2020)

Would it be possible to detect other universes? Some physicists believe gravitons from one universe might leak into another. Perhaps gravity from another universe is fooling physicists into thinking dark matter exists. But, if our universe is contained in a multiverse, let's hope that multiverse is expanding rather than imploding!

9: First World

What was created first?

The Bible starts with a very poetic phrase that is easily remembered, even by millions who've never read it; "In the beginning God created the heavens and the earth." (*NIV*, Gn 1:1) Does "heavens" refer to the cosmos? No. Bible references to the earth and heavens are about *this* planet and the sky in which *its* "fowls of the air" fly. (Hebrew *shamayim*=skies, see Genesis 1:26 and 2:19)

A recent Jewish translation says, "*When* God began to create heaven and earth..." (*JPS bible (Jewish Publication Society)*, Gn 1:1) The use of "when" limits the topic to the time Earth was born rather the beginning of the *entire cosmos* and makes it clause dependent on what directly follows, "the earth was without form, and void" — the real beginning of the narrative.

So, although Genesis mentions Earth and its skies first, it cannot be the first created thing, else it would have floated aimlessly in space without any sun to orbit! Astronomical models all say the universe was created before the stars, some of which exploded to provide the materials from which the Sun formed. Last of all, the planets accreted from the swirl of matter around the sun. Even Jupiter, the sun's largest planet, was created before earth. (*Jupiter is the Most Ancient Planet in the Solar System*, by Andrew Masterson, CosmosMagazine.com, 12 Jun 2017, DL 20 Jul 2020)

The presence of a tiny bit of nickel (atomic #27) in our sun strongly suggests that it was not created first. Fusion stars cannot create elements heavier than iron (atomic #26), so the sun got its heavier-than-iron elements from the explosions of earlier stars. More than half the Milky Way's stars are older than our 4.5-billion-year-old sun.

Older still is Sagittarius A-star (Sgr A*), the supermassive black hole (SMBH) at the heart of the Milky Way! It was created in the very early universe. A black hole warps spacetime in on itself, leaving no escape path for anything, not even for light – thus the moniker *black hole*, which is much easier to say than *gravitationally collapsed cosmological object.*

The astrophysical creation sequence looks like this:

1. **The Universe** – from unknown conditions, perhaps from a meta-universe or multiverse.
2. **The Central Black Hole Sgr A*** – from smaller black holes, gas, plasma, and/or other stars.
3. **The Milky Way Galaxy** – from gas, dust, and other objects.
4. **The Solar System** – from large clouds of hydrogen and helium gas in the galaxy, mixed with the debris from super novae.
5. **Planets and moons** – from gas, dust, and exploded star debris swirling around the sun.

So, after the cosmos, Sgr A* is the first creation critical to life in our galaxy. In fact, it powered the galaxy's formation. (cosmotography.com/images/supermassive_blackholes_driv e_galaxy_evolution_2.html, DL 13 Oct 2018)

Has God mentioned Sgr A*?

Bound into the holy Book of Abraham is Facsimile (facs) No. 2, a sketch of an ancient Egyptian hypocephalus, a circular papyrus amulet representing the circles and cycles of the cosmos. The amulet, which was created by one or more priests for a guy named Sheshonq, was found under the head of a mummy circa 1818 in Thebes by a tomb raider shortly after Napolean conquered Egypt.

At the amulet's center is the figure of the Egyptian creator god. (*The Complete Gods and Goddesses of Egypt*, Richard H. Wilkinson, 2003, p95) What does a stick figure creator god on a piece of papyrus have to do with Sgr A*?

Joseph Smith said this same stick figure also represented the first star created in the cosmological order to which Earth belongs, "Kolob, signifying the *first creation*, nearest to the celestial, or the residence of God". (Abr, Facs2, Fig1) The text to which this facsimile belongs tells of God showing a star hierarchy to Abraham in which Kolob is the God Star.

Why would God be talking about astronomy with Abraham? The scripture isn't really about the order of stars in the physical heavens. It's about the Creator! It uses the central star in the system as a figure for Jehovah. Like Kolob, Jehovah is the first creation. He, like Kolob, is at the top of the hierarchy. He, like Kolob, is nearest the throne of God. He, like Kolob, is the engine that created everything else. And, just as Kolob governs the movement of nearby stars gravitationally, which in turn govern other stars gravitationally, and so on through the entire galaxy, Jehovah governs the entire creation through a hierarchy of influence.

"Kolob is the greatest of all... If there be two things, one above the other... a planet or a star may exist above it... [just as when] there are two spirits, one being more intelligent than the other; there shall be another more intelligent than they; I am the Lord thy God, I am more intelligent than they all." (*PGP*, Abr 3:16-19) Thus the Lord used the physical hierarchy of the stars in the galaxy to illustrate the spiritual hierarchy of intelligent beings with Himself at the apex.

In his Egyptian Alphabet and Grammar (notes related to work on the Book of Abraham), Joseph Smith said that Kolob is "the *eldest* of all the Stars" and the "*first beginning* to the bodies of *this creation*". (*EAG (Egyptian Alphabet and Grammar)*, pp28-32) The term "this creation" probably refers to *this galaxy*, which is likely the scope God meant when He told Abraham, "I have set this one (Kolob) to govern all those (stars) which belong to the *same order* as that (planet) upon which thou standest." (*PGP*, Abr 3:3) No structure larger than a galaxy has central gravitational governance. And, in Abraham's day, humans could not see any stars beyond those in the Milky Way. The farthest Milky Way star discernible with the naked eye is V762 Cas, 16,308 lightyears away. That's not even halfway across the galaxy.

Four million times larger than the sun, Sgr A* is the most massive star in the galaxy. God said, "Kolob is the greatest." (*PGP*, Abr 3:16)

Its name is pronounced "Sagittarius A star". So, do physicists consider Sgr A* to be a star? Well, black holes *of stellar mass* form when massive stars collapse at the end of their life cycle but physicists believe that collapsed stars grow too slowly by the accretion of swirling material to have become *supermassive* black holes in the short time since the universe began. And Sgr A* *is* a monster. The formation of such humongous black holes likely included many mergers of super-massive objects like giant stars and collapsed stars (predecessor black holes). So, Sgr A* is probably an gravitationally imploded mess of primordial stars and stellar flotsam from supernovas. The asterisk (*) was appended by co-discoverer Robert Brown to denote how exciting Sgr A* because that is the notation sometimes used to denote molecules or atoms in an excited state; e.g., Fe* or He*.

Jehovah pointed out that Kolob shares governing power with other great stars. Likewise, He delegates power to lesser but noble and great intelligences. Sgr A* shares the gravitational hub with a dozen or so smaller black holes and a huge swarm of shining stars. There are about 10 billion stars in the central bulge, each tugging at each other and at all the other stars in the galaxy gravitationally, holding the galaxy together as it whirls.

In 1930, when the idea of galaxies sprinkled across the universe was still in its formative stages, a CJCLDS author named B. H. Roberts wrote that all stars orbit Kolob; "There are many great stars, called 'governing ones,' one of which rises preeminent above the others 'nearest the throne of God,' and is called 'Kolob.' This great star constitutes the grand center of our universe, around which all other great stars revolve..." (*CHC (Comprehensive History of the Church)*, bk2, p128-129) By "center of our universe" Roberts meant, as he had earlier stated, "the great centre of that part of the universe to which our planetary system belongs". (*New Witness for God*, Roberts, bk1, 1898, p444) Roberts' source, Joseph Smith, was way ahead of his time!

Other authors and speakers in The Church of JESUS CHRIST of LDS have since gone on record to agree with Roberts. For example, Apostle J. Reuben Clark said, "In reference to this *hub of stars…* we might well recall what God said to Abraham, 'These are the governing ones; and the name of the great one is Kolob.'" (*What Was This Jesus*, BYU Selected Speeches, 1951)

Does science accept Kolob?

Of course, secular scientists do not identify Sgr A* as Kolob. But their curiosity about the central black hole has remained extremely high ever since its existence was signaled in 1931 by radio waves emitted from a particular point that turned out to be the center of the galaxy. It had only been about ten years since astronomers realized that the band of light across the night sky, admired by humans for millennia, was actually an edge-on view of a huge swirl of stars.

In 1974 the central radio source was identified as a star and in 1982 it was named Sgr A*. It is about 26,000 light-years away, hiding behind a shroud of glowing gas and dust near the galactic center. Scientists are still working to get a better peek at Sgr A* using special telescopes that see various types of radiation, monitoring flares caused by in-falling objects, and gathering data on the movements of stuff nearby. The veil obscuring Sgr A* is likely what Joseph Smith meant when he wrote; "God has said, '*Let this be the centre for light, and let there be bounds that it may not pass.*' He hath set a cloud round about in the heavens." (*EAG*, p25)

Humans will not be sending a probe to Sgr A* any time soon. The fastest man-made devices are now travelling away from the sun at 50,000 MPH. At that speed, it would take 3 trillion years to cover the 152,840,995,060,000,000 miles to the galactic center. And, even if a ship were to arrive safely there, it would be ripped apart by black hole gravitational forces before it arrived.

To locate Sgr A* in the night sky, a person in the northern hemisphere should wait until summer and then find the teapot astarism (star pattern) in the Sagittarius constellation in the southern sky. Sgr A* is just upper-right off the spout where the steam would spray.

Isn't Kolob crazy talk?

Where did Joseph Smith get the word *Kolob*? He knew no Egyptian. That he tried to capture some knowledge of Egyptian in the form of notes is true, but this was done as part of his good faith effort to understand the manuscripts God helped him translate. In the process, Egyptian terms were transliterated to English. Among these were *kokob* and *Kolob*.

Names usually have meanings. One professor of ancient scripture, Michael D. Rhodes, wrote that Kolob "most likely derives from the common Semitic root *QLB, which has the basic meaning of 'heart, center, middle.' In fact, the Arabic form of this word, *qalb*, is in the Semitic names of several of the brightest stars in the sky including Antares, Regulus, and Canopus". (*Review of Books on the Book of Mormon,* p124)

For example, **Antares**, the star at the heart of the constellation Scorpius, is still known in the Mideast as *Kalb al Akrab,* meaning "heart of the crab". (constellation-guide.com/antares, DL 20 Feb 2019) Standardization of Arabic transliteration replaced *kalb* with *qalb* for *heart* (*kalb* now means *dog*) but the name of the star has not changed.

In Hebrew, the word for *heart* is לֵב, and for *in* is ב and together (קלב) they translate to English as *Calab*, meaning *in the heart.* (translate.google.com, DL 1 Jun 2018) So, *Kolob* makes a great Semitic name for the galaxy's heart star.

What about *kokob* which Smith translated as *star*? (*PGP,* Abr 3:13) As it turns out, there is a star 16 degrees from Polaris is named "Kochab" from the Arabic ال كوكب al-*kawkab* (also transliterated *Kaukab*) or Hebrew כוכב, *kōkhāv* (also transliterated *kōkhābh*), all of which mean "star". (*Dictionary of Modern star Names*, by Kunitzsch, Paul & Smart, Tim. 2006) The Hebrew plural suffix "im" makes it *kokhabhim* or "kokaubeam" as Joseph Smith spelled it.

Joseph Smith had naught but three years of formal schooling and was neither a linguist nor a scientist. So, his 1830 telling of a central body gravitationally governing the motion of other stars which governed the motion of yet other stars in the local order came almost a century before the Milky Way Galaxy was discovered by science. And his use of Semitic-based Egyptian names and their meanings is impressive considering that the work done by Champollion to decipher the Rosetta Stone would not be available as an Egyptian-French dictionary until ten years after Joseph Smith published his translation of Egyptian papyri. Joseph Smith never learned French.

Why create Kolob first?

Once upon a time, astronomers thought that supermassive black holes arrived on the cosmic scene rather late in the evolution of the universe. But now they see that the opposite is true; "In the evolution of stars and galaxies, supermassive black holes were around at their beginnings and played a major role in shaping them." (*When giants warped the universe*, by Graham Phillips, cosmosmagazine.com, DL 31 Mar 2017)

Almost every galaxy has a central SMBH that drove its formation and its star creation. (*Simulations Show How Growing Black Holes Regulate Galaxy Formation*, from Carnegie Mellon University, sciencedaily.com, DL 16 Nov 2008) In the vernacular of Joseph Smith, Kolob was created first to govern the later formation of kokaubeam in the swirling stellar order we call the Milky Way Galaxy.

Why is Kolob a black hole?

The Abrahamic writings from Egypt, as translated by Joseph Smith, say Kolob is a star. But couldn't it be some kind of huge star other than a black hole? No. Why not?

Firstly, only a very massive star could be the chief gravitational governor or orbital anchor of all other stars, large and small. Black holes are the most massive type of star.

Secondly, only a black hole blasts sufficient energy out to push enough material away to allow for shining stars to be created in a galaxy; "Simulations demonstrate that self-regulation can quantitatively account for observed facts associated with black holes and galaxies." (spaceflightnow.com/news/n0502/17blackholes, DL 16 Jul 2020)

Thirdly, Kolob was the "first creation" and therefore is the eldest star, much older than the galaxy itself, just like Sgr A*.

Fourthly, Kolob was set to govern our entire order of stars for the duration. Only a black hole outlives its galaxy, unlike any other type of star. The larger a fusion star, the sooner it runs out of fuel and explodes – not a good governing star! The most massive known star, R136a1, at 265 solar masses, puts off about 7 million times more energy than our sun but has a life expectancy of only 4 million years, whereas the larger a SMBH is, the longer it lives. Even if Sgr A* accumulated no more mass ever, it would last 1.34×10^{87} years. Black holes will outlast every other type of celestial object.

In translations of Egyptian papyri, Joseph Smith said that *Enish-go-on-dosh* was "the power of attraction" by which a star influences other stars. (*PGP*, Abr Facs2, Fig5) He was quoted in the Church's newspaper as saying, "Were I an Egyptian, I would exclaim Jah-oh-eh, Enish-go-on-dosh, Flo-ees- Flos-is-is" translated as, "O the earth! The power of *attraction* and the moon passing between her and the sun." (centerplace.org/history/ts/v4n24.htm, DL 9 Jan 2020) And, in his Egyptian language notes, he said that Earth is "under the government of an other or the second of the fixed stars which is called *Enish-go-on-dosh*, or in other words *the power of attraction it has with the earth*… Kolob signifies the highest degree of power in government, pertaining to heavenly bodies". (*EAG*, p29-30)

So, the physical governing power of Kolob is gravity which serves as an analog for the governing power of Christ which is the intelligence or "light that proceedeth forth from the presence of God to fill the immensity of space… which is the law by which all things are governed, even the power of God". (DC 88:11-13) In this light there are no shadows!

Is Kolob a portal to Heaven?

What does it mean that Kolob is the star *nearest unto the throne of God*? (*PGP*, Abr 3:2) Surely God's celestial home isn't lurking in the center of our galaxy!

Kolob belongs to the "same order" as earth. (*PGP*, Abr 3:3) The entire Milky Way Galaxy belongs to this order. But Kolob is not near God's home in the same sense as a person's office might be next to her boss's. God's celestial home belongs to a higher order that may be described as above or outside the physical universe, perhaps overlaid in some paranormal way.

Imagine a 3D person standing above a 2D realm watching stick people going about their business. Intervention would be as easy as drawing a new stick figure or changing the landscape – miracles to the stick people. Likewise, God stands above our 3D realm and observe us going about our business and He can intervene with precision at multiple locations in our 3D universe.

Is Kolob a portal from our 3D world to God's world? In the Disney film *The Black Hole*, future travelers encounter hell as they pass through a black hole but find heaven on the other side of it. Such fiction sounds farfetched but serious scientists have speculated that a black hole could exit into other universes as white holes. Travel through such holes would not be a problem for spiritual beings and things. People who have residual memories of premortality describe passing through a dark tunnel on their way to Earth and people who recount near-death experiences describe entering a dark tunnel leading to a white light in some kind of heaven, though not the origin of the 19th-century railroad metaphor.

The standard NDE account includes passage through a tunnel that is very similar to a wormhole:

1. The spirit leaves the body and rises up.

2. It then enters a dark tunnel or passageway off the planet.

3. At the end of the tunnel is a heavenly place of light and brilliance and angelic beings.

4. Beyond that is a portal to a higher heaven. Disembodied spirits who go past that cannot return to mortality.

Of course, if the "dark tunnel" were a physical wormhole, it would have to be lightyears away from earth to avoid detection by scientists. Perhaps people perceive distances differently after they are dead. More likely is that the tunnel is a spirit structure. One NDE account has an angel telling the disembodied spirit that he helped to build the tunnel in his premortal life. Perhaps the tunnel leads to Kolob where a halfway house heaven exists to receive the spirits of the freshly deceased. And perhaps Kolob is where the portal to God's celestial home lies.

Surely there is no better place than the firstborn physical world for the angelic operations center where deceased Earthlings take their first step back toward God the Father?

How can Kolob be a "fixed" star?

The *Egyptian Alphabet and Grammar* paper by Joseph Smith and his scribes classifies Kolob as one of twelve "*fixed* stars." Of course, all stars move. So, how are they *fixed*?

Anciently, stargazers distinguished objects that were obviously moving, such as moons, asteroids, comets, the sun, and planets, from objects that seemed not to move. The word "planets" comes from the Greek πλανῆται (planētai) meaning *wanderers.* "Fixed stars" is an archaic astronomical term that refers to bright objects in the sky that appear stationary with respect to the rest of the starry background.

Nowadays, in the age of high technology, the motion of almost any star can be detected by tracking the object over a period of just a few short years. For example, in 2018, after analyzing twelve years of data from high-powered X-ray telescopes, astronomers announced the discovery of twelve black holes directly orbiting Sgr A* inside the galactic hub. (nature.com/articles/nature25029, DL 29 May 2018)

The first book on the topic was "Book of Fixed Stars" by Abd al-Rahman al-Sufi, written around 964AD. Finally, in 1718AD, Edmund Halley proved that even a so-called fixed star actually has proper motion. So, although *every* star moves, the Book of Abraham is astronomically 100% correct to call Kolob a fixed star!

What is Kolob time?

God told Abraham that Kolob is "the last pertaining to the measurement of time". (*PGP*, Abr Facs2, Fig.1) Also, "Kolob was after the manner of the Lord, according to its times and seasons in the revolutions thereof; that one *revolution* was a day unto the Lord, after his manner of reckoning, it being *one thousand years.*" (*PGP*, Abr 3:4)

Does this mean that Kolob rotates slowly? Not at all. On average, celestial objects spin faster the more massive they are. The largest planet in our solar system is Jupiter and it spins faster than all its sibling planets. Saturn is the next largest and has the second fastest spin. Extremely massive black holes have acquired so much angular momentum from billions of years of swirling in-falling objects that their spins are often more than a thousand times per second. The ergosphere (just outside the event horizon) of galaxy NGC 1365's central black hole is spinning at 84% of light-speed. (phys.org/news/2014-02-fast-black-holes.html, DL 1 Jun 2018)

The astronomic definition of a revolution is *the orbit of an object around another object* – its *year* – not *rotation*. But if Kolob is at the center, it would be odd to speak of its orbit.

So, how should the 1,000:1 ratio be understood? Perhaps God meant that one Kolob year *seems like* a day to Him but *seems like* a thousand years to humans. In other words, what seems like forever to humans is but a moment in the mind of God. Joseph Smith said, "The great Jehovah contemplated the whole of the events connected with the earth… the past, the present, and the future were and are, with Him, one eternal 'now.'" (*TPJS*, p220) And, Alma said, "Time only is measured unto men." (*BOM*, Alma 40:8)

The Kolob passages should be understood in the context of God's purpose in sharing Egyptian astrology with Abraham; "I show these things unto thee before ye go into Egypt that ye may declare all these words." (*PGP*, Abr 3:15) God was not sending Abraham to the Egyptians to correct their false astrology. Rather, He was sharing Egyptian astrology with Abraham so he could preach the true God and doctrine in a manner that would resonate with the Egyptians.

The message for the Egyptians was, in essence; If there be two heavenly objects, one above another in its cycles, there will be yet another above that all the way up to Kolob at the top. Likewise, if there be two spirits, one more intelligent than the other, there shall be another spirit more intelligent than both of them, and so on all the way up to God who is more intelligent than all of them. (*PGP*, Abr 3:18-19 paraphrased)

That the astronomy described by Abraham is Egyptian is borne out by the fact that the cycles of Kolob and other heavenly objects were drawn on a circular papyrus amulet by an *Egyptian priest* for a man named Sheshonq, not by Abraham! (*PGP*, Abr, Facs2) Abraham's Kolob time figures match ancient Egyptian astrology rather true astronomy.

Whence came the Kolob papyri?

One of the tomb raiders who joined in the plunder of Egypt following its conquest by Napoleon was an Italian adventurer by the name of Antonio Lebolo. Sometime around 1821AD, he found a huge mummy cache in a shaft at Sheikh Abd el-Qurna on the side of the Nile opposite the city of Thebes.

Lebolo turned most of the mummies over to his boss Drovetti to be sold but kept eleven for himself, shipping them back to his home in Italy. When Lebolo died, his estate shipped the mummies to New York where an American named Michael Chandler acquired them. Chandler took the curiosities on the road, exhibiting them around New England to people willing to pay a small fee.

Over time, Chandler sold all but four mummies. In 1835 he arrived in Kirtland, Ohio where the restored Church of JESUS CHRIST was headquartered. Having heard rumors of Joseph Smith's translation of ancient characters, Chandler asked him for his opinion. After examining some of the characters, Joseph said that the scrolls contained writings of Abraham and Joseph, along with a tale of an Egyptian princess named Katumin.

Joseph Smith wanted to buy the papyri but Chandler wanted to sell them with the mummies as a package. So, Smith made the purchase with $2,400 donated by friends (about $60,000 in 2018 dollars).

The papyri had been found in the breast wrappings of one of the mummies. After Joseph Smith's murder by a mob, his widow gave some papyri to her maid and sold the rest. For decades it was thought that all of the papyri had been acquired by a museum in Chicago and destroyed in the fire that burned the city down in 1871. Skeptics who had obstreperously attacked Smith's work as total fantasy were somewhat taken aback in 1966 when several fragments of the papyri were found in the Chicago Metropolitan Museum archives. Another fragment was found in church's archives where it had been filed with a copy of the Egyptian Alphabet and Grammar.

What do Egyptologists say?

The Kolob Hypocephalus, aka Facsimile No. 2 of the Book of Abraham has gotten a lot of attention, in part because its place in Latter-day Christianity renders it controversial. Egyptologists have offered their translations of the figures and, to the disappointment of many, the rediscovered papyri fragments have nothing to do with the Book of Abraham. However, Smith's translation of the Kolob hypocephalus has some themes in common with secular interpretations, as is

248

evident from a look at figures 1 and 4 on the amulet. Terms common to both interpretations are underlined:

Explanation Figure 1

▶ Fig.1 – Smith: The image of a ram-headed person flanked by baboons at the center represents the star Kolob which governs other stars, a metaphor for the supremacy of the <u>Creator God</u>. (*PGP*, Abr 3:17-19) Through the <u>Son, the Father</u> created all things. (Eph3:9) Kolob was <u>created first</u> and is last in the <u>measurement of time</u>.

▶ Fig.1 – Egyptologists: The image represents Khnum, <u>Creator God</u>. Khnum is also called Chnum-Re or Amun-Re, conveying the <u>father-son</u> concept. The three horizontal lines denote "Nu" or "Nun" or the <u>primeval creation</u> waters. The baboons (Figures 22 and 23) on each side of Khnum were associated with stars and constellations. The lunar disk headdresses on each represent Thoth or Djehuty, a moon god associated with the <u>regulation of time</u>.

♦ What was it about a stick figure with an ant-like head between two baboons that gave Smith the correct themes about <u>father and son creator gods</u>, the <u>beginning of creation</u>, and the <u>governance of time</u>? Statistically, there are too many correlations for Smith to have guessed blindly.

Explanation Figure 4

▶ Fig.4 – Smith: The bird in a boat represents the expanse of the sky and signifies <u>1,000</u>, the revolution ratio of Kolob to Earth, the 2nd position star having the same <u>revolution</u> rate as Kolob.

▶ Fig.4 – Egyptologists: The bird represents Sokar, who is sometimes equated with Osiris, both of whom are Gods of the dead. Whether Sokar or Osiris, the god is sometimes shown in the ship of the dead. The Coffin Texts discuss the ship, "He takes the ship of <u>1000</u> cubits from end to end and sails it to the stairway of fire." (*Coffin Texts 162*, II, ca. 2134-2040 BCE, pp403-4) And the sarcophagus of princess Anchenneferibre features a description of "Osiris in his ship of a <u>thousand</u>". (*Die Religiösen Text auf dem Sarg der Anchneneferibre*, C. E. Sander-Hansen, 1937, pp36-37) The same sarcophagus describes "Khabas in Heliopolis." (ibid.) Khabas (Egyptian h3-b3=s) refers to the stars of the sky and means, "A thousand is her souls." (*Wörterbuch der Ägyptischen Sprache*, Adolf Erman and Hermann Grapow, Akademie Verlag, 1971, bk4, p562, entry 7) In the festival of Sokar, a high priest would place the Sokar-boat on a sledge and pull it around the sanctuary as part of a procession symbolizing the <u>revolution</u> of the sun and other <u>celestial bodies</u>.

♦ What was it about a bird in a boat that suggested <u>stars,</u> <u>revolutions,</u> and <u>1,000</u> to Smith's mind in line with Egyptology? How did Smith come up with a number as specific as 1,000 from such a picture? How could he know that the figure had anything to do with stars or revolutions of celestial bodies?

How did Joseph Smith know that anything on the hypocephalus had a religious theme? He recorded his translation in English eight years before Champollion's Egyptian dictionary was published in French, which he could not read. He had promised to provide more from Abraham but was murdered by a mob in mid-1844 while awaiting trial in Carthage Jail. (*The Story of the Book of Abraham*, H. Donl Peterson, 2008, p146)

Egyptologists are scientists and therefore cannot cite revelation as a source for interpreting ancient hieroglyphs. They must employ secular understandings to ancient religious figures rich in paronomasia, riddling, cryptography, and other types of ancient wordplay. Smith's translation of the Kolob hypocephalus is not scripture; it's simply an inspired prophet's best understanding of the intended meanings. The deeper meaning of the hypocephalus would only be apparent to a Hebrew priest adapting Egyptian figures to convey religious themes.

What did Smith mean by "translate"?

Artist depictions of Joseph Smith often show him at a table upon which the golden plates lie open. His early attempts to decipher the characters could have looked something like that. However, over time, Joseph switched from trying to work out what the characters meant, to using the Urim and Thummim stones that came with the record, to using a stone of his own that he felt comfortable carrying in his pocket. In these later attempts, God sent Him the English translation without the necessity of opening the ancient record.

When Smith later produced the Book of Abraham, he used the same spiritual gift from God as with the Book of Mormon. Although he had put a great deal of effort into understanding Egyptian, the final text is pure revelation. The English text of Abraham's words does not correspond with any ancient manuscript known to exist now or to have been in his possession at the time.

But God did not need Joseph to have any type of ancient record in hand. For example, Joseph received an English translation of words written by John 2,000 years ago without a scrap of manuscript. (See DC 7)

As a prophet, Joseph Smith did not need to be an expert on ancient languages. The fact that he had some misconceptions about several ancient artifacts that passed through his hands does not at all diminish any of the revelations God gave him.

10: The council

Who are the creation Gods?

Genesis Chapter One says Elohim created Heaven and Earth. *El*, the Hebrew word for God, is found in names like Emmanuel (God with us) and Israel (God contends). The Hebrew name for God is *Eloah* (אֱלוֹהַּ, Strg#433). The suffix -*im* makes it plural, *Elohim (*אֱלֹהִים, Strg#430=*gods, rulers, angels)*, just as the plural of cherub is cherubim. This is important because every instance of "God" in the Genesis Chapter One is *Elohim.*

Countless pundits have opined that Elohim is simply a majestic or royal plural form denoting reverence. But the translators rendered *elohim* as "gods" in 240 other places in the Old Testament, e.g. "Thou shalt have no other *gods* before me." (Gn 20:3) So why not in Genesis Chapter One? According to one expert, the first sentence of Genesis should be, "In the beginning *gods* created the heaven and the earth." (*From The Alpha and the Omega,* Jim A. Cornwell, ch1, 1995)

Genesis also uses plural pronouns for God here and there. Elohim said, "Let *us* make man in *our* image, after *our* likeness" (Gn 1:26) In tempting Eve to eat the forbidden fruit, Satan promised her that she would "be as the *gods*, knowing good and evil". (Gn 3:5) And, after Adam and Eve tasted the fruit, God said, "The man is become as one of *us*, to know good and evil." (Gn 3:22) In response to Babel's tower project, God said, "let *us* go down, and there confound their language."

There are also a handful of places in the OT where Hebrew verbs used with God were conjugated in a way that indicate the subject was plural *Gods*. And, in Abraham's version of the creation, *Gods* plural did the work. (*PGP*, Abr 4)

The "Gods" include The Father, The Son, The Holy Spirit, and all the hosts of heaven who helped with creation as apprentice gods.

Who helped with creation?

Who were the *others* with God? A 1st century translation says, "And the Lord said to the *angels* who ministered before Him... Let *us* make man in *Our* image." (*Targum of Palestine*, Gn 1) So, in addition to the obvious companions, the Word and the Holy Spirit, all the angels were with God in the beginning.

"God himself, finding he was in the midst of spirits and glory, because he was more intelligent, saw proper to institute laws whereby the rest could have a privilege to advance like himself." (*TPJS*, p354) So, since God's spirit children wanted to become more like Him, so, like any good father and mentor, God involved them in His creation plans.

God told them, "*We* will go down, for there is space there, and *we* will take of these materials, and *we* will make an earth whereon *these* may dwell." (*PGP*, Abr 3:25) His meeting with the spirits is known as "...the Council of the Eternal God of all other gods before this world was..." (DC 121:32)

Early Christians compared the Divine Council to the Egyptian pantheon, Ogdoad. In a 3rd century text, Jesus said, "Before the foundation of the world... the whole multitude of the Assembly came together upon the places of the Ogdoad." (*Second Treatise of the Great Seth VII*, as quoted in *The Nag Hammadi Library*, by Apostle Horn, p257)

An early Jewish Midrash Rabbah says, "We took counsel with the souls of the righteous... The supreme King of Kings, the Holy One, blessed be He, sat with the souls of the righteous with whom He took counsel before creating the world." (*Midrash Rabbah*, Freedman & Simon, Gn 1:59; 8:7) The Midrash also says, "With the Almighty King of kings, the Holy One, blessed be He, dwelt the souls of the righteous with whom He decided to create the world." (ibid., Ruth 2:3)

And, according to *Bereshith rabba* teachings, Adam, Noah, Abraham, Enoch and Moses were all there "with God before the creation of the world" and consulted with God on the Creation. (*Analysis of the Book of Abraham*, Rabbi Nissim Wernick, p24) Also, the Midrash says that God counselled with the righteous spirits yet to be born when creating the world, "With the souls of the righteous among them he took counsel and created the world." *(A Theological Commentary to the Midrash: Genesis Rabbah*, Jacob Neusner, 2001, p27)

Involving His children in His works always was and still is the Father's modus operandi. By allowing His children to participate in His work, God prepares them for life on Earth and molds them to become more like Him.

Is the Divine Council biblical?

The psalmist wrote, "Elohim standeth in the *assembly of the gods*; in the midst of the judges He judgeth… *Ye are gods, and all of you children of the Most High*." (Septuagint (LXX), Ps 82:1-6) Some translations use different terms for the meeting and attendees. For example, "God has taken his place in the *divine council*; in the midst of the *gods* he holds judgment." (*ESV*, Ps 82:1) So, the children of the Highest God are the gods.

When defending himself against charges of blasphemy from Jewish leaders for claiming to be God, Jesus cited the ye-are-gods scripture. (Jn 10:34-35) His spirit children were still immature, but God still included them in His council of *gods*.

The psalmist said, "The heavens shall praise thy wonders, O Lord; and thy truth in the *assembly of the holy ones*. For who in the heavens shall be compared to the Lord? And who shall be likened to the Lord among the *sons of God*? God is glorified in the *council of the holy ones*." (*LXX (Septuagint)*, Ps 89:5-8)

Disappointed with profane mortal prophets, God asked, "Who among them has stood in the *council of the Lord* to see and to hear his word? ... But if they had stood in *my council*, they would have proclaimed my words to my people." (Jer 23:18, 22) Mortals are usually not attendees to the council.

The prophet Micaiah told the king he had a vision in which he; "saw the Lord sitting on his throne, and all the host of heaven standing by him on his right hand and on his left. And the Lord said, Who shall persuade Ahab?" In response, several spirits volunteered with various plans and God chose one for the job. (1 Kg 22:19-22) That's a council! However, only two plans were proposed in the creation council that we know.

The story of Satan and Job features a recurring assembly of the sons of God with their Father. (Jb 1:6; Jb 2:1) Job's friend asked, "Were you brought forth before the hills? Have you listened in the *council of God*?" (*ESV*, Jb 15:7-8)

Emanuel Tov said the Song of Moses in Deuteronomy 32:8 originally "referred to an *assembly of the gods*...[but] the scribe of an early text... did not feel at ease with this possibly polytheistic picture and replaced 'Bene El,' (sons of El), with 'Bene Yisrael.'" (*Textual Criticism of the Hebrew Bible*, Emmanuel Tov, 1992, p269)

The previously quoted Bible passages referencing the Divine Council are only what survived ancient monotheistic censorship. A case in point was the effort by King Josiah to suppress pagan beliefs in myriad gods. Archeological and textual evidence shows that, during the last few decades of the 7th century BC, a group of Hebrew scribes, court officials and priests edited and redacted the holy records to remove polytheistic references while King Josiah waged a campaign to destroy the shrines and priests of idolatry.

Michael Heiser told the Society of Biblical Literature in May, 2011, "El-Elyon (Most High God) and Yahweh were the main deities, positioned in the pantheon as father and one son among many in an Israelite pantheon or, as it is more commonly referred to, a *Divine Council*... But as time went on... the redaction of the Deuteronomistic History dealt with those other gods and the gods of *Yahweh's own council*, by downgrading them to angels." (*Divine Plurality in the Dead*

Sea Scrolls, quoted by mormonchristian.webs.com/god.htm, DL 4 Jun 2017)

What do scholars say?

"In the beginning was the Word." These first words in John's gospel should have been translated, "In the beginning was the Council," according to Hugh Nibley, Professor of Ancient Scripture. (*Ancient Documents and the Pearl of Great Price,* p6)

Michael Heiser's doctoral dissertation in Hebrew Scriptures and Ancient Semitic Languages was based on his extensive study of an impressive number of ancient tablets and Mideastern texts in which the Divine Council is mentioned. He concluded, "The term *Divine Council* is used by Hebrew and Semitic scholars to refer to the heavenly host, the pantheon of divine beings who administer the affairs of the cosmos." (thedivinecouncil.com/HeiserIVPDC.pdf, DL 2 Aug 2017)

Paul Sumner presented a thesis to the faculty of the religion division of Pepperdine University in which he said, "Working through *innumerable hosts of angelic servants*, God creates and rules the physical universe, as well as the world of men... The existence of the *Divine Council* is witnessed to by various literary genres of the Hebrew Bible. It is mentioned in historical, narrative and poetic passages, prophetic visions, Temple liturgy, apocalyptic visions... The concept and imagery of the Divine Council is thus woven throughout the pages of the Hebrew Bible. In the Hebrew Bible, a few select men gain access to the Divine Council. Such visits or 'throne visions' are for the purpose of giving the prophet a message to announce to his people." (vdocuments.mx/the-divine-council-in-the-hebrew-bible-chapter-2-of-hebrew-.html, DL 19 Oct 2018)

Ferrar Fenton "translated into English directly from the original Hebrew" the first words in Genesis as "By headships..." (*The Holy Bible in Modern English*, 1903, Gn 1:1, fn1) Plural heads only makes sense in the context of a council of gods! And, among the heads is one Head God, identified in a large handful of ancient texts as the premortal Messiah. (fairmormon.org/conference/august-2004/the-king-follett-discourse-in-the-light-of-ancient-and-medieval-jewish-and-christian-beliefs#en49, DL 17 Apr 2018)

The prophet Joseph Smith said the first verses of the original Hebrew Bible meant, "The Head of the Gods brought forth the Gods" and created heaven and earth. (*TPJS*, p371) The prophet explained, "The head God called together the gods and sat in *grand council*... and contemplated the creation." (*TPJS*, pp348-349) In the same discourse, Joseph Smith also phrased it this way; "In the beginning the head of the Gods called a council of the Gods; and they came together and concocted a plan to create the world and people it." (*TPJS*, pp348-349)

So, what threw the KJV translators off? The Bible's first Hebrew word is *bereshith (בְּרֵאשִׁית,* "In the beginning"*)*, which, by the way, is what Israel named its 2019 lunar lander.

Joseph Smith explained that *"be"* (Heb בְּ=*in*) was added to the original text by an uninspired copyist trying to add literary value, something like adding "Once upon a time" to a children's story. If that *"be"* prefix is removed, the remaining word *reshith* (ראשית) means "beginning" or "chief" — its root, *rosh*, is translated "head" or "chief" or "captain" or "ruler" 455 other places in the Bible. (Str#7225 & #7218)

Furthermore, Joseph Smith said that *reshith* was originally two words, *rosh* (head) and *shith* (bring/brought), that were incorrectly merged into one word by a clumsy scribe. (*WJS*, 7 April 1844, p399, nt107)

Then, with the remaining words, *bara* (organize), *elohim* (gods), *hashamayim* (heavens) and *haarts* (earth), the correct Genesis 1:1 translation should be "The Head (God) brought forth and organized gods, and heaven, and earth." With Smith's corrections, the existence of a creation council is obvious!

Some experts argue that *bara* is only applicable to the one male deity, so Elohim, though having the plural suffix, must be singular masculine — a sort of royal plural as a sign of respect. However, it also appears in many other places in the OT where the actor or actors are human. So, it depends on the context *as interpreted by the reader.*

Why convene a council?

The primary purpose of the Divine Council was to prepare the creation plan, engineer every process, design every life form, preordain every player, and sequence all events; "The Gods took counsel among themselves and said: Let us go down and form man in our image, after our likeness; and we will give them dominion over the fish of the sea, and over the fowl of the air, and over the cattle, and over all the earth, and over every creeping thing." (*PGP*, Abr 4:26)

The Divine Council also designated the blessings that would accrue to spirits who lived worthily in mortality. Joseph Smith taught, "All Blessings that were ordained for man by the Council of Heaven were on conditions of obedience to the Law thereof." (*Scriptural Items*, F. D. Richards, WJS, 16 July 1843, p232)

God summarized some of the Council's planning work for Joseph Smith as he languished in Liberty Jail; "If there be bounds set to the heavens or to the seas, or to the dry land, or to the sun, moon, or stars-- All the times of their revolutions, all the appointed days, months, and years, and all the days of their days, months, and years, and all their glories, laws, and set times, shall be revealed in the days of the dispensation of the fulness of times-- *According to that which was ordained in the midst of the Council of the Eternal God of all other gods before this world.*" (DC 121:30-32)

"God, who created all things by Jesus Christ," used the creation event as the beginning of the "eternal purpose which He purposed in Christ Jesus our Lord". (Eph 3:9-11) Jesus is "Alpha and Omega, the beginning and the end". (Rv 22:13) The creation started with Him and the plan will end with Him.

Jesus, in turn, delegated responsibilities to His archangels. He called Michael as His first assistant in the creation and his chief captain against evil. Epiphanius of Salamis (ca. 310-403AD) wrote in his *Hexaemeron* that Michael was appointed to the position previously held by Satan when the latter fell from grace. (darkchamberz.com/michael-who-is-like-god-archangel, DL 14 May 2026)

Joseph Smith said that other spirits were also called to be archangels in premortality and leaders in mortality; "At the general & grand Council of heaven, all those to whom a dispensation was to be committed, were set apart & ordained at that time, to that calling. The Twelve also as witnesses were ordained." (*WJS*, 12 May 1844 discourse, recorded by Samuel W. Richards, p371)

They spent six periods drawing up the blueprint for creation; "The *Gods plan the creation* of the earth and all life thereon—Their *plans for the six days* of creation are set forth." (*PGP*, Abr 4 chapter synopsis) The works started after the six planning periods; "The Gods said: We will do everything that we have said... On the seventh time we will end our work, *which we have counseled.*" (*PGP*, Abr 4:31; 5:2)

Were you in the council?

The most valiant spirits were given key roles in the creation project. God stood with the most "noble and great" spirits and said, *"These will I make my rulers."* (*PGP*, Abr 3:22-23) But what about the less spectacular spirits?

The Prophet Joseph Smith explained, "Every man who has a calling to minister to the inhabitants of the world was ordained to that very purpose in the Grand Council of heaven before this world was." (*TPJS*, p365) Referring to the run-of-the-mill layperson, Apostle Russell Nelson said, "A council in heaven was once convened in which we participated." (*The Creation*, CJCLDS semi-annual Conference Report, Apr 2000) So, every person was called to fill an important role and accepted it in the premortal Grand Council.

"We made vows, solemn vows, in the heavens before we came to this mortal life... We made covenants. We made them before we accepted our position here on the earth… We committed ourselves to our Heavenly Father, that if He would send us to the earth and give us bodies and give to us the priceless opportunities that earth life afforded, we would keep our lives clean." (*Be Ye Therefore Perfect*, SLC Institute of Religion devotional address, 10 Jan 1975, p2)

Roy Mills, not affiliated with a church that believes in a premortal existence, never lost his memory of it; "The angels in heaven showed us what we needed to grow spiritually and worked with us in choosing life experiences that would teach us the things we needed to learn." (near-death.com/paranormal/pre-existence.html, DL 10 Jul 2016) Some spirits intentionally choose lives with more difficulties. Why? Because enduring difficulties well and suffering for others builds more character and brings more blessings than an easy life. The One who suffered the most is the Greatest.

Others who have premortal memories say that the system that prepares spirits for mortality is well organized and that the decisions are mutual. Respect for individual agency is the prime directive. Some even recall raising their hand to support the plan.

Some NDE spirits are shown details of future events so that, when experienced in life, a sense of déjà vu would reassure them they are on the plan. Spirits are given missions for life and receive divine help to choose well the surmountable challenges that they will face on Earth. Spirits train and rehearse to improve the chances that they will make good choices in life.

Couldn't God have done it all without you? The same question could be asked of His plan for today. Couldn't God visit all the sick and give food to all the hungry and help all the poor? Yes! So, why does God want your help when He could do it all single-handedly?

There are good reasons God doesn't work solo. He called the Twelve Apostles to make leaders of them and they grew His church for years after He ascended to heaven. Serving others with love makes us more like Him. Jesus showed others how it was to be done; "He that believeth on me, the works that I do shall he do also; and greater *works* than these shall he do." (Jn 14:12)

Creation work was the same way. As members of the Divine Council, all of the spirit children of God participated the planning and execution of the creation and volunteered for important earthly missions.

Where did the Council take place?

What location did God choose as the site for the creation command and control center? It must be a place...

> ➢ convenient to the creation project site.
> ➢ inaccessible to mortals.
> ➢ Appropriate for all God's spirit children, including those with rebellious tendencies like Lucifer and his gang (not God's celestial home where no unclean or evil thing is dwells).

A poem attributed to Joseph Smith refers to "the *council in Kolob*". (*T&S (Times and Season)*, bk4, p82, based on DC 76) God descended to the physical realm where space and materials existed for the creation of the first world. And, if that first world was Kolob, as the Abraham papyri assert, and if Kolob is Sgr A*, then it is indeed the headquarters of the Divine Council. (*PGP*, Abr 3:3)

What happened to the Council?

Daniel saw "a watcher and an holy one came down from heaven." The holy one foretold the fate of King Nebuchadnezzar using a tree allegory and then stated, "This matter is by the decree of the watchers, and the demand by the word of the holy ones: to the intent that the living may know that the most High ruleth in the kingdom of men..." (Dn 4:13-17) That the watchers and holy ones come from the Divine Council is evident from the fact that the decree was deemed the will of the most High God.

Isaiah was permitted, while still in the flesh, to attend the council where the Lord sat on a throne surrounded by holy angels and said, "Whom shall I send, and who will go for us?" Isaiah answered, "Here am I; send me." (Is 6:8) Then the Lord gave him a message for the wicked.

So, the Divine Council is still operational. It continues to watch over Earth. How long will it remain intact? Until creation is complete! Until the last new baby has arrived! Until the last new plant has sprung up! Until the last new mountain thrusts forth. Until everyone is done using their agency to embrace God's plan, ignore it, or rebel against it.

A modern example of the Council in operation was related by Heber J. Grant. As he was riding a horse in Arizona in a contemplative frame of mind, He had a vision: "I seemed to see…a Council in Heaven… the Savior was present, my father was there, and the Prophet Joseph Smith was there. They discussed… and decided that the way to remedy the mistake that had been made in not filling these vacancies [in the Quorum of the Twelve] … was to send a revelation." (*CR*, Apr 1941, pp4-6)

Why was God the Father able to rest from His labors on the seventh day? Because Christ's Divine Council assumed responsibility for the project and day-to-day operations at the conclusion of the sixth day.

11: Rise of Evil

Who are the spirit devils?

As Woodward's dictum states, "When you hear hoofbeats, think horses not zebras." When devils are mentioned, think rebellious spirit children of God rather than grotesque red creatures with horns, arrow-point tails, and tridents in hand. Such characterizations tend to put devils in the same fantasy realm as Santa and the Easter Bunny.

The Bible calls devils "unclean spirits" because they are the "angels that sinned" and thus soiled their own souls. (2 Pt 2:4) Their sin is compounded because they know that Jesus is the Son of God. (Mk 3:11) Despite their bad choices, they are still children of the "Father of Spirits". (Hb 12:9)

Why did God create devils?

How could a perfectly good God create a perfectly bad Lucifer? He didn't. He perfectly organized Lucifer's spirit using pre-existing less-than perfect intelligence. A cake made by the best chef in the world may turn out to be less than perfect if some of the ingredients are wrong.

So, the answer is in the raw ingredients. God started with ingredients that existed from eternity. "The intelligence of spirits had no beginning." (*TPJS*, p353) Lurking somewhere in the mix of intelligence from which God organized Lucifer was a tiny bit of pride and despotism that precipitated his eventual turn to the dark side.

Apostle Parley P. Pratt said, "It may be inquired, why God made one unequal to another, or inferior in intellect or capacity. To which I reply, that *He did not create their intelligence* at all. It never was created, being an inherent attribute of the eternal element called spirit, which element composes each individual spirit, and which element exists in an infinitude of degrees in the scale of intellect, in all the varieties manifested in the eternal God, and thence to the lowest agent, which acts by its own will." (*JD*, bk1, 10 Apr 1853, p258)

So, God is innocent of creating evil because the ingredients for all imperfect spirits existed eternally. Before God made them into spirits, the angels that fell had made choices as intelligences that sowed the seeds of their own sedition. God had to allow them their agency to choose darkness over light.

Couldn't God have selected intelligence from chaos in such a fashion that none of the spirits created would fail Him? The Lord said, "Every spirit of man was innocent in the beginning." (DC 93:38) So, just as God wouldn't judge mortals for what He knows they will likely do in the future, neither will He judge intelligences or spirits for deeds they haven't yet done. The fact that many spirits turned to evil strongly suggests that primal intelligences had already voluntarily organized themselves into discrete but diverse clusters and striven toward The Light before they thus qualified for transmogrification into spirits at the hand of God. Striving toward light at one point is no guarantee against future failure.

Christ's pre-spirit intelligence had made better choices than the pre-spirit intelligence of anyone else. For this reason, He was the first to be organized into a spirit body. And, after being organized into a spirit, Jesus still progressed further and faster than all others with perfect choices.

But surely God knew which spirits would go bad in advance. Couldn't He have just not created those? God does not deny choice makers opportunities based on decisions they haven't yet made unless He knows their choices would throw a big monkey wrench into His plans for others. In the case of devils, the struggle between good and evil that resulted from creating Lucifer and his minions was an important part of His plans for others! It is necessary "that there is an opposition in all things. If not so, … righteousness could not be brought to pass, neither wickedness, neither holiness nor misery, neither good nor bad". (2 Nephi 2:11)

Why did God allow evil?

Did God allow evil just so He could have something from which to save us? No. Evil is an integral part of the path forward! If God had only created children from intelligence that would never make mistakes, there would have been very little progression out of the chaos from which all spirits come.

Lucifer was a shining star at the beginning and there were no devils to tempt him and yet he turned dark. Spirits have the capacity to make sinful choices without any outside help. This capacity for evil is inherent in the intelligence from which each spirit was made. Satan did not invent evil. Agency has always existed and with it the ability to make bad choices.

If the glory of God is the bright type of intelligence, then the darkness of evil is the dull kind. Not all the intelligence in the primeval pool of urstoff is good. Just as spirits can decide to exercise evil control over other spirits, intelligence can decide to exert dominion over other entities.

During one of the annual training sessions with the angel Moroni at the stone box containing the golden plates, Joseph Smith was shown a vision contrasting "the glory of God and the power of darkness… the two powers". (*Latter Day Saints' Messenger and Advocate,* Oct 1835, p198) The power of darkness is dull intelligence that resists light. Lucifer's dark choices made him the personification and master of darkness.

Knowing Satan would rebel, God incorporated him into the divine plan. And, instead of sending him immediately to the abyss when he rebelled, God allowed Satan to tempt premortal spirits and later earthlings for a wise purpose. Without Satan, humanity would have remained naïve.

God knew that evil would cause a good portion of His children to stumble. Selecting a savior was part of His plan even before creating His spirit children. The goal was to help them become eternally happy and evil-proof.

In remarks given at the Nauvoo, Illinois temple site, Joseph Smith said, "The design of God before the foundation of the world was that… through faithfulness we should overcome &… obtain glory honor power and dominion for this thing is needful, inasmuch as the Spirits in the Eternal world glory in bringing other Spirits in Subjection unto them, Striving continually for the mastery." (*Martha Jane Knowlton Coray Notebook*, Church Archives, 21 May 1843)

Devils are so persuasive that, without the power and protection of glorified resurrected bodies, all disembodied people would eventually fall prey to Satan. As the ancient American prophet Jacob put it, without the resurrection "our spirits must have become like unto him, and we become devils, angels to a devil, to be shut out from the presence of our God, and to remain with the father of lies, in misery, like unto himself". (*BOM*, 2 Ne 9:9)

The consequences of evil help people learn and grow spiritually as demonstrated by cases where God allowed the buffetings of Satan to chasten transgressors in the church. (1 Tim 1:20; 1 Cor 5:5; Alma 37:15; DC 82:21)

Even the child Jesus "learned obedience by the things which He suffered". (Hb 5:8) For the sake of all His children, the Father made "the captain of their salvation perfect through sufferings". (Hb 2:10) Jesus underwent temptation at Satan's hands at the end of His forty-day fast but passed the test and was stronger for having done so. It was all part of the Father's plan who knew from the beginning that Christ would triumph over Satan in the end. (1 Jn 3:8)

God's objective in promoting intelligences to spirithood and beyond is to help them ultimately achieve eternal immunity from the powers of darkness. Obtaining bodies is a huge step toward achieving eternal glory beyond the clutches of power-mongering evil entities.

Where does Satan get his power?

A commonly asked question in scripture study circles is, "Where does Satan get his power?"

Some say God lends him power sufficient to tempt and torment humans to help them grow in character. But that would make God an active accessory to evil works.

Others say Jesus stripped Satan of his power in Eden. But they are very wrong. During an angry rant by Satan, Moses "began to fear exceedingly". (*PGP*, Mo 1:20) And, circa 1489BC, Satan scared the wits out of the archangel Michael who, "when contending with the devil... durst not bring against him a railing accusation". (Jude 9:1)

The New Testament blamed devils for blindness, dumbness, madness and other ills. Was it a cultural tendency to blame every malady on devils or did Satan actually make people sick? There was a "woman which had a spirit of infirmity... whom Satan hath bound, lo, these eighteen years". (Lk 13:11) Since the NT does not diagnose her medical problem, psychosomatic illness can't be ruled out.

Paul taught, "Satan himself is transformed into an angel of light." (1 Cor 11:14) How can Satan appear as an angel of light if he and his devils were bound in "chains under darkness"? (Jude 1:6) Does Satan use borrowed light? Who would lend him light? It's more likely that he stole it. In other words, in addition to his spirit minions, myriad intelligences have succumbed to the powers of his persuasion and do his bidding to the extent that God allows.

Jacob said, Satan only "transformed himself *nigh* unto an angel of light." (*BOM*, 2 Ne 9:9) The prince of darkness has little light of his own. As Moses put it to Satan, "where is thy glory, for it is darkness unto me?" (*PGP*, Mo 1:15) Satan's power is dull intelligence.

Satan is still using the power he acquired before he became the devil. He competes with the Holy Ghost for influence over our spirit minds! We choose. Devils are masters of manipulation, deception, illusion, and imitation. Their telepathic abilities are their primary weapons.

Can Satan answer prayers intended for God? Yes, but only with feigned love and assurance. His attempt to interlope when Joseph Smith prayed was frightening and engendered despair. Then God dispelled the darkness with light and love. God has placed limits on Satan. For example, "Instituted before the creation, the Devil could not come in sign of a dove." (*WJS*, 29 Jan 1843, p160) The dove was an emblem or Token of Truth (WJS, 29 Jan 1843, p163)

In summary, Satan's powers come from elemental intelligence loyal to his person from the beginning, myriad higher-order intelligences seduced by him, evil spirits that follow him, and humans who desire wickedness. The spirits in the premortal rebellion granted him power by listening to him. People on planet earth give him power the same way. Scripture says that if everyone obeyed God, Satan would be bound; "Because of the righteousness of his people, Satan has no power." (*BOM*, 1 Ne 22:26)

Can Satan read your mind?

After thousands of years of observing human behavior, Satan and his minions are quite adept at guessing what is going on inside people's minds based on their situation and their behaviors. Then, to the extent they are susceptible or receptive, he can plant suggestions in their minds. (e.g., Jn 13:2, Acts 5:3)

However, Satan's knowledge is finite. He cannot read human thoughts unless a person intentionally directs them to him. For example, even if Satan binds a person's tongue, as he did to Joseph Smith in the Sacred Grove, the victim can use thoughts to command the devil in the Savior's name to depart.

And, during demonic possession, the demon merges with the victim's consciousness, effectively generating thoughts in their mind and controlling their intellect and emotions.

How did God choose leaders?

God called prophets and priests based on their good choices. He chose the "noble and great ones" in the Divine Council to be, as He said, "my rulers". (*PGP*, Abr 3:22-23)

God said, "They who keep their first estate shall be added upon; and they who keep not their first estate shall not have glory in the same kingdom with those who keep their first estate; and they who keep their second estate shall have glory added upon their heads for ever and ever." (*PGP*, Abr 3:24-26)

The second estate is mortality after which judgment and glory are bestowed. The first estate was premortality where those spirits who chose to be valiant for God instead of following Satan were made "elect" angels. (1 Tim 5:21) These were blessed with the privilege of Earth life.

Who volunteered first?

"Father told us that after we were born on the earth, we would not remember our life in heaven… that some of us would choose evil and would not be able to return to him… we would all make mistakes… that *we would need a savior to atone for our sins* so that we could repent and receive forgiveness." (*The Great Council in Heaven*, Liahona, Apr 1984)

To be clear, the Father did not ask the council what to do or whether anyone had a plan. Jesus and Satan did not both put forth plans of their own. Rather, the Father presented the plan, using previously created worlds as examples, and explained that mortals make mistakes and that there would be a need for the role of Savior and Light of world; "Whom shall I send?" (*PGP*, Abr 3:27) And His Firstborn answered, "Here am I, send me… Father, thy will be done, and the glory be thine forever." (*PGP*, Abr 3:27, Mo 4:2)

Then Lucifer volunteered, "Here am I, send me... I will be thy son, and I will redeem all mankind, that one soul shall not be lost, and surely I will do it; wherefore give me thine honor." (*PGP*, Mo 4:1) Christ made the Father's plan His own but Lucifer thought He had a better one!

What was Lucifer thinking?

Satan had everything – almost. He was once an angel of great leadership. But his ambitions took a dark turn, as poetically lamented by Isaiah; "How you have fallen from heaven, morning star, son of the dawn! You have been cast down to the earth...! You said in your heart, 'I will ascend to the heavens; I will raise my throne above the stars of God (spirits/angels of God); I will sit enthroned on the mount of assembly (head of the Divine Council)... I will make myself like the Most High." (Is 14:12-14 NIV; parens added)

Lucifer was an "angel of God who was in authority in the presence of God, who rebelled against the Only Begotten Son... was thrust down from the presence of God... And was called a Perdition, for the heavens wept over him... and sought to take the kingdom of our God and his Christ." (DC 76:25-28) When the Father chose Jesus for the coveted leadership role, Satan chose insurrection!

Satan was thinking he could capitalize on his spirit siblings' fear of making mistakes as mortals whose memories would be lost when they took earthly bodies. That was scary. So, he "boasted of himself saying 'send me I can save all, even those who sinned against the Holy Ghost' and he accused his brethren and was hurled from the council for striving to break the law". (*WJS*, Joseph Smith, as recorded in George Laub Journal, 7 Apr 1844, p362)

But the Father said, "I will send the first." (*PGP*, Abr 3:27) "And there was Light." (Gn 1:3) "That was the true Light, which lighteth every man that cometh into the world." (Jn 1:9) "The second (Satan) was angry, and kept not his first estate; and, at that day, many followed after him." (*PGP*, Abr 3:28)

Then "Michael and his angels fought against the dragon," wrote John, "and the great dragon was cast out, that old serpent, called the Devil, and Satan, which deceiveth the whole world: he was cast out into the earth, and his angels were cast out with him". (Rv 12:7) "And his tail drew the third part of the stars of heaven (spirits/angels of God), and did cast them to the earth." (Rv 12:4)

When the disciples of Jesus came to him rejoicing, "Even the devils are subject unto us through thy name," Jesus recalled a premortal memory, "I beheld Satan as lightning fall from heaven." (Lk 10:18)

So, in a nutshell, having seen God's glory, Satan coveted it and tried to bargain for it with God. In exchange for saving everyone, he said, "Give *me* thine honor." (*PGP*, Mo 4:1) When his offer was rejected, his thinking was clouded by anger, and he tried to create His own kingdom by convincing others to follow him. He made decisions without all the facts; "He knew not the mind of God." (*PGP*, Moses 4:6)

Why was Satan's plan infeasible?

Satan said, "I will be *like* the most High!" Wasn't that a noble aspiration?

Satan's sin was not that he wanted to emulate God. He wanted a *shortcut* to deification and omnipotence. He wanted to "ascend into heaven" and sit on a "throne above the stars (spirit children) of God". (Is 14:13) He wanted to be the head of the Divine Council; "I will sit also upon the mount of the congregation." (Is 14:14) He wanted control of everything. He "sought to take the kingdom of our God and his Christ". (DC 76:28) And he wanted to change God's plan and take all the credit and glory; "I will redeem all mankind, *that one soul shall not be lost*, and surely *I* will do it." (*PGP*, Mo 4:1)

Would Satan's plan have worked? Not without reversing the arrow of causation or somehow depriving all of individual agency. Satan could not have forced everyone to be righteous. Even people bound in straight-jackets and strapped to a gurney with their mouths taped shut can think nasty thoughts. People who are forced by confiscatory taxation to help other people do not get the good feeling that comes from freely giving, nor do they grow spiritually as a result.

But Satan was not too worried about the truth. He "was a liar from the beginning". (DC 93:25) If Satan had been given power to force everyone to behave a certain way, he would have violated the eternal Law of Agency, and we would have had a devil as head of the Divine Council and god of creation.

Would his plan to "*redeem* all mankind" have worked if he just kept them all ignorant of right and wrong, naive like Adam and Eve in Eden? No. Naïveté would have prevented them from ever being like God; "It is impossible for a man to be saved in ignorance." (DC 131:6)

Could he have saved them despite their bad choices? No. Even God "cannot save them in their sins... no unclean thing can inherit the kingdom of heaven". (*BOM*, Alma 11:37) Transgression separates people from God's presence; "There was no means to reclaim men from this fallen state, which man had brought upon himself because of his own disobedience; Therefore, according to **justice**, the plan of redemption could not be brought about, only on conditions of repentance.... for except it were for these conditions, **mercy** could not take effect except it should destroy the work of **justice**. Now the work of **justice** could not be destroyed; if so, God would cease to be God." (*BOM*, Alma 42:12-13)

If perfect *mercy* and *justice* are the biggest parts of what defines God, Satan's plan would have never led to divinization for anyone. Violating the eternal **Law of Justice** by allowing boundless sinning with unconditional mercy and no consequences would have destroyed his legitimacy as a savior.

His kingdom would have been a hell and the Father would have ceased to be God for having chosen Satan. Satan's plan was to take away agency with the idea that everyone would then remain innocent and not require mercy. But that kind of innocence means ignorance; "A man is saved no faster than he gets knowledge, for if he does not get knowledge, he will be brought into captivity by some evil power." (*TPJS*, p217) God's plan was to imbue his children with the knowledge and glory that would immunize them against evil powers forever. Lucifer is not the only evil power out there!

Just as physical laws rule out the possibility of a perpetual motion machine, Eternal Law prohibits "Get Out of Jail Free cards" for sinners. The Law of Justice requires consequences commensurate with choices. Christ was willing to suffer the consequences so that God could remain both just *and* merciful whereas Satan did not picture himself making any sacrifice at all.

Jesus agreed to submit to the amnesia that accompanies mortal birth, relying on His divine character to keep His life sin-free and qualify to be the proxy sacrifice that would satisfy the Law of Justice for the sins of all. How did Lucifer plan to obtain a body? Was he willing to be born a helpless mortal baby with no memory of his premortal greatness nor any knowledge of his position as dictator-god of the world?

We can't know all that Satan was thinking but he must have been smart enough to know his plan had flaws. He peddled a counterfeit salvation and sold it to a third of the spirits as a risk-free path to glory, the first in a long series of whoppers that earned him the moniker, "father of lies". (Rv 12:4; Jn 8:44)

What made Christ's plan feasible?

Christ's plan was in fact the Father's, the **Plan of Mercy**. It leverages a "loophole" in the **Law of Justice**. To wit, the punishment for transgressions can be borne by someone other than the guilty party, provided the following three conditions are met:

1. The proxy is guiltless.

2. The proxy is willing.

3. The sinner repents.

Jesus believed He could live a guiltless life and was willing to pay the price for everyone's sins; "God himself atoneth for the sins of the world, to bring about the plan of mercy, to appease the demands of justice, that God might be a perfect, just God, and a merciful God also." (*BOM*, Alma 42:15)

Justice would be done. Mercy would be extended. Agency would be preserved. And blessings would be given to the penitent. Thus, Jehovah's **Plan of Mercy** complied with the **Law of Justice** *and* conformed to all other eternal laws.

How could heaven have warfare?

"There was a war in heaven." (Rv 12:7) Hold on! As General William Tecumseh Sherman said, "War is hell!" Hell in heaven doesn't make sense! Why try to go there? For some answers, let's review John's vision of the precreation drama in the realm of spirits:

"There appeared a great sign in heaven *in the likeness of things on the earth*: a woman clothed with the sun, and the moon under her feet, and upon her head a crown of twelve stars." (*JST*, Rv 12:1) The woman is not on earth. She is the precreation church of God, as is stated later in this scripture.

"And the woman, being with child, cried, travailing in birth and pained to be delivered. And she brought forth a *man-child, who was to rule all nations* with a rod of iron; and *her child was caught up unto God and his throne.*" (*JST*, Rv 12:2-3) The man-child is the Christ who was caught up to the Father and made King of creation and ruler of the Kingdom of God henceforth. So, the heaven from which Christ was caught up was the precreation spirit realm rather than the celestial heaven that is the Father's home base.

"And there appeared another sign in heaven--and behold, a great red dragon, having seven heads, and ten horns, and seven crowns upon his heads. And his tail drew the third part of the stars of heaven and did cast them to the earth. And the dragon stood before the woman which was delivered, ready to devour her child after it was born." (*JST*, Rv 12:4) Satan and his minions threatened to destroy God's church and the new Kingdom of Christ. "Third part" is not the fraction 1/3. Rather, it indicates that there were three groups in this spirit heaven; 1) God's chosen leaders, such as the twelve apostles of His church, 2) Satan and his minions, and 3) everybody else.

Satan's forces posed such a threat that the church of God withdrew to a refuge prepared by the Father until it could be sufficiently strengthened to overpower Satan's army of devils; "The woman fled into the wilderness, where she had a place prepared of God, that they should feed her there [a while]." (*JST*, Rv 12:5)

Then the church of God, led by the archangel Michael, exerted all their influence and power to convert Satan's group and everybody else; "There was war in heaven; Michael and his angels fought against the dragon; and the dragon and his angels fought against Michael; And the dragon prevailed not against Michael, neither the child, nor the woman, which was the church of God, who had been delivered of her pains and brought forth the kingdom of our God and his Christ. Neither was there place found in heaven for the great dragon, who was cast out--that old serpent called the devil... and his angels were cast out with him." (*JST*, Rv 12:6-8)

Solomon acknowledged various heavens when he prayed to God, "The heaven and heaven of heavens cannot contain thee; how much less this house (temple) that I have builded." (1 Kg 8:27)

As Joseph Smith elucidated, "[Christ] took the liberty to go into other heavens." (byustudies.byu.edu/content/volume-5-chapter-22, p426, DL 18 Aug 2018)

Jesus said, "We will go down, for there is space there, and we will take of these materials, and we will make an earth." (*PGP*, Abr 3:24) So, the project began, in a heaven below the Father's heaven of heavens. Satan aspired to ascend to the highest heaven's throne. His rebellion could not have happened in the highest heaven because a strict zero-tolerance policy is in place there; "There shall in no wise enter into it any thing that defileth." (Rv 21:27) "No unclean thing can enter into his kingdom." (*BOM*, 3 Ne 27:1) "He who is not able to abide the law of a celestial kingdom cannot abide a celestial glory." (DC 88:22)

Certainly, Satan and the rebels who followed him were not clean and had not learned to obey celestial law. So, the war in heaven occurred in a lower heaven where the Divine Council was being held. Where was the Divine Council held? A poem by Joseph Smith based on a vision from God tells us; "From the *council in Kolob*, to time on the earth... unto [the faithful]

I will show my pleasure & will." (*Times and Seasons*, bk4, p82, Nauvoo. Ref. DC 76)

What were the weapons of war?

"Michael and his angels fought against the dragon." (Rv 12:7-9) Michael and Satan were immortal spirits at the time. So how would they hurt each other and with what weapons?

The same weapons are used on Earth by the forces of good and evil. Michael's angels used loving messages and persuasion. Satan's forces used fear and intimidation.

This was a struggle for hearts and minds much like a hotly contested election year in the USA; "The contention in heaven was—Jesus said there would be certain souls that would not be saved; and the devil said he could save them all, and laid his plans before the grand council, who gave their vote in favor of Jesus Christ. So, the devil rose up in rebellion against God, and was cast down, with all who put up their heads for him." (*TPJS*, p357)

Lucifer's loyalists focused on the risk of failure on Earth under Christ's plan. He labelled it a gamble. He pointed to the fact everyone would be making important choices in a state of total amnesia, having forgotten all they had learned about God as premortals. He pointed out that Earthlings would be making choices of eternal consequence, such as whom to marry, without any memory of premortality and while buffeted by the vagaries of mortality. He said under Jehovah's plan only a few would be exalted and all others would go to hell. He said that most would not even get a chance to hear about the true God in mortality. He promised that his plan avoided all these negatives and would provide glory for everyone in the end.

Spirits had to choose one campaign or the other. Friendships were broken. Mourning ensued for all those deceived by Satan. Since spirits knew each other's thoughts telepathically, those with opposing opinions were no longer comfortable associating. So, those who favored God gathered to His group and those who favored Satan to his group.

This war of wills was the way spirits who were not ready for life on Earth filtered themselves out.

Which spirits were undecided?

Premortal spirits were probably distributed all along the wicked-righteous continuum, from the prince of darkness through all shades of gray to the King of light. But, at some point in the struggle for souls, each spirit made an up or down vote on Jesus as their Savior; "There were no neutrals [at the conclusion of] the war in heaven. Each took a side either with Christ or with Satan." (*DOS (Doctrines of Salvation)*, bk1, p65) Failure to follow Jesus was a vote for darkness; "He that is not with me is against me." (Mt 12:30)

The Light of Christ shone in the darkness but "the darkness comprehended it not." The word "comprehend" is translated from the Greek καταλαμβάνω meaning to "seize tight hold of". (Jn 1:5) Those who don't hold onto the Light are the darkness. Paul wrote to his converts, "You were once darkness, but now you are light in the Lord." (Eph 5:8) Satan and his followers let go of the Light and therefore remain darkness.

Who kicked the devils out?

With telepathy for communication, spirits cannot keep secrets from each other. And, every spirit's luminosity reveals its character to other spirits. So, it was expectable that spirits with predominantly dark thoughts and dark auras withdrew from Christ's holy presence; "Every one that doeth evil hateth the light." (Jn 3:20)

So, the evil spirits were relieved to be released from the brilliance of Christ's presence; "God spared not the angels that sinned, but cast them down to hell (on Earth), and delivered them into chains of darkness (cut off from God)." (2 Pt 2:4) "The angels which kept not their first estate (premortal life with God), but left their own habitation (spirit world) he hath reserved in everlasting chains under darkness (hell) unto the judgment of the great day." (Jude 1:6)

When Michael fought with the devil over the fate of Moses' body, He told Satan, "The Lord rebuke thee." (Jude 1:9) What does such a rebuke mean? Well, to heal a boy who was possessed, "Jesus *rebuked* the devil, and he departed out of him". (Mt 17:18) Likewise, Michael and those who followed Jehovah invoked the *power of God* to cast the evil ones out of their midst.

Satan and his followers wanted to leave and craved something vengeful to do. So, they waited on Earth for the arrival of creatures sufficiently sentient to succumb to temptation or demon possession. They even achieved a degree of success practicing their evil arts on animals as evidenced by their use of a serpent to seduce Eve; "And now the serpent was more subtle than any beast of the field which I, the Lord God, had made. And *Satan put it into the heart of the serpent, for he had drawn away many after him...* to destroy the world." (*PGP*, Mo 4:5-7)

Why do devils seem free?

The scriptures say that God put Satan in chains. (2 Pt 2:4) So, why are devils still wreaking havoc around the globe?

The manacles on devils are "chains of darkness". (Rv 20:2) He is not bound so strictly that he can't cause any trouble at all. Total binding of devils *forever* will not happen until after Christ's millennial reign and the little season of strife that follows. (Rv 20:2-10)

The power of devils is limited by their inability to accrue enough intelligence (spirit power) to effect physical miracles and intervention. And God prevents them from tempting us more than we can bear. (1 Cor 10:13)

But God indeed allows evil spirits to exercise their agency within the bounds of His plans. For example, "God sent an evil spirit" to deal mischief between Abimelech and the men of Shechem. (Jdg 9:23) An "an evil spirit from the LORD terrorized" Saul. (1 Sm 16:14) And, when God asked his spirit children for a volunteer to persuade Ahab to head into battle "there came forth a spirit, and stood before the LORD, and said, *I will persuade him. And the LORD said unto him, Wherewith? And he said, I will go forth, and I will be a lying spirit in the mouth of all his prophets. And he said, Thou shalt persuade him, and prevail also: go forth, and do so".* (2 Kg 22:20-22) While the wording attributes everything to God, as was the custom, these were not true angels of light, rather evil spirits who sought permission to harass evil men. God had given the afflicted men plenty of chances to repent before relaxing His protection over them from devils.

The case of Job was different because he was righteous. God knew Job would withstand Satan's arrows so He temporarily removed the protective "hedge about him, and about his house, and about all that he hath on every side". (Jb 1:10) God allowed Satan to do what he had long wanted to do, afflict him for his righteousness. (Jb 1:8-12)

The Lord allowed the devil in Judas to lead him by the nose, saying, "That thou doest, do quickly." (Jn 13:27) His betrayal of Christ played right into God's plan. All evil serves God's purpose. It's not a tragedy that evil exists. It's only a tragedy when it is not annulled through repentance.

Does God hate Satan?

If God hates sinners, then, he must hate Satan. Right? Some preachers say that God indeed hates sinners and back their assertion with several scriptures, for example; "The wicked and the one who loves violence [God's] soul hates." (Ps 11:5) Also, "The LORD hates… one who sows discord among brethren." (Prv 6:16-19) There are about three more less direct statements like this, all in the Old Testament.

But, if God hated sinners, He would hate everyone; "All have sinned, and come short of the glory of God." (Rm 3:23) And, if God hates everyone, what do the scriptures mean that say, "God is love." (Jn 4:8) And, "God showed his great love for us by sending Christ to die for us *while we were still sinners*. (*NLT (New Living Translation)* Rm 5:8)

The common phrase is true; "God hates the sin but loves the sinner." The love/hate question is reconciled by acknowledging both of God's roles; as the God of Justice he must judge the sin and convict the sinner but as the God of Mercy he loves each of His children unconditionally; "God sent not his Son into the world to condemn the world; but that the world through him might be saved." (Jn 3:17)

Satan's sins are indeed more egregious than any human's. But where should the limit on God's love for his children be set? Joseph Smith said this in poem, "An angel of light… was thrust down to woe… *And the heavens all wept*." (*Times and Seasons*, bk4, #6) Or as the canonized version states, "the heavens wept over him." (DC 76:26) So God loves Satan but hates his rebelliousness and the suffering he must endure.

Are evil spirits lost forever?

Hell is the absence of God's light and love, thus the term *"outer darkness"*. (*BOM*, Alma 40:13-14) Only those condemned to it will know its true horrors; "And the end thereof, neither the place thereof, nor their torment, no man knows." (DC 76:45)

Has God lost His desire or ability to help these wayward spirit children? Is there no hope for them? Don't they still have agency to make choices? What if they regret their rebellion and want to reform to the light? Might not something less than an eternity of extreme sorrow and misery satisfy the demands of justice or rehabilitate them?

Through the modern-day prophet Joseph Smith, God explained: *"It is not written that there shall be no end to this torment*, but it is written endless torment… that it might work upon the hearts of the children of men… For, behold, I am endless, and the punishment which is given from my hand is endless punishment, for Endless is my name. Wherefore-- Eternal punishment is God's punishment. Endless punishment is God's punishment." (DC 19:6-12)

So, God keeps the door open for those who have chosen darkness to change their attitudes and get some degree of relief from the hell they suffer. They are still His children.

Whom did you support?

A small airplane crashed and occupant Ranelle Wallace suffered severe burns over more than 60% of her body. While the medics tried to resuscitate her, her spirit left its body and she was given a glimpse of her premortal decision to come to earth:

"Then we raised our right arms, just as we might in a court of law, and we made a sacred covenant with God that we would do all in our power to accomplish our missions on earth. And I felt the tremendous honor of making this covenant before our Heavenly Father. We vowed, in effect, to become partners with him in bringing about goodness on earth." (*The Burning Within*, Ranelle Wallace, 1994, pp106-107, 115-116)

All of those and only those who chose Christ in premortality have the privilege of coming to earth. If you're reading this book, you're one of them.

What about premortal sins?

During the spirit world war, Lucifer's gang could not possibly have been the only spirits making mistakes. Lucifer was at the evil end of the spectrum and Jesus was at the righteous end. All the rest of the spirits fell somewhere between these two extremes. The spectrum must have included spirits who seriously entertained following Satan but ultimate put their hands up in support of Christ.

Justice is being meted out for the devils. But how is the Law of Justice being satisfied for the spirits who were guilty of selfish or unholy thoughts before deciding to follow Jesus?

The apostles of Jesus believed people could sin before coming to Earth. When they saw the beggar, who was blind from birth, they asked Jesus, "Who did sin, this man or his parents, that he was born blind?" (Jn 9:1-2) Did Jesus correct them by asserting that nobody sinned before birth? No. Instead, He declared that birth defects are not punishment for premortal sin. (Jn 9:3)

Part of preparing for Earth life was to repent of any sins committed in premortality. "The Lord cannot look upon sin with the least degree of allowance." (DC 1:31) So, spirits who vacillated a bit before choosing Jesus, spirits who let fears cloud their minds, spirits who harbored hatred for a dissenting brother, and spirits who had any unkind thoughts committed sin and needed to repent before being born on Earth.

The Law of Justice is not limited to the comparatively fleeting moment a person is in mortality. So, why wouldn't mistakes made before and after mortality cause an individual to fall short of the glory of God? Justice requires "not a sacrifice of man, neither of beast, neither of any manner of fowl; for it shall not be a human sacrifice; but it must be an infinite and eternal sacrifice". (*BOM*, Alma 34:10) An eternal sacrifice would address mistakes retroactively and proactively to infinity both ways.

Thankfully, Christ's atonement reaches, not just back to Adam, but all the way back into the premortal spirit world; We needed and received Christ's grace as spirits when "called us with an holy calling... according to his own purpose and *grace*, which was given us in Christ Jesus *before the world began*". (2 Tim 1:9) Grace would not be needed unless spirits in the beginning fell short.

Of course, Satan and his followers proved that spirit children of God can sin. The premortal struggle between the forces of good and evil did not result in purely binary outcomes (thoroughly good spirits or thoroughly bad ones). There were undoubtedly imperfect thoughts among those who ultimately chose to follow Christ. Some veered toward Satan a little before changing their minds and following Christ after all.

Will wrong thoughts and deeds by such spirits ultimately bar them from God's presence? The Lord said, "I... cannot look upon sin with the least degree of allowance." (DC 1:31)

God made all His children innocent before the Spirit World War but they didn't all stay that way; "Every spirit of man was *innocent in the beginning*; and God having redeemed man from the fall, *men became again, in their infant state, innocent before God.*" (DC 93:38) Premortal sins are swept away by Christ's atoning sacrifice by the time a spirit is born into Earth life, making that spirit innocent *again.*

"Little children are redeemed from the foundation of the world through mine Only Begotten." (DC 29:46) "Little children are holy, being *sanctified through the atonement of Jesus Christ.*" (DC 74:7) The retroactive nature of Christ's atonement was part of the plan from the beginning.

In 1853, Apostle Orson Pratt wrote, "All the spirits when they come here are innocent... if they have ever committed sins, they have repented and obtained forgiveness through faith in the future sacrifice of the Lamb... spirits must be forgiven and become innocent before they can even come here." (*The Seer*, 1993 Seagull Book & Tape, pp146-147)

So, Christ's atonement can be applied to any mistake in the eternal past or in the future for as long as there is time to repent. It is "an *infinite* atonement". (*BOM*, 2 Ne 9:7)

Like those in fiction who, by some misfortune, suffer total amnesia and then define themselves anew without any baggage from their past, those coming to Earth are given a "veil of forgetfulness" that provides the clean slate they need to decide afresh what kind of individual they want to be!

12: Spiritual Creation

Why is mortality in the plan?

"What is the meaning of life?" It's a query commonly bandied about as *the* unanswerable question. So, apparently a lot of people still don't know the answer. The meaning of life and the purpose of creation are one and the same. Earth was created to provide a physical theater for spiritual improvement. It's a training ground where humans gain knowledge, mature spiritually, experience parenthood, overcome hardships, and test their own love for God and His children. Well-handled experiences can mold godlike character while badly-handled ones can lend wisdom.

But Earth is not just for humans. Every form of life has its path to glory along with inanimate creations. John the revelator "saw a new heaven and a new earth: for the first heaven and the first earth were passed away". (Rv 21:1) "For after it hath filled the measure of its creation, it shall be crowned with glory, even with the presence of God the Father... for it... transgresseth not the law." (DC 88:19, 25)

So, "earth abideth the law of a celestial kingdom" and will traverse its pathway to that glory, eventually shining like "a sea of glass mingled with fire" where Christ and the righteous will spend eternity. (DC 88:25; Rv 15:2) All God's thoughts are focused on mentoring and advancing His creations. He said, "This is my work and my glory—to bring to pass the immortality and eternal life of man." (*PGP*, Mo 1:39)

Who wrote the creation story?

Not everyone who knows and loves the Bible venerates Genesis as holy writ. A class of historical and literary scholars known as *higher critics* claim Genesis was compiled from various sources rather than authored by Moses:

1. The "Priestly" source (Gn 1:1 – 2:3, the first creation account)

2. The "Jahwist" source (Gn 2:4 – the second creation account through the story of Cain and Abel)

Sometimes called Critical Theory, the idea is that much of the Genesis creation story is actually a tapestry of texts from Assyrian and Babylonian myths woven together by Hebrew priests and scribes in the fourth and fifth centuries before Christ.

These higher critics base their view on the diversity of writing styles and name usage in the original Hebrew text that usually suggest multiple authors. However, such differences in style might also have resulted from diverse copyists tackling different portions of the original text over the course of thousands of years. It's also possible that different portions were received and documented in disparate ways such as from a vision, from God's voice, or by working with scribes to record the essence of what Moses taught. It's even possible that the pieces of Moses' works found in today's Genesis came from separate lines of preservation. If so, editing and copying by different groups would have introduced a variety of styles before the portions were brought back together as part of the Torah.

Higher critics point to the fact that the first chapter references Elohim for God while the second uses Jehovah. But, it's more likely that the two different names are used purposefully to emphasize theology, the first chapter describing the works of the Council of Gods while the second chapter credits accomplishments to Jehovah as the lone God.

The question is answered by the writings of Moses himself, as revealed afresh to the prophet Joseph Smith — God said to Moses, "Behold, I reveal unto you concerning this heaven, and this earth; write the words which I speak." (*PGP*, Mo 2:1) The inspired version of the first two chapters of Genesis follow, firmly establishing them as having come directly from the mouth of God to the mind of Moses.

Can Genesis and Science agree?

Genesis Chapter One seems at odds with the scientific version of the physical creation of earth and its biome, a mismatch that has led many to abandon religion altogether, especially young people. Here are a few examples of contradictions that arise from misunderstandings of Genesis:

♦ Science says Earth was created from the detritus of exploded stars. But Genesis says Earth was created from nothing.

♦ Science says the sun was created first and then Earth was formed from leftover material. But Genesis starts with the earth already created and the sun appearing three days later.

♦ Science says the stars were created long before Earth. But Genesis says the lights in the skies did not appear until after Earth's plant life appeared.

♦ Science says Earth is 4.5 billion years old. But Genesis says God created it only six 6,000 years ago.

♦ Science defines a day as one complete rotation of Earth on its axis with respect to the sun. But Genesis counts three days before the sun existed.

♦ Science says that the first life was aquatic. But Genesis says that aquatic life appeared two creation days after land life appeared.

♦ Science says that there are about 9 million species of animals on Earth. But Genesis says that Adam named them all in a single day (more than 104 species per second).

Can such contradictions be reconciled? If so, how?

Some Christian groups accuse scientists of trying to destroy people's faith by spinning data and fudging evidence. But, as scientists continue to improve their craft and amass more data supportive of their assertions, it's time for Christians to reevaluate their interpretation of Genesis.

A few Christians (and some Jews) have conceded to the higher critics' view that Hebrew scribes cobbled Genesis together from Babylonian and Assyrian myths. Still others have concluded that the Genesis creation story is purely allegorical. But these are uncomfortable assessments of scripture for staunch believers.

The best clues for solving the Genesis creation mystery come from ancient Egyptian papyri, temple ritual, and the second chapter of the Bible. Genesis Chapter One is not about the Big Bang, cosmic evolution, or the gravitational collapse of a nebula to form the solar system. Read on!

Why do Chapters 1 & 2 conflict?

An oft-asked question is, "Doesn't Genesis Chapter Two contradict Genesis Chapter One?" The usual response from preachers, pastors, pundits, and apologists is something like, "Genesis Chapter Two is a more detailed account of the creation of mankind previously mentioned in Genesis Chapter One, focusing on Adam." (godandscience.org/apologetics/genesis2.html, DL 17 Jun 2018) So, does Genesis Chapter Two simply elaborate on Day Six?

These first two chapters are as different as chalk and cheese. So, applying Occam's razor – The explanation that makes the fewest assumptions is that Genesis Chapter Two is different from Genesis Chapter One because *it is about different events*. Stay tuned!

Why was light created twice?

On Day Four, God said, "Let there be lights…" (Gn 1:14) The Day Four lights were the sun, moon, and stars. (Gn 1:16) But, God had already created light back on Day One; "Let there be light." (Gn 1:3) So, what was the Day One light?

Young-earth creationists say the light of Day One was all the photons strung across 94 billion lightyears of observable space, all in such a way that they are still giving astronomers the impression that the universe is billions of years old.

Really? Did He also fill the universe with CMB radiation on Day One, setting its wavelength and temperature to exactly the values predicted by the Big Bang model?

Another theory is that the Day One light was the invention of the photon; "God said, Let there be photons and let the photons be divided from the absence of photons..." Ludicrous! (Compare Gn 1:3-5) In a take-off from this idea, the Berkeley University store sold T-shirts that say, "In the beginning, God said, The four-dimensional divergence of an antisymmetric, second rank tensor equals zero, and there was Light, and it was good." Funny, but not what Genesis means!

Another theory is that Day One's light was the flash of the Big Bang. (*Genesis and the Big Bang*, Dr. Schroeder, 1992) A lot of people like this idea. But it has a timing problem because Genesis starts with Earth already created, "formless and empty", meaning the Big Bang is already more than 9 billion years in the rear-view mirror by the time God says, "Let there be light." The initial flash of the Big Bang, stretched into the microwave frequency range, remains to this day, one of the main pieces of evidence for the Big Bang.

Hmm! What if Day One isn't about physical light at all? The creation narrative in the book of John begins with the premortal Jesus who "was the Word... *That* was the true Light". (Jn 1:1, 9) At the beginning of creation, Jesus was chosen as "Light of the World". (Jn 8:12) Jesus is God and "God is light". (1 Jn 1:5) Jesus is the "true Light that gives light to everyone". (*NIV* Jn 1:9) Let's substitute Him into Genesis; "God the Father said, Let there be a Light for the world: and Jesus took the job. And the Father saw that the Light was good." (Compare Gn 1:3-4) That works!

What does it mean that "God divided the light from the darkness"? (Gn 1:4) It means Christ and His angels of Light fought the devil and his angels and cast them out. (Rv 12:7-9) John's creation narrative supports this interpretation; "The light shineth in darkness; and *the darkness comprehended it not*." (Jn 1:5, DC 6:21) Satan's gang rejected the Light; "The way of wickedness is *darkness*." (Prov. 4:19) The *good* angels "comprehended the *light*". (*PGP*, Abr 4:4) Jesus said,

"Here is the agency of man, and here is the condemnation of man; because *that [Light] which was in the beginning* is plainly manifest unto them, and they receive not the light." (DC 93:30-32; 37:4)

Dr. Michael S. Heiser, PhD in Hebrew, said, "We have dependent clauses... two full verses leading up to... verse 3. And that is the... first creative act... It's verse 3." Light! (youtube.com/watch?v=KmFE-qjoksc, DL, 20 Sep 2025)

Why the order of days?

Hebrew writers employed a rich variety of literary devices to enhance the poetic and mnemonic value of their text, often through creative repetition. A close look at the Creation Week reveals an interesting pattern:

A - Day 1: light – divided from the darkness
B - Day 2: waters – an expanse in the midst of the waters
C - Day 3: earth – dry land divided from the waters
a - Day 4: lights – sun, moon, and stars
b - Day 5: waters – life begins in the waters
c - Day 6: earth – land life emerges

Spiritual design can be done in almost any order but every project requires light and leadership before it can begin. In the case of creation, the Light of the World is both.

How were days defined?

Genesis says, "The evening and the morning were the *first day*." (Gn 1:5) Hold on! There were still two more creation days before the sun was created! How could any days cycle without the sun? How could there be evenings and mornings? Why is "evening" first and "morning" last in the phrase?

Was Moses writing about his own days? Did he get the creation story in seven parts, one each day for a week? No. Genesis Chapter One is the result of God's command to Moses; "Write the words which I speak." (*PGP*, Mo 2:1)

The word "day" is translated from the Hebrew *yom* (Str#3117), which is also used at the end of Creation Week to summarize the *entire week* as; "the *day* that the LORD God made the earth and the heavens." (Gn 2:4) It is translated as "*time*" 65 other places in the Bible, as in time period. In fact, the Book of Abraham's creation saga uses *time* instead of *day* for these periods. *Day* is still used to mean *time* in our *day*.

So *yom* means time period, era, epoch, or eon! "These are the generations of the heavens and of the earth when they were created, in the *day* that the Lord God made the earth and the heavens." (Gn 2:4) Problem solved! A primordial era, epoch, or eon can indeed take place without the sun.

But, if a literal day is *not* meant, then why does Genesis use "evening" and "morning" to delineate a 12-hour night? Ancient Hebrews, even during Christ's time, didn't even count the hours between evening and morning. The day started with sunrise and ended at sunset. The hours of daylight were divided up into 12 hours. This resulted in different lengths for an hour, depending on the season. The priests were responsible for calculating the hours, called halachic times, to determine the time for the soundings of trumpets.

Evening and morning is an example of a Hebraism in which the word for *evening* ('ereb Str#6153) denotes obscurity or chaos and the word for *morning* (boqer Str#1242) denotes order or clarity, thus conveying the concept of progress each period. (Abraham ben Meir Ibn Ezra, Genesis 1:5) This would also explain why "evening" has always marked the start of the Jewish or Hebrew day.

Many latched onto the idea that each creation day was actually 1,000 years, as mentioned in several scriptures; "One day is with the Lord as a thousand years." (2 Pt 3:8) Also, "A thousand years in thy sight are but as yesterday." (Ps 90:4) That God was not providing an astronomical ratio is evident from the next part of that Psalm verse, "or like a watch in the night." So, these two scriptures are similes not equations! To children, a year seems like eternity, to the elderly a year seems like a day, but to God time is irrelevant.

290

Furthermore, 1,000-year periods are woefully insufficient to resolve the scripture-science conflict. The 2,000 years in Days Five and Six in which life emerged and flourished are a mere blip in geologic time, about 4 billion years shy of what the record of natural processes in the earth's crust say.

If God could create everything in just one week, why not in one day, or one hour, or one second, or in no time at all with the command, "Let there be everything!"

For perspective, let's step back and look at one of God's commands with which every parent is familiar, "Be fruitful and multiply and fill the earth." (*ESV*, Gn 1:28) Thousands of years later, this process is still underway. It was not and cannot be compressed into an instant or a day. Each human baby takes about nine months. And to fill the earth took thousands of years and many generations of humans. God creative processes unfold at natural rates and always have. So, there's no reason to think that creation processes were unnaturally accelerated or compressed in the beginning.

Why was darkness on the deep?

"Darkness was upon the face of the deep. And the Spirit of God *moved* upon the face of the waters." (Gn 1:2) "Moved" is translated from the Hebrew *rachaph*, the primitive root of which means "to brood." (Str#7363) Indeed, the Book of Abraham says, "The Spirit of the Gods was *brooding* upon the face of the waters." (*PGP*, Abr 4:2) Only Smith got that right!

What does it mean to brood upon the waters? It could mean that the Divine Council was contemplating creation. The Father would not have needed to ponder. Creation activities are second nature to Him. But the spirits on His Divine Council were learning on the job. God laid out the general plan and let the Divine Council to work out the details.

The brooding of the Spirit of Elohim could also refer the pervasive intelligence that was the creative power guiding the blossoming of God's work on our planet, as a hen incubates her eggs and covers her young as they develop.

Why was there still no man?

After the six creation periods in Genesis Chapter One, Genesis Chapter Two states, *"There was not a man to till the ground."* (Gn 2:5) What happened to him? Wasn't man created on Day Six?

"The LORD God made… every plant of the field *before it was in the earth*, and every herb of the field *before it grew."* (Gn 2:5) The activities of the six-day creation in Genesis Chapter One were all *before any life was physically on the earth*. God told Moses, "I, the Lord God, created all things, of which I have spoken, *spiritually*, before they were naturally upon the face of the earth." (*PGP*, Mo 3:4-5)

So, Genesis Chapter One is six days of planning, organizing, ordaining and foreordaining premortal spirits to callings. Genesis Chapter Two tells a brief version of the physical creation to set the stage for the two-tree drama in the Garden of Eden.

As these are primarily activities by the Divine Council, it should not be a surprise that Genesis Chapter One doesn't match the science. How unfortunate that thousands have ditched the Bible because its first chapter disagrees with science!

Genesis Chapter Two continues, "God had not caused it to rain upon the earth." (Gn 2:5) No rain, no life! Being solely a physical phenomenon, rain is used to denote when things got physical; "There went up a mist from the earth, and watered the whole face of the ground. And the LORD God formed man of the dust of the ground." (Gn 2:6-7)

Speaking of the transition between Genesis Chapter One and Genesis Chapter Two, President George Albert Smith said, "This [Genesis Chapter One] was all a spiritual creation. Then follows the physical creation [in Genesis Chapter Two]." (*CR*, Apr 1946, p183)

What was the spiritual creation?

The *spiritual* creation in Genesis Chapter One was not just a *spirit* creation. Spirits, at least human spirits, had already been created long before Genesis begins. The first meeting of the Divine Council of spirits was held just before Genesis begins.

In Abraham's version of Creation Week the Divine Council said "we will do" this and "we will do" that – not "we have done" this or that. God told Ezra, "When I prepared the world, which was not yet made... no man spake against me. For then every one obeyed." (2 Esdras 9:18-19)

So, the Divine Council contemplated, conceptualized, organized, defined, designed, planed, and blueprinted the entire creation in advance of the physical implementation. And at the end of the sixth day, the Gods said, "We will do everything that we have said... We will rest on the seventh time from all our work which we have *counseled*... And thus were their *decisions* at the time that they *counseled* among themselves to form the heavens and the earth." (*PGP*, Abr 4:31; 5:1-3) Young's Literal Translation says that God then, "ceased from all His work which God *had prepared for making*". (*YLT*, Gn 2:2-3)

The spiritual creation also included the construction of spirit environments such as gardens, buildings, cities, paths, and portals – things described by that people who return from NDEs. One person who died three times, asked his spirit guide how the portal or tunnel got there and was told, "You built it." So, premortal spirits constructed gateways and causeways for easily accessing Earth from spirit realms, probably through something analogous to hyperspace or interdimensional voids.

Where's proof of planning?

How strong is the evidence that the six days of Genesis Chapter One were spiritual activities rather than the physical creation? The sequence of creation events rules out a physical creation:

♦ The planning for the sun (Day 4) was done after the planning for plant life (Day 3) which can't survive without the sun in the physical world.

♦ The planning for land animals (Day 6) was done after the planning for birds (Day 5) even though in the physical world, land animals came first.

♦ The planning for land animals (Day 6) was done after the planning for whales (Day 5) even though in the physical world, land animals came first.

♦ The planning for the sun and stars (Day 4) was done after Earth was planned (Day 0) even though Earth was already planned as of the first verse of the Bible.

Project managers have more leeway sequencing planning, designing, and organizing tasks. When constructing a building, the floor layout can be designed before the foundation layout. But, during the physical construction, foundation must put in place before the floor and walls.

Speaking of the six days of creation, President Joseph Fielding Smith said, "Abraham gives an account of planning in heaven for this earth and its inhabitants, before the work of building." (*DOS*, bk1, p75) This was also taught in CJCLDS Sunday School classes as outlined in the 1985 *Gospel Doctrine Teacher's Supplement*. Whereas the Old Testament reading schedule in the January 1, 1994 edition of *Church News* lumped the Genesis and Abraham creation weeks together and characterized them as planning by the Divine Council, the conclusion is that the creation days in Genesis Chapter One also refer to planning, organizing, and foreordaining.

Abraham saw a vision of the premortal realm and asked, "O Eternal, Mighty One! What is this vision and picture of the creatures?" And God replied, "This is my will for those who exist in the *divine world-counsel...* whatever I had determined to be, was *already planned beforehand* in this picture-vision before you, and *it has stood before me before it was created."* (*Apocalypse of Abraham*, 22:33, Translation #2)

Elder Orson Pratt taught: "We used to read the first and second chapters of Genesis… but did not distinguish between the *spiritual* work and the *temporal* work of Christ… there was no flesh upon the earth until the morning of the seventh day." (*JD*, bk21, p200)

So, on Day Seven, the Divine Council physically carried out the creation plans; "And the Gods came down and formed these the generations of the heavens and of the earth… *According to all that which they had said concerning* **every plant of the field before it was in the earth, and every herb of the field before it grew**; for the Gods had not caused it to rain upon the earth *when they counseled to do them*, and had not formed a man to till the ground." (*PGP*, Abr 5:4-5)

In the Jewish *Apocalypse of Abraham*, a book dated to around 100AD but believed to contain narratives from much earlier sources, God told Abraham, "This is my will with regard to those who exist in the divine world-counsel… I gave commandment to them through my Word. And it came to pass whatever I had determined to be, was already planned beforehand… and it *stood before me ere it was created.*" (marquette.edu/maqom/box.pdf, p43, DL 15 Feb 2020)

How does Abraham tell it?

Both Abraham and Moses follow the same seven-day narrative but with slightly different verbiage but the two accounts are virtually identical in sequence and substance. Abraham's account is a bit more verbose. He references "Gods" plural, a more literal translation of *Elohim*, and *time* instead of *day* for creation periods. Here are some ways that Abraham's text adds perspective:

EPOCH ONE – Light and dark: Abraham said, "[the Gods] *comprehended* the light." (*PGP*, Abr 4:4) Righteous spirits supported Jehovah as the Light that would power creation, govern its workings, and redeem them from error.

EPOCH TWO – Atmosphere and clouds: Abraham uses the word "ordered" to describe God's instructions for Nature.

EPOCH THREE – Seas, land and plants: "Let us *prepare* the earth *to bring forth grass.*" (*PGP*, Abr 4:11-13) The verbiage clearly indicates preparation prior to execution.

EPOCH FOUR – Sun, moon and stars: Abraham said, "The Gods *organized* the lights." (*PGP*, Abr 4:14-19) Earth's days, years, and lunar months were designed.

EPOCH FIVE – Water Creatures: "Let *us [the Gods] prepare* the waters to bring forth… [life]." (*PGP*, Abr 4:20-23) They planned for the waters to bring forth aquatic life.

EPOCH SIX – Land Creatures: "The Gods *organized* the earth to bring forth" living things; "The *Gods took counsel among themselves…* to *organize* man… in the image of the Gods... We *will* cause them to be fruitful and multiply… *We will do everything that we have said, and organize them; and behold, they shall be very obedient.*" (*PGP*, Abr 4:24-31)

EPOCH SEVEN – Rest and all things physical: "And thus *we will* finish the heavens and the earth, and all the hosts of them.… On the seventh *time we will end our work, which we have counseled; and we will rest* on the seventh time from all our work *which we have counseled.* And the Gods concluded… that on the seventh time they would rest from all their works which they (the Gods) *counseled among themselves to form*; and sanctified it. *And thus were their decisions at the time that they counseled among themselves to form the heavens and the earth.*" (*PGP*, Abr 5:1-3) After six days, it was time for the Divine Council to stop counseling among themselves and execute the plans in the physical reality; "And the Gods came down and formed these the generations of the heavens and of the earth." (*PGP*, Abr 5:4)

What did planning involve?

God said He "laid the foundations of the earth". (Jb 38:4) Of course, Earth glides through space rather than sits on a physical foundation. So, God was obviously speaking about planning, designing, and preparing for the physical creation.

The Divine Council designed the solar system such that Earth would be perfectly suited for humankind. The geology was designed to support a diverse biome sufficiently symbiotic to flourish. The tides, weather, and ecosystems needed to be suitable for the generation and support of life.

Planning resolved many questions: What should the earth look like? What would provide warmth to it? How warm should it be? How long would its days be? How much water should it have? How would water get there? How would mountains be raised? How would life be sustained? Who would go to Earth first? Who would lead the people at various times? Who would restore truth when it had slipped away?

Spirits accepted life missions commensurate with their abilities and befitting God's purposes. God called "noble and great ones," saying, "These I will make my rulers." (*PGP*, Abr 3:32) He called Michael the archangel to be Adam. (DC 27:11) God said, "I placed on earth a second angel… I called his name Adam." (*Secrets of Enoch*, 30:12-13) The angel Gabriel was called to play the role of Noah. (*HC*, bk3, p386)

"Every man who has a calling to minister to the inhabitants of the world was ordained to that very purpose in the Grand Council of heaven," said Joseph Smith. (*TPJS*, p365) He also taught, "At the general and grand council of heaven, all those to whom a dispensation was to be committed, were set apart and ordained." (*WJS*, 12 May 1844, recorded by Samuel W. Richards, p371)

God told Jeremiah, "Before I formed thee in the belly I knew thee and ordained thee a prophet unto the nations." (Jer 1:5) Solomon said, "As a child I was by nature well endowed, and *a good soul fell to my lot*; or rather, being good, *I entered an undefiled body*." (*RSV*, Wisdom of Solomon, 8:19-20) And Paul referred to divinely foreordained saints of his day; "He [God] hath chosen us in him before the foundation of the world." (Eph 1:4)

"The Order & Ordinances of the Kingdom were instituted by the Priesthood in the council of Heaven before the World was," said Joseph Smith. (*Scriptural Items*, WJS, 11 Jun 1843, recorded by Franklin D. Richards, p215) Smith also spoke of the "Blessings that were ordained for man by the Council of Heaven." (ibid., p232)

NDE veteran Sarah Hinze, author and speaker, said that every person has a "unique mission... to accomplish during his specified time on earth". (sarahhinze.com/characteristics-of-prebirth-experiences, DL 18 Jun 2018)

And Julie Aubier, whose NDE converted her from atheism, explained why we left the premortal spirit realm for Earth, "A lot of us came as benevolent souls to answer a call." (the-formula.org/why-we-are-here-julie-aubier, 18 Jun 2018)

So, there is significant scriptural, circumstantial, and anecdotal evidence that God involved His spirit children in selecting a site for creation, designing it, educating project participants, architecting construction, assigning tasks, appointing leaders, designating life missions, and every other planning activity necessary to ensure optimal outcomes.

Did God get tired?

"On the seventh day God ended his work which he had made; and he rested on the seventh day from all his work which he had made." (Gn 2:2-3) Why would God need rest take a break just as the physical creation was beginning?

God did not heave a sigh, wipe His brow, and sit down in an easy chair; "*On the seventh day he finished his work* and sanctified it and *also formed man* out of the dust of the earth." (DC 77:12) God rested from the spiritual creation when the physical creation began! "The defense rests" doesn't mean it stops paying close attention to the proceedings.

The creation week serves as a pattern; "Six days shalt thou labor... but the seventh day is the Sabbath." (Ex 20:9-10) God did *spiritual* work for six days and then *physical* work on the Sabbath whereas humans do *physical* work for six days and then *spiritual* work on the Sabbath. Jesus asked, "Is it lawful to do good on the Sabbath?" (Mk 3:4) Then he healed the sick and taught the people about the Sabbath – *spiritual* work! We are still in creation Day Seven, a continuation of Genesis Chapter Two. Joseph Smith called it "the Sabbath of creation" which would be "crowned with peace" in its 7th millennium at the end of which the salvation of man will be complete. (*TPJS*, p13)

The end of the spiritual creation also marked the point at which the Bible switches from "Elohim" to *Jehovah* (YHVH/יְהוָה) as the God in charge. (Gn 2:4) This was the point at which the Father transitioned all remaining God responsibilities from Himself to His Son.

Since the Gods had planned, organized, and ordered everything in the first six periods, on the seventh, they simply "watched those things which they had ordered until they obeyed". (*PGP*, Abr 4:18) There is no need to push planets along their courses nor coax any aspect of Nature along; "[He] hath given a law unto all things, by which they move in their times and their seasons; And their courses are fixed, even the courses of the heavens and the earth, which comprehend the earth and all the planets." (DC 88:42-43)

Airplanes fly on auto-pilot by following signals. In similar fashion, every scintilla of the creation is tuned into the Light of Christ that fills "the immensity of space" and is "the light which is in all things, which giveth life to all things, which is the law by which all things are governed, even the power of God". (DC 88:12-13) This Light sustained all Nature without the slightest flicker even while Christ was in the womb, while he ministered as an earthling, and while He suffered the mind-numbing agony of His sacrifice. All Nature is like the earth, filling the measure of its creation without additional work by God. (DC 88:25) He rests. Natural law is programmed such that all creation rolls along without intervention from God.

God did not *need* to rest on the 7th day, rather He had completed the premortal creation work and could switch to watching the development and evolution of His plans by His Son; "On the seventh day God ended *his* work." (Gen 2:2)

13: Physical Creation

Where is the physical creation account?

The Archbishop of Canterbury, Stephen Langton, divided the Bible into chapters in 1227AD but, because he cut Genesis Chapter One short by a verse, the summary statement for the spiritual creation slipped into the Chapter Two, "Thus the heavens and the earth were finished, and all the host of them." (Gn 2:1) Thus ended the Divine Council's planning week.

The next verse starts the physical creation day; "On the seventh day God finished His work." (Jewish Publication Society Bible (JPS) Gn 2:2) The Hebrew word *kalah*, from which "finished is translated", means *accomplished* or *completed*. (כָּלָה Str#3615) Then the physical creation narrative begins with the pronouncement, "These are the generations of the heavens and of the earth when they were created." (Gn 2:4) From that verse on, the Hebrew switches from *Elohim* as "God" to *Yahweh Elohim* as "Lord God". So, *Elohim* rested but *Yahweh Elohim* continued on with the physical creation.

Yahweh told the angels in the council, "Yonder is matter unorganized... Let us go down" — meaning they left project headquarters, descended into the physical realm, and actually created something tangible. They returned to headquarters and reported, "We have been down... [and completed the work requested.]"

Neither of the physical creation accounts paralleling Genesis Chapter Two provide any information that can be compared to the scientific version. (Compare Moses 3 & Abraham 5.)

However, there is one more inspired physical creation story! President Joseph Fielding Smith said, "The account of the creation... as given in the temple, is the creation of the physical earth." (*DOS*, bk1, p75) Indeed, the pre-2023 temple narrative showed a very strong correlation to science's physical creation sequence. It was divided into six time periods, but the sequence and content of days was different from scripture (G=Genesis, M=Moses, A=Abraham):

<u>Epoch One</u>
– G/M/A: Light
Temple: Earth
Science: Earth 4.5 BYA (Billion Years Ago)

<u>Epoch Two</u>
– G/M/A: Clouds, Oceans
Temple: Land, Seas, Rivers
Science: Land, Seas, Rivers, Rain 4.2 BYA

<u>Epoch Three</u>
– G/M/A: Land, Seas, Plants
Temple: Sun, moon, stars & planets *appear*
Science: Sun, moon, stars & planets *appear* 4.1 BYA

<u>Epoch Four</u>
– G/M/A: Sun, moon, stars & planets
Temple: Vegetation
Science: Vegetation & other non-animal life 2.4 BYA

<u>Epoch Five</u>
– G/M/A: Birds, Fishes
Temple: Animals, Birds, Fishes
Science: Animals, Birds, Fishes .58 BYA

<u>Epoch Six</u>
– G/M/A: Animals, Mankind
Temple: Mankind
Science: Mankind .0002 BYA

Earth was shrouded by asteroid impact dust and hot gases during what is aptly called the *Hadean eon*. Then, during **Epoch Three**, the sky cleared sufficiently to allow the sun, moon, and stars to be seen from the planet's surface. With that acknowledged, the scientific sequence of creation aligned nicely with the temple sequence before it was modified to match the scriptural versions. The six days in Genesis, Moses, and Abraham align with each other but not with science because they do not describe physical creation stages.

When Joseph Smith produced revised versions of Bible text, he tended to err on the side of mimicking the Bible. So, it's quite unlikely that he *accidently* deviated from the Bible when producing the original temple sequence way back in the 1840s before science was popularized for laypeople!

Whence came life?

"Spirit and element, inseparably connected, receive a fulness of joy." (DC 93:33) To enable His spirit children to achieve the ultimate happiness, God set about creating a physical environment suitable for their physical bodies.

There are two beginnings in the creation story that remain shrouded in mystery, 1) the beginning of the universe and, 2) the beginning of life. Science may never have the tools to solve the first mystery and may always lack the data to solve the second. But, at least in the case of the second mystery, it's plausible that scientists might stumble across the set of conditions that jump-started life billions of years ago.

Scientists have tracked life's chain of causality back more than four billion years but are not sure where, when, or how the first living cell came to be. They have created an evolutionary or phylogenic tree of life that organizes the taxonomic branches within the ancestral domains:

1) Archaea – Single-celled with no cell nuclei.

2) Eukarya – Single- or multi-celled organisms with membrane-bound cell nuclei.

3) Bacteria – Single-celled organisms with cell walls made of peptidoglycan and membrane-bound cell nuclei capable of producing spores for reproduction.

4) Viruses – Acellular infectious agents containing double strand of Ribonucleic Acid (RNA) or DNA with a protein coat, dependent on host cells for

reproduction and therefore not considered *alive* by most scientists.

Here are the most oft-mentioned theories of how life came to be on Earth:

♦ Panspermia – Tiny extremophile forms of life exist throughout the universe and rained down on planets, seeding them with the beginnings of life. From such a seed, evolution produced more complex forms.

♦ Biopoesis – Natural chemical reactions produced organic molecules that, under suitable conditions such as simmering deep-sea vents or warm clay crystal matrixes or primitive tropical pools energized by sunlight and perhaps lightening, combined to produce replicating structures. From these, single-celled life forms and eventually earth's more complex ones evolved.

Do Science and Genesis disagree?

Nowhere in the Genesis does God say, "Let there be earth!" Likewise, Genesis says nothing about the beginning of the universe, the formation of the galaxy, or the swirling cloud of stuff that collapsed slowly in on itself to create the solar system.

Nothing in the Bible rules out evolution. Anyone using the Bible to argue against evolution is twisting scriptures.

However, science strongly disagrees with mainstream Christianity's false interpretation of Genesis, a fact that underscores the importance of viewing Chapter One as creation planning and Adam as the first patriarch and prophet *of the Jews* rather than the first Homo sapiens on Earth.

Genesis and science simply don't have enough subject matter in common to conflict with each other.

What makes up everything?

The ancient Greeks thought everything was made of four things: Earth, Fire, Water, and Air! That seems ludicrous to the modern mind. But the Greeks also theorized that a smallest piece of matter must exist that could not be cut into smaller pieces, the atom (*a*=not, *tom*=cut). They were right about that although they did not have the tools to prove it.

Atomic theory found evidentiary support around 1800AD when John Dalton used it to explain why certain elements chemically reacted with each other in predictable ratios. For example, $NaOH(aq) + AgNO_3(aq) ==> NaNO_3(aq) + AgOH(s)$ is an equation showing that the weights of the resulting compounds could be balanced back to the weights of the original compounds. Weighing the pre-reaction materials and comparing them to the post-reaction materials indicated that individual atoms were separating and recombining in compounds in predictable ratios.

Around 1900AD, while researching radioactivity, Ernest Rutherford discovered that certain atoms decayed, proving that the name *atom* was a gross misnomer. And electrons were discovered. They seem to be uncuttable!

About 1911, Rutherford shot alpha particles through thin gold foil and, based on sharp angles of deflection, concluded that atomic nuclei must have positive charges. That protons were responsible for the positive charge was deduced in 1919 and neutrons were discovered in 1932.

As it turned out, protons and neutrons were not uncuttable. That they were made of even smaller particles was established in 1968 by analyzing the detritus of particle collisions. Ultimately protons and neutrons were found to consist of six types of quarks and that was the end of the atomic divisibility quest. Quarks seem to be uncuttable!

The current Standard Model of fundamental particles has seventeen uncuttable particles (6 flavors of quarks that make up protons and neutrons, 6 types of leptons including electrons, and 5 types of bosons including the Higgs particle responsible for mass). These fundamental particles are responsible for every known thing in the physical universe (it is not known what makes up dark matter and dark energy).

A fundamental particle is a packet of energy or excitation expressed as a wave in its own universally present quantum field (e.g. electron field, quark fields, gluon field, neutrino fields, boson fields, et cetera). Field excitations travel as waves but collide like particles. For example, a photon travels like a wave in the ubiquitous electromagnetic field until it collides with something. Then it vanishes and out pops one or more particles behaving like waves in their respective fields.

Why are there so many fundamental fields? Physicists would like to find a single fundamental field that underlies all other fields. But that Holy Grail of physics has eluded them experimentally. As physics Nobelist Niels Bohr said, "Everything we call real is made of things that cannot be regarded as real." At the smallest scales, there are only quantum field energy levels.

Ancient Hebrew literature describes the primordial "light" of the Infinite ("Ohr Ein Sof") that underpins all existence; "Before the emanations were emanated and the creations were created, there was a supernal, simple light filling all of existence... There was no beginning and no end; rather, all was one simple light." (*Eitz Chayim*, Rabbi Chaim Vital, ch1) This idea matches the primordial urstoff called intelligence now serving as the spiritual light of which all existence consists. Matter may be the natural state of physical things but, since all physical things are structure from primordial light, it is a temporary and unstable form of things. The famous physicist of the 19th century believed the force of an intelligent Mind is "the matrix of all matter". (en.wikiquote.org/wiki/Max_Planck, DL 25 Sep 2025) Said he, "I regard consciousness as fundamental. I regard matter as derivative from consciousness." (ibid.)

One school of Hindu philosophy teaches that a primordial field of consciousness exists as the fundamental component of all reality that "is formless and has no boundaries and is the only thing that existed before the universe was created".

(researchgate.net/publication/266618399_Unified_field_of_
consciousness, DL 1 Feb 2019)

Brigham Young said, "The life that is within us is a part of an eternity of life, and is *organized spirit...* The matter composing our bodies and spirits has been organized from the eternity of matter that fills immensity." (*JD*, bk7, p285) "Fills immensity" refers to the universal intelligence or spirit urstoff that pervades the universe. (DC 88:11-13) Matter, energy, forces, and spirits, are created from this same primal essence.

This ubiquitous metaphysical light constitutes *the* universal field from which all other fields derive. It cannot be created or destroyed; "The intelligence of spirits had not beginning, neither will it have an end." (*TPJS*, p353) It can act and/or be acted upon. (*BOM*, 2 Ne 2:14) It is the fundamental component of all reality and the correct basis for a Theory of Everything (TOE).

Of what is Earth made?

In 1841, the poorly educated modern prophet Joseph Smith said, *"Earth has been organized out of portions of other Globes that has been Disorganized."* (*WJS*, from McIntire Minute Book, 5 Jan 1841, pp60-61) The idea that Earth is made of stardust seemed quite strange at the time. It wasn't until the 1920's that astronomer, physicist, and mathematician, Arthur Eddington, proposed that, up through iron, the elements found on earth are the product of fusion in stars and that natural elements heavier than iron are created by star explosions.

Eddington's theory hit a slight speed bump when physicists realized that fusion stars can only "forge" elements lighter than iron. So, where did heavy elements like cobalt, nickel, copper, zinc, silver, gold, and lead come from? Supernovae are primarily responsible for cooking elements heavier than iron. When stars begin to run out of fuel, they can explode with energies that cram heaver elements together.

The solar system, including its planets, assembled from star explosion detritus. As Carl Sagan said on his 1980 television show Cosmos, "We're made of star stuff." How prescient of the prophet Joseph Smith to say that Earth came from stars when the prevailing opinion his day was that stars were created after Earth, as per Genesis Chapter One. (*Joseph Smith and the Creation,* William Stokes Lee, 1991, p154)

How good is Bible science?

Some say the Bible describes the expansion of the universe when it says God "stretcheth out the heavens as a curtain, and spreadeth them out as a tent". (Is 40:22) But it's more likely this was a view of the skies as perceived by the ancients.

In the late 1800's, Lady Elizabeth Blount authored the *Earth not a Globe Review* in which she asserted that any person who believed in a spherical earth could not be a Christian because the idea contradicted the Bible. She supported her flat-earth argument with scriptures that mention the "four corners of the earth" or the "ends of the earth". (Is 11:12; Rv 7:1; 20:8; Jb 38:13) However, such verbiage is idiomatic like "the four winds" (N, S, E, W).

Does the reference to "pillars of the earth" prove the world sits on pillars? (1 Sm 2:8) No. What would the pillars stand on? Or is it pillars all the way down? Ludicrous! Do scriptures such as "The world is firmly established; it cannot be moved" mean the earth doesn't rotate? (Ps 93:1) Obviously not.

Some anti-flat-earthers have tried to use the Bible to prove the earth is a globe with passages like the "circle of the earth". (Is 40:22; Prv 8:27) However, a circle can be flat. Others use the passage that says God "hangeth the earth upon nothing" to prove the Bible agrees with astronomy. (Jb 26:7) However, flat-earth models also show it hanging on nothing.

Obviously, the Bible has been used to support whichever position the debater prefers. A person can know the truth by direct observation of Earth's shadow on the moon, solar eclipses, ships disappearing below the horizon, Foucault pendulums rotating everywhere except on the equator, and road grid corrections where surveyors had to shave triangular pieces off the widths of "square" sections of land due to surface curvature.

The Greek philosopher Eratosthenes (276-194BC) calculated Earth's circumference quite accurately using a stick stuck in the ground at different latitudes and a bit of geometry on the resulting shadows when the sun was at its highpoint. Indirect observations of Earth such as photos from space, airplane flights, and bouncing earthquake waves prove it is not flat. If God's purpose were to use the Bible to state the shape of Earth, He would have revealed to a prophet the perfect answer, that Earth is an oblate spheroid or ellipsoid.

God did not weave science into His dealings with people. The Bible's authors were only familiar with very small regions and their language reflected their limited perspective. They lacked the modern means necessary view the world globally and the solar system was a mystery to them. The prophets were neither scientists nor cartographers.

How old is Earth?

Aristotle thought the earth had always existed. But Bible believers figured they could establish the age of the earth from scripture. Some counted the years from Adam's expulsion from Eden and made some estimates:

♦ 5530BC – The oldest extant Bible-based Christian estimate of Earth's age was recorded by Theophilus (115-181AD) in his apologetic writings to Autolycus. (*Theophilus to Autolycus*, bk3, ch24-25)

♦ 5490BC – Early Syrian Christian's year of Creation.

♦ 4004BC – Irish Christian theologian James Ussher's year of Creation, calculated in 1650AD.

♦ 3760BC – Hebrew calendar year of Creation beginning in 15th century AD.

It's obvious from the diversity of speculation that the Bible was not designed to chronicle the ages of Earth's epochs. And all these estimates were based the unproven assumptions that, 1) Earth was created in a literal six-day workweek and, 2) Adam's fall and banishment from the garden happened immediately thereafter. But the Bible specifies neither the length of each creation epoch nor how much time Adam spent in Eden before transgressing.

All these guesses are demolished by the mutually-reinforcing chronologies found in Nature. A wide variety of radiometric dating methods prove the ancientness of Earth's crustal layers and the ages of meteorites agree with ages in Earth's geology. Collectively, scientific methods have produced age estimates clustered closely around 4.54 billion years.

But, anyone with no more than a superficial knowledge of Nature must know that 6,000 years is insufficient to explain common things seen in the earth. Layers upon layers of coal are proof of millions of years of plants dying and metamorphosing under heat and pressure. The same is true for large deposits of oil, gypsum, cement and myriad other layered materials.

And the top of the highest mountain in the world is made of limestone, a sedimentary rock composed largely of the minerals calcite and aragonite that were once in the skeletons of marine organisms such as coral and mollusks. Everest did not rise from the sea overnight. It took tens of millions of years for the limestone to form and tens of millions of years more for it to be thrust from the sea floor to 29,000 feet by torpid tectonics. The evidence in the earth for its extreme ancientness would fill tons of tomes.

It's not a conspiracy by scientists to destroy faith. It was God's choice to put the preponderance of evidence in Nature rather than scripture. It was not God's intention to have people of faith argue with scientists about the age of the earth.

Not all Christians bought into the young-Earth theories. For example, in the mid-1800s, LDS Apostle William Phelps said, "Eternity, agreeable to the records found in the catacombs of Egypt, has been going on *in this system* almost 2,555 millions of years..." (*Times and Seasons*, bk5, #758)

While off by a factor of about two, Phelps' calculations were billions of years closer to modern science's estimate than the 6,000 years that other religions were teaching at the time.

But there is one scripture that might provide an estimate of elapsed time between Adam's fall and the end of Christ's millennial reign. It says that the earth will see "seven thousand years of its continuance… [and] in the beginning of the seventh thousand years will the Lord God sanctify the earth". (DC 77:6-12) This aligns with classical Hebrew sources that say the Messiah will come no later than 6,000 years from Adam's banishment from Eden. This belief is based on the pattern established in the Bible; "In six days God made the heavens and the earth, the oceans and all therein, and He rested on the seventh day." (Ex 20:11) Each thousand years of Biblical history is a "day" with the 7th being the Messiah's millennial reign.

Why is seven so special?

The number seven appears throughout Hebrew scripture and culture. The menorah has seven branches. Jericho's walls fell after seven laps of marching Israelites. Naaman washed seven times in the Jordan to be cured of leprosy. A comprehensive list would be very long, especially if we were to include multiples of seven. But here are the main ones:

Sevens represent the phases of creation.

I. The <u>spiritual</u> creation was six planning periods followed by a seventh as a Sabbath. (Gn 1)
II. That Sabbath contained six periods of <u>temporal</u> creation followed by a seventh as a Sabbath.
III. That Sabbath contained six periods of <u>temporal</u> history followed by seventh as a Sabbath.
IV. That Sabbath contains Christ's return and reign in peace until which all <u>spiritual</u> work is finished. (DC 77:12)

Perhaps this creation pattern is what God meant when He said, "For by the power of my Spirit created I them; yea, all things both spiritual and temporal-- *First spiritual*[I], *secondly temporal*[II], which is the beginning of my work; and again, *first temporal*[III], *and secondly spiritual*[IV], which is the last of my work." (DC 29:31-32)

A similar cascade of sevenths will recur in the last days leading up to Christ's return in glory:

I. The Lamb will open seven seals representing the seven thousand years of God working His gospel with humans.
II. In the seventh seal, angels will sound seven trumps.
III. In the seventh trump will be poured out seven vials.
IV. In the seventh vial, amid thunder and earthquake, God will announce, "It is done". (Rv chapters 5-17)

Was all land one continent?

"God said, Let the waters under the heaven be gathered together unto one place, and let the dry land appear: and it was so." (Gn 1:9) Many sincere preachers have interpreted this to mean that there was just one huge continent and that it was split apart in the days of Peleg; "For in his days was the earth divided." (Gn 10:25)

Some say the continents split apart while Noah's flood was on the earth. But Peleg was born three generations after the flood! And, since Peleg comes from the Hebrew for *Division*, his parents obviously named him for a division that was underway at the time of his birth.

But, if the supercontinent had split and rushed to opposite sides of the globe within Peleg's lifetime, the resulting tsunamis, earthquakes, superstorms, and years of famine-inducing volcanic winter would have resulted in a mass extinction event that would have been recorded by every *surviving* culture with a written language. And yet, no such cataclysmic event appears in any record from Peleg's time, not in the Bible, and not in any legend.

The entire focus of the chapter in which Peleg is on the dispersion of Noah's descendants as they were "*divided* in their *lands*; every one after his tongue, after their families, in their nations". (Gn 10:5) The scripture means, "The land was divided *among the nations.*"

What does Genesis mean by the waters being gathered unto one place? "As the mantle cooled, land would have gradually appeared as the oceans became deeper." (metro.co.uk/2008/12/31/early-earth-was-covered-in-water-274995, DL 30 Oct 2018) God said, "Let the dry land appear." (Gn 1:9) "One place" means off the land!

Here's another misunderstood scripture; "The land of Jerusalem and the land of Zion shall be turned back into their own place, and the earth shall be like as it was in the days before it was divided." (DC 133:24) Does this mean the continents will be pasted back together at Christ's second coming? No, for the same reasons they couldn't have separated quickly – tsunamis, earthquakes, volcanoes, superstorms, etc. God promised never again to wipe out entire nations by flood. Here are three plausible interpretations:

When Christ returns in glory, *Zion*, the City of Enoch, which was taken up into heaven, will return to Earth from heaven with Him and become part of *Zion*, the *New Jerusalem*. (*PGP*, AOF (Articles of Faith), #10) The Lord told Enoch, "Truth will I cause to sweep the earth as with a flood, to gather out mine elect from the four quarters of the earth, unto a place which I shall prepare, an Holy City… and it shall be called Zion, a New Jerusalem… Then shalt thou and all thy city meet them there… And there shall be mine abode, and it shall be Zion." (*PGP*, Mo 7:63-64)

Another plausible interpretation is the Land of Jerusalem means the area associated with Jerusalem while the Land of Zion" means the entire original kingdom as it was under Solomon, before it was divided into the northern kingdom of Israel and the southern kingdom of Judah. In this case, the remaining area east of the Jordan River that was under Solomon's rule will change ownership from the Jordan back to Israel when Christ returns.

A third plausible interpretation is that the division was not physical and that the reunification is political and spiritual under Christ's reign; "The Lord, even the Savior, shall stand in the midst of his people, and shall *reign over all flesh*." (DC 13:24) Enoch's City, the New Jerusalem, and the Old Jerusalem have all been called Zion. The Enoch's City has been called the New Jerusalem. (Rv 21:2) The gathering place for latter-day saints in America will be called the New Jerusalem also. So, it's confusing. But, all of these will be of one faith and one kingdom under Christ. Everyone on Earth will effectively be in one place, the Kingdom of Jesus Christ.

That's not to say that there has never been a supercontinent. Scientists say the first one was Ur, 3.1 BYA, followed by Rodinia 1.1 BYA, and Pangaea 300 MYA (million years ago), which split down the middle during the age of the dinosaurs to form the land masses of our two hemispheres. The Atlantic Ocean between them has been widening for the last 100 million years.

Today's seven continents are still slowly floating around on the partially viscous mantle of earth, the fastest ones grinding along at almost three inches per year. As a result, the Global Positioning System must periodically update itself to remain accurate. The average continent moves slower than the most sluggish snail, about as fast as fingernails grow. So, the current configuration of the major continents has been essentially unchanged for tens of thousands of years.

How was Earth made habitable?

4.5 BYA, Earth was still over 8,000 degrees Fahrenheit from the energy of objects smacking into it. Scientists call it the Hadean eon (hellish).

4.3 BYA, water from icy comets precipitated steam and thunderstorms. The atmosphere was sweltering and poisonous. It took millions of years for the surface of the earth to cool sufficiently for water vapor to condense into rain. Then oceans formed and waves pummeled volcanic islands.

3.5 BYA, granite floated up through the heavier basalt crusts and gradually became the most dominant rock on the planet's surface. The first strands of self-replicating molecules formed from which the first life forms evolved.

2.2 BYA, oxygen from stromatolites, rocks formed by microbial mats, gradually diluted the noxious gases in the air, making it more transparent and less toxic.

1.5 BYA, continents grew to cover a quarter of the earth's surface. Convection in earth's mantle drove the land masses together, forming the supercontinent Rodinia 1.1 BYA.

.7 BYA, Rodinia blocked warm ocean currents from the equator, allowing the polar icecaps to expand down to the equator, resulting in a snowball Earth.

.5 BYA, Rodinia broke up and released internal planetary heat trapped by the ice, resulting in a great thaw. Life exploded into myriad multicellular species. Plants crept onto dry land followed by insects, amphibians, and reptiles, protected from ultraviolet solar radiation by the ozone layer.

.25 BYA a mantle plume eruption in Siberia caused a global catastrophe that killed 95% of all life on earth. Continents clustered again to form the supercontinent Pangea and dinosaurs appeared.

.18 BYA oxygen levels increased dramatically as plant diversity surged in response to rising carbon dioxide levels, enabling dinosaurs to attain enormous sizes.

.065 BYA over 70% of all species died in the aftermath of one or more asteroid or comet strikes combined with several massive waves of super-volcanism in India. These events filled the atmosphere with particulates that plunged the planet into darkness and precipitated many years of winter-like conditions. The reign of the dinosaurs came to an end.

From then until now, the earth has experienced less extreme ice ages and warming cycles. Compared to previous millennia, conditions on Earth have been quite stable for the last 8,000 years.

Who knew Earth was moving?

Peoples of the Old Testament, like most folks of their day, viewed the earth as being flat, surrounded by ocean, sitting on pillars in a watery abyss, and totally stable; "The world also is stablished that it cannot be moved." (*ERV (English Revised Version)*, Ps 96:10)

Then around 500BC, Greek philosophers speculated that the earth was spherical. Around 400BC, Greek philosopher Philolaus proposed that the earth rotates to create day and night. Around 300BC, the Greek genius Eratosthenes noted that a vertical pole in Egypt cast no shadow right at mid-day. So, using a little geometry on shadow angles to the north, he arrived at a circumference for Earth of 25,000 miles, which is only 99 miles over the actual value!

The idea that Earth orbits the sun had been proposed by Aristarchus in the 3rd century BC but, in 1543AD, Copernicus put his name on the more complete heliocentric theory that solved the mystery of why planets appear to go backwards against the backdrop of stars and why Venus has moon-like phases.

About 6BC, an ancient American Nephite wrote, "It is the earth that moveth and not the sun." (*BOM*, Hel 12:15) But, astronomers of the Mayans and Incas always thought Earth was at the center of all things. (thoughtco.com/ancient-maya-astronomy-2136314, DL 6 May 2022)

Humankind has now gone from thinking that Earth was stationary and flat to learning that Earth's equatorial surface revolves at 1,000 MPH. The entire planet orbits the sun at a mean velocity of 66,600 MPH. The sun orbits the galactic center about once every 237 million years at a mean velocity of 492,000 MPH. And the galaxy is moving with respect to the CMB at about 825,000 MPH. Wheeeee!

Is this God's only earth?

In the 1997 movie *Contact*, young Ellie asks, "Dad, do you think there's people on other planets?" Her father answers, "I don't know, Sparks. But I guess I'd say if it is just us... seems like an awful waste of space." Even if *every* star's planets all had life, there would still be "an awful waste of space" out there by his reckoning, what with all the interstellar gaps and intergalactic voids. But God sees it differently.

As Jehovah said, "There is no space in which there is no kingdom." (DC 88:37) So, God doesn't waste space. Who views the 14.6 million square miles of uninhabited Moon real estate as a waste of space? Vast and rugged barrenness can be beautiful and awesome!

Scientists disagree widely as to how many habitable planets there might be in our galaxy. Anywhere from 1% to 50% of the 100 to 700 billion stars might have a planet with some kind of life. NASA's Chandra X-ray Observatory has found evidence of planets in other galaxies. So, taking the estimate that there are at least two trillion galaxies in our universe, the probable number of habitable planets boggles the mind. (bigthink.com/starts-with-a-bang/how-many-galaxies, DL 6 May 2022)

Hence arises the Fermi paradox: Given the extremely high probability that there is intelligent life that started sooner and advanced farther than we have, why haven't ETs visited Earth already? Is anyone out there? Where are all the alien tourists or conquerors? Do they find us or our planet inhospitable or repulsive? Or are they still on their way here after millions of years of star hopping? Or do they simply lack the necessary technology for interstellar travel, like endless propulsion, radiation shields, and force fields to prevent specks of dust or rocks from destroying their spaceships at high speeds?

God said, "Worlds without number have I created… But only an account of this earth, and the inhabitants thereof, give I unto you… There are many worlds that have passed away by the word of my power. And there are many that now stand, and innumerable are they unto man… And as one earth shall pass away, and the heavens thereof even so shall another come; and there is no end to my works." (*PGP*, Mo 1:33-38) Do the people on other worlds look anything like us or are they small and green with huge bulbous heads and slanted eyes? Are they even carbon-based?

Professor Nibley wrote, "Those other worlds are uniform and similar, as in 'the other worlds we [the Gods] have heretofore formed.' Everything follows the same pattern… all made of the same substance." (nibleys-commentary.com, bk2, p453, DL 2 Aug 2017) So, we must assume that if there are children of God among the life forms on those earthlike planets, they were also created in His image.

God showed Moses a vision in which he "beheld the earth, yea, even all of it… also the inhabitants thereof… And he beheld many lands; and each land was called earth, and there were inhabitants on the face thereof." (*PGP*, Mo 1:27-29)

Enoch praised God, "Were it possible that man could number the particles of the earth, yea, millions of earths like this, it would not be a beginning to the number of thy creations." (*PGP*, Mo 7:30) Then God told Enoch, "I can stretch forth mine hands and hold all the creations which I have made; and mine eye can pierce them also, and among all the workmanship of mine hands there has not been so great wickedness as among thy brethren." (*PGP*, Mo 7:36)

Only people who've heard the gospel can be charged with wickedness. So, God was talking about peopled earths.

"Millions of earths like this" unequivocally refers the type of habitable exoplanets scientists are seeking. So far, of the thousands of exoplanets found, only about 12% appear to be in habitable zones. Even if only 1 of 10 of those in habitable zones harbor intelligent life, the number would be in the billions in our galaxy alone, trillions in the known universe.

The only Bible-believing church that teaches the existence of other planets inhabited by children of God is The Church of JESUS CHRIST of Latter-day Saints. Per Pew Research, about 65% believe there is Extraterrestrial Intelligent life. Muslims quote the Quran to support their belief in ETI; "It is God who has created seven heavens and similar number of Earths." (Quran 65:12)

Surprisingly, the idea of alien worlds was broached during Christianity's early years. Father Origen Adamantius (184-253AD) said, "Not then for the first time did God begin to work when He made this visible world; but as, after its destruction, there will be another world, so also we believe that others existed before the present came into being." (*De Principiis*, bk 3, ch5, sec 3)

When did Day Seven end?

Conspicuously missing from Genesis is any mention of "morning and evening" for Day Seven. So, when did Day Seven end, or did it?

Carl Sagan cited a story from the Talmud wherein God told Adam and Eve that He had left the creation unfinished and that humans, over long periods of time, would help in completing it. (*Pale Blue Dot*, Sagan, 1997, ch21) It's true! Creation continues! And humans are accountable for Earth's care until all the babies are born and all the plants have sprouted and all the other creation events are accomplished.

As long as physical creation processes continue to unfold, Day Seven will remain *in progress*. Tectonic plates grind along. New volcanoes create new mountains. Crustal layers continue to raise mountains higher. New strains of drug resistant bacteria continue to evolve. New babies are born. New bird hybrids are being classified as distinct species. New stars and nebulae continue to pop up. The list goes on.

Creation's Day Seven will continue until we see "the end of wickedness on earth, and the *Sabbath of creation* crowned with peace". (*TPJS*, p13) It will be the millennial Sabbath of peace that follows the Sabbath of Creation. "As God made the world in six days, and *on the seventh day he* finished his work… even so, in the beginning of the seventh thousand years will the Lord God sanctify the earth, and complete the salvation of man, and judge all things." (DC 77:12)

In the Sabbath of creation, Christ will reign a thousand years. Day Seven will end when the earth dies! "For after it hath filled the measure of its creation, *it shall be crowned with glory… yea, notwithstanding it shall die*, it shall be quickened again… and the righteous shall inherit it." (DC 88:17-26)

14: Creation of Life

What is life?

Dictionary definitions of life are circular:

Life: The property, quality, state or condition of *living*, distinguishing a *living* organism from a *dead* organism.

Living: Possessing *life*.

Dead: No longer *living*.

But science must have a good definition of life. In fact, science has *various* definitions and no single universally accepted one. The best ones go something like this: "The state of a material complex or individual characterized by the capacity to perform certain functional activities including metabolism, growth, and reproduction…" or "signaling, self-sustaining metabolism, and reproduction."

What about the "live" cancer cells from Henrietta Lacks that researchers have been culturing and using ever since she died in 1951? More than 50 million tons of Henrietta's cells have been grown in labs. Are her cells alive? Which definition of life is appropriate depends on context.

Are viruses alive? Most scientists say no. Why not? Because they cannot metabolize, grow, or reproduce on their own – they need a host. But, in normal conversation, even virologists use the words *living, killed,* and *dead* when discussing vaccines. Says one, "Viruses remain dormant when they aren't *living* inside a host organism." (*Can Viruses be Killed?,* Madhavi Deshpande, ScienceABC, 2015)

The use of the word "living" belies the truth of viruses. To quell vaccine fears, scientists say the viruses are *dead*. If they can die, then they must have been alive, right? Even the definition of *biological* warfare includes any kind of germ, including viruses. Nobody uses the term *virological warfare* to avoid lumping viruses in with a term whose root means *life*!

Viruses share a common ancestor with living cells. Some have even contributed beneficial genes to humans. Some viruses have more genes than the average bacteria. Giant viruses contain thousands of genes, can get infected by other viruses, can slurp up free-floating RNA or DNA, can transfer genes in and out of other virus or other cells, and can have genes that are not found in any cell-based living organism.

The scientific line between "alive" and "not alive" is being blurred as stranger and larger viruses are discovered. And, if probes find live viruses on Mars, some headlines will undoubtedly read something like "Life Discovered on Mars!" So perhaps the phylogenetic tree of life will eventually include a domain for Viruses. The fact that animated things evolved from inanimate things means that any precise boundary between the two will be somewhat arbitrary.

Biologists organize living things into three *domains*:

♦ Archaea – Microorganisms that are somewhat different from modern bacteria in composition and chemical structure, often inhabiting extreme environments such as hydrothermal vents or animal digestive systems.

♦ Bacteria – Single-celled spherical, spiral, or rod-shaped organisms that have cell walls but no organized nucleus and no organelles.

♦ Eukaryota – Organisms whose cells have membrane-bound organelles and nuclei.

These domains branch out into kingdoms, phyla, classes, orders, families, genera, and species to form the phylogenic tree of life that reflects biologists' understanding of evolutionary relationships. People use various mnemonics to remember the levels, such as, "Do Keep Ponds Clean Or Frogs Get Sick" (DKPCOFGS) or "Dapper Kings Play Chess On Fine Green Silk."

When God breathed a spirit (Heb *neshamah*: breath or spirit) into Adam's body, "man became a living soul". (Gn 2:7) A living soul is a spirit in a body that's breathing. "The spirit and the body are the soul of man." (DC 88:15) Death is the separation of spirit and body along with loss of vital signs.

If animals, plants, and microbes are also alive, then they too must have spirits, despite the teachings by many clerics that they don't, or that their spirits cease to exist when their bodies die, and that they therefore cannot go to heaven. But, John saw creatures in heaven. (Rv 4:6) How did they get there without a soul?

The doctrine of The Church of JESUS CHRIST of LDS closely resembles animism, the religious belief that all things, whether deemed alive or inanimate, are imbued with spirit essence. Joseph Smith taught, "They are figurative expressions, used by the Revelator, John, in describing heaven, the paradise of God, the happiness of man, *and of beasts, and of creeping things, and of the fowls of the air.*" (DC 77:2) So, heaven is chock full of all kinds of life! All life forms of life are candidates for heaven.

Even the earth is alive in a sense. Enoch heard the spirit of the earth lament, "Wo, wo is me, the mother of men; I am pained, I am weary, because of the wickedness of my children. When shall I rest, and be cleansed from the filthiness which is gone forth out of me? When will my Creator sanctify me, that I may rest, and righteousness for a season abide upon my face?" (*PGP*, Mo 7:48) Earth's spirit is intelligent. It is displeased with the wickedness it senses on its surface; "The earth laid accusation against the lawless ones." (Enoch 7:6) So plaintive was the earth's cry that Enoch had pity on it and prayed, "O Lord, wilt thou not have compassion upon the earth?"

Ezra said, "Ask the earth, and she shall tell thee, that it is she which ought to mourn for the fall of so many that grow upon her." (2 Esdras 10:9) And, in response to the crucifixion of its Creator, Earth trembled. (Mt 27:51)

That the earth is alive is supported by the prophecy that it "shall pass away". (DC 45:22) It can't die unless it lives. "Notwithstanding it shall die, it shall be quickened again, and shall abide the power by which it is quickened." Earth will be exalted to celestial glory because it "abideth the law of a celestial kingdom". (DC 88:25-26)

"Our way of grouping things in categories of 'alive' and 'not alive' is only from our limited physical point of view," said a man who died and was revived, "I saw that everything is alive." (nderf.org/Experiences/1mohammad_z_nde.html, DL 21 Jun 2018) Myriad others have returned from death to confirm that all things are indeed alive and aware.

So, it behooves us to treat all of God's creations kindly because our actions with everyone *and everything* leave indelible impressions on spirit memories that will endure beyond mortality.

How was life jumpstarted?

When evolution was still controversial in schools, kids were asking, "Did we come from scum or climb from slime?" Biologists would like to know the recipe for creating life so that they can get a handle on the process that created the Last Universal Common Ancestor (LUCA).

In 2009, a team synthesized RNA molecules that self-replicated in a concoction of raw materials. In 2010, a team synthesized DNA molecules and substituted them into an existing cell's nucleus and watched it multiply. In 2018, a team created a flotilla of pulsating cell-like blobs with part of the machinery needed for self-replication. Biologists have the list of ingredients but lack the correct directions for assembly. Bottom-up synthetic biologists believe they will succeed in jumpstarting the first fully artificial cells around 2028. (nature.com/articles/d41586-018-07289-x, DL 13 Jun 2020)

If, as science teaches, Nature created life by happenstance, why can't today's geniuses copy the trick? While there are dozens of laboratories that have been working on synthesizing life for several decades, Earth had zillions of natural laboratories over the course of billions of years. Natural laboratories included ocean vent environs, hot springs, warm freshwater ponds, and perhaps other conditions scientists haven't identified. It is believed that atoms chemically bonded to form amino acids which then formed the proteins that catalyzed the creation of ribonucleic acid

(RNA) which has the ability to replicate. The rest of the story is evolution.

But a few scientists wonder whether life on Earth could have been seeded from space (panspermia). So, they are using telescopes to look for organic molecules of the type that might have facilitated the beginnings of life on earth. If there is life throughout the universe, panspermia could be the most common way Nature jumpstarts it on habitable planets. Stephen Hawking speculated, "Life could spread from planet to planet or from stellar system to stellar system, carried on meteors." (*Origins Symposium*, 2009)

Could aliens have seeded earth? Even atheist activist Richard Dawkins entertained that idea, provided the aliens were products of evolution. Scientist Albert Rosenfeld mused, "I'm somehow not surprised at the idea that someone out there put us here. And if such a magical, mysterious, and powerful intelligence exists that is utterly beyond human imagining, can you give me a good reason why I shouldn't call it God?" (As quoted in *CWHN (Collected Works of Hugh Nibley*, bk1, ch4. p82)

How did life get sapient?

The progression from dust to apex predator was durative but wondrous for Homo sapiens. Earth had to be just the right distance from the sun, and contain just the right proportions of various versatile molecules needed for life, such as carbon, water, et cetera. It needed a moon that would lull the tides in and out at just the right sweep to support stromatolites, the source of atmospheric oxygen that animals would need.

Environments here and there around the globe had to change in such a way as to steer evolutionary diversity toward more intelligent species. Evolution usually selects for intelligence but sometimes for more beastly qualities. Course corrections often occurred in the form of local or mass extinctions, redirecting the evolution toward the emergence of modern species.

The *Jurassic Park* movie unknowingly overstated the intelligence of velociraptors when it portrayed them as quickly learning to open doors. In actuality, they owed their success to instinctual ferocity. Velociraptors had feathers, did not hunt in packs, and were actually no larger or smarter than an average domestic turkey. By contrast, T-rexes were surprisingly intelligent and ruled the planet for a long time. But God did not create the earth to be ruled by large lizards.

Convenient to His plan, an asteroid and some super-volcanism intervened to partially reset evolution. The dinosaurs, from the titanic T-rex to the tiny Tethyshadros, were destroyed. After the Cretacious-Tertiary (K-T) extinction event, the smartest surviving animals were diminutive mammals, tiny rodents scurrying around in the underbrush looking for edible seeds, leaves, bugs, and grubs. Since then, mammals have ruled the earth. But no other animal had been as dominant as humans.

A more subtle but equally important miracle was the massive climate change in East Africa from tropical to arid several MYA that drove the primates out of the trees, making bipedalism a key survival adaptation. Hands that were already quite dexterous from arboreal life could now carry modified sticks and stones as tools.

One advancement in technology led to another that improved the odds that the most intelligent individuals would survive. The brain-to-body mass ratio of humans is now tied with mice for first place among mammals.

How do scientists date stuff?

A mountain of fossil and archeological evidence contradicts young-Earth interpretations of the Bible. How did scientists arrive at such old ages for Earth? Are modern dating methods reliable? How can scientists be 100% certain that the Creation Museum in Kentucky is wrong to portray humans living alongside dinosaurs?

Scientists actually have quite a variety of trustworthy dating methods. Here's a partial list:

♦ **Dendrochronology**: Tree rings, caused primarily by weather-induced changes in growth rates and also seasonal temperatures, serve as annual tick marks in the wood. Climate changes and forest fires also affect the rings. Such changes affect entire forests, enabling researchers to synchronize the rings where generations overlapped by matching outer patterns from earlier trees with the inner patterns in later trees. Dendrochronologists can compile tree ring histories that stretch across thousands of years. The oldest tree in the world is a clonal colony of quaking aspens in Utah that have shared a single root system for 80,000 years. The oldest individual tree is a Great Basin Bristlecone Pine at 5,064 years, predating Noah's flood by about 1,000 years.

♦ **Radiocarbon**: Earth's atmosphere is circa 0.04% CO_2. About one in every ten trillion carbon atoms is the isotope carbon-14 (C14) rather than the more stable form (carbon-12). All non-aquatic organisms have the same C14/C12 ratio as that found in the atmosphere until they die. Since organisms stop absorbing carbon when they die, the ratio decreases from the point of death as the C14 decays. So, the lower the C14/C12 ratio is, the older a specimen is. Half of the C14 decays every 5,700 years. It works quite well for land-based organism's younger than 40,000 years, after which there is not enough C14 left for accuracy.

♦ **Potassium-Argon**: The K-Ar radiometric technique uses the same decay-rate principle as the C14 method. Scientists know from analysis of fresh magma that about 2.5% of the earth's crust is potassium and that about $1/10,000^{th}$ of that is K40. So, knowing the rate of decay of K40 to Ar40 (K-Ar), they can measure the K40/Ar40 ratio to determine the number of years since the magma cooled.

♦ **Rubidium-Strontium**: This method is similar to the K-Ar dating method. Rubidium decays to Strontium at a known rate. The lower the ratio, the older the material. Allowances must be made if the material has been exposed to liquid.

♦ **Uranium-Lead & Uranium-Thorium**: This method is also similar to K-Ar dating. Uranium-238 decays to Lead-206 or to Thorium-230. Allowances must be made if the material has been exposed to liquid.

◆ **Accelerator Mass Spectrometry**: AMS works on the same principle as the Carbon-14 method but works on a much wider range of elements simultaneously. Since AMS is especially useful for small sample sizes, it was used in 1988 by three laboratories to test the Shroud of Turin with minimal damage to it other than to its reputation as the genuine article.

◆ **Thermoluminescence**: The electron count in objects such as obsidian arrowheads and potsherds increases with exposure to the constant radiation from sources such as the sun, Earth's interior, and radioactive elements in the environment. When such an object is heated by fire, the trapped electrons are released, effectively resetting the "clock" to zero. So, scientists can reheat any artifact that has been on an ancient fire, such as a stone weapon, a tool, or an earthen vessel, and count the electrons that pop out. The more electrons released, the more time has passed since the item was last heated. This method is reliable to 500,000 years.

◆ **Electron Spin Resonance**: ESR measures how many electrons from radiation have been trapped in a crystalline material over time. The more electrons trapped, the older it is. But rather than heating it to release electrons, scientists expose a specimen to a varying magnetic field while probing it with a laser and take a paramagnetic resonance measurement to determine the level of electron saturation. Then, by comparing that to the yearly radiation absorption rate, the lab can estimate the number of the years the specimen has been trapping electrons. This works especially very well for ancient teeth like those from Neanderthal jaws.

◆ **Paleomagnetic Analysis**: When rocks are molten, any iron oxide particles they contain align themselves with the earth's magnetic field and remain frozen in that position when the rock cools. Successive layers are a record of magnetic field changes and/or tectonic plate orientation. By noting the orientation of iron oxide particles and matching it to other known orientations, scientists can gauge the age of the rock.

♦ **Amino Acid Racemization**: Amino acids in living organisms are molecularly structured in what is called the L-isomer configuration. After death, the amino acids slowly restructure to a D-isomer configuration at a predictable rate through a process known as racemization. This method can be accurate to millions of years depending on conditions. Allowances must be made for moisture and temperature.

♦ **Biostratigraphy**: The deeper a layer of rock is, the older will be the fossils in it. Layers exposed in the Grand Canyon range from about 200 million years old near the top to nearly 2 billion years old at the bottom. The progression of evolution is on display with excellent consistency from the newest layers containing modern forms to the oldest layers containing primitive forms or none at all.

♦ **Optically Stimulated Luminescence**: Minerals trap more and more electrons from ionizing radiation and cosmic rays over time as long as they are unheated and in the dark. The more electrons trapped, the longer it's been since the minerals have been exposed to sunlight or heat. Exposure to sunlight or heat from campfires or kilns clears the trapped electrons, like resetting a stopwatch to zero. Lab technicians can expose the material to light or heat and, by measuring the amount of light released thereby, use known rates of electron absorption to calculate how long it's been since those materials were last heated or exposed to sunlight.

♦ **Tephrochronology**: Layers of volcanic ejecta (tephra) can be accurately dated using argon-argon (Ar-Ar) dating. Then any layer sandwiched between layers thusly dated must have a date in between that of the layer above and that of the layer below.

The usual reaction from young-earth proponents to scientists' war chest of dating tools is to focus on a single type of test in isolation and point to scenarios that might yield less accurate results. But scientists commonly use more than one method to reconstruct a history.

How accurate is dating?

Young-earth proponents who point to large accuracy ranges in archeological or geological dates, or highlight discrepancies between results from different types of tests, usually ignore the fact that even the low ends of the ranges still prove that the dinosaurs and the earth are far older than such creationists believe them to be.

For example, creationist Henry Morris cited K-Ar dates on specimens from an 1801 lava flow in Hawaii that ranged from 160 million to 2.96 billion years. (*Scientific Creationism*, Morris, 1985) But even taking the low-end date blows young-earth dates away. Scientists can usually explain wide ranges. The Hawaii case was unusual because the lava sample contained xenoliths for which the K-Ar method is useless. (talkorigins.org/faqs/dalrymple/radiometric_dating.html, DL 1 Aug 2021)

When different dating methods are used, the results are almost never identical. There is no single ideal method that produces accurate results for every kind of sample, in every context, for every chronology. So, scientists employ multiple applicable methods to a particular sample, consider the factors that may influence accuracy, and calibrate the estimate accordingly.

Most dating methods yield results fairly consistent with other dating methods and with widely validated timelines compiled by science. Never has a mammal fossil been found in the Cambrian or Paleozoic strata nor has a dinosaur fossil ever been found in the Devonian or Pleistocene. So, although dates are usually expressed in ranges, they are sufficiently accurate to provide a good picture life's history on Earth.

Must fossils kill faith?

Faith and fact should agree, otherwise there are flaws in the faith and/or facts. And deep-seated fear that their faith is flawed is the reason that young-Earth fans feel threatened by fossil evidence.

To skirt the fossil-faith conflict, some Bible apologists manufacture alternative scenarios that only people in scientifically illiterate circles can swallow. For example, a large number of defenders of their faith float the idea that all the fossil-rich strata were laid down during the Noahic global flood, ignoring the fact that most fossils are millions of years older. They also ignore the principles of hydrological sorting that indicate the layers took millions of years to accumulate. Sedimentation over millions of years is quite different from mud laid down quickly in a single catastrophe.

Another apologist theory is that there were two creations. All ancient organisms whose fossils have been found belonged to the first creation. They died out for some reason, perhaps from one or more widespread catastrophes. Then God started over as described in Genesis. Since God is all-wise, mass extinction events would likely be part of His plan to guide evolution to His desired variety of life forms.

Another theory is that ancient fossils lived and died and were fossilized on other worlds that broke up and fell to Earth. However, the 8,000-degree accretion heat of early molten Earth would have totally destroyed them all 4.6 BYA. If they fell later, atmospheric friction would have melted them.

Even if the fossils had landed unscorched and intact on Earth's surface by some miracle, they would have had random orientations instead of positions consistent with the gradual layering over the course of earth's geological history. Dinosaur eggs would be found in upside-down nests as often as right-side up and piles of dinosaurs would be found in the absence of water evidence as often as not. But the opposite is true. Fossils lie situated consistent with the strata in which they are found, layer upon layer testifying of countless cycles of life and death, from sophisticated forms near the surface to more primitive ones in deeper layers.

To avoid conflict with fossil evidence, Bible believers have only to embrace Genesis Chapter One as a series of premortal planning periods and Genesis Chapter Two as a physical creation story that sets the stage for the drama of Adam and Eve deciding whether to maintain status quo or try forbidden fruit. Since there is no precise creation sequence in Genesis Chapter Two, it can hardly conflict with Earth's history as told by scientists.

Do fossils refute the idea of an Adam?

The belief that Adam is responsible for bringing death into the world is based on an interpretation of the scripture; "For as in Adam all die, even so in Christ shall all be made alive." (1 Cor 15:22) If Adam brought death into the world, how can fossils be indicating that cycles of life and death had been underway for almost 4 billion years? And, if Adam didn't bring death, why would a Savior be needed to restore life? The science seems to threaten the very core of Christian faith!

One way to resolve this apparent conflict between scripture and science is to acknowledge that God can apply justice or mercy retroactively. To wit, penitent people who lived prior to Christ's resurrection were forgiven of their sins by virtue of his atoning sacrifice because it was retroactive. Similarly, other life forms that lived prior to Adam and Eve could have been doomed to die as a result of the fall from Eden because it was retroactive. God's foreknowledge baked these kinds of retroactivity into the plan from the beginning!

A better interpretation is to view Adam's primary punishment as spiritual death rather than physical death. By getting themselves ejected from the Garden of Eden, Adam and Eve were immediately separated from God. That's spiritual death! Paul was talking about spiritual death in the previous quote, and also when he wrote, "By one man sin entered into the world, and *death by sin*; and so death passed upon *all men*, for that *all have sinned.*" (Rm 5:12)

James wrote, "Sin, when it is finished, bringeth forth death." (Js 1:15) The apostles could not have been talking about physical death! People sin without dying and Christ died without sinning. Adam's transgression didn't kill him physically. It brought spiritual death upon him.

It was the banishment from paradise that took Adam and Eve off life support. Consequently, *they died of natural causes, like all the other creatures outside the Garden of Eden.* The only physical death caused by their transgression was their own. But, so as not to get ahead of the story, an in-depth look at the forbidden fruit and Adam's physical death will have to wait until the chapter on "The Fall".

Does Genesis prohibit evolution?

Pope Pius XII declared his acceptance of evolution in 1950, Pope John Paul II in 1996, and Pope Francis in 2014. If these Catholic pontiffs found nothing in their Bibles that rules out evolution, then it's probably not there.

And, there are many scientists who see nothing in evolution that contradicts scripture. The modern apostle, chemist, geologist, and author of *Jesus the Christ* wrote, "The oldest… rocks… reveal the fossilized remains of once living organisms, plant and animal. The coal strata… are highly compressed and chemically changed vegetable substance. The whole series of chalk deposits and many of our deep-sea limestones contain the skeletal remains of animals. These lived and died, age after age, while the earth was yet unfit for human habitation." (*The Earth and Man*, Talmage, Deseret News, 21 Nov 1931)

The Bible doesn't say *how* life came to be, only that God ordered it and was obeyed. If God chose evolution as the means by which Earth brought forth life, who should argue?

Many Christians believe that the text of Genesis rules out evolution where it says, "The earth brought forth grass, and herb yielding seed *after his kind*, and the tree yielding fruit, whose seed was in itself, *after his kind*: and… every living creature… *after their kind*, and every winged fowl *after his kind*, and… cattle, and creeping thing, and beast of the earth *after his kind*." (Gn 1:12, 21, 24)

If a "kind" equates to a "species", then there are somewhere between 2 million and a trillion kinds. (livescience.com/54660-1-trillion-species-on-earth.html, DL 30 Oct 2017) How is a distinct species identified? For organisms that reproduce sexually, a species is the largest group in which two individuals can produce fertile offspring. For species that don't rely on sex to reproduce, definitions consist of a comparison of DNA, morphology, and ecological niche similarities, which often requires categorization judgment calls.

There are sixteen instances of the *after-kind* idiom in the KJV that have nothing to do with creation. For example, it was used in Noah's day; "Whatsoever creepeth upon the earth, *after their kinds*, went forth out of the ark." (Gn 8:19) And it was used in Moses' day; "Ye shall not eat... the vulture after his kind, and *every raven after his kind...*" (Dt 14:12-14) The latter is better translated, "any kind of raven". (*NIV* Dt 14:14)

So, the term "after their kind" cannot mean that every generation of a particular species must look exactly like the previous one. If it meant a species can't change, then the global flood apologists would have to stop saying that jaguars, lynxes, ocelots, leopards, and tabby housecats all come from one cat couple that Noah took into the ark.

The *after-kind* idiom is translated very differently in various versions of the disembarkation of animals from the ark: "one kind after another," "kind by kind," "as families," & "in groups of their own kind." So, the idea that *after-kind* was part of an anti-evolution law is a misinterpretation of the text. And, the fact that every child looks different from its parents is proof that God did not freeze genetic variability.

But the *kinds* mentioned in Genesis are much broader categories than species: grass, trees, fowls, fishes, whales, cattle, and beasts. There are about 12,000 grass species, about 80,000 tree species, about 10,000 bird species, about 34,000 fish species, 89 species of cetaceans, over 1,000 breeds of cattle, and "beasts" includes *all animals other than humans.* (Wordnik) So, there is a lot of room for evolution within each *kind*! There is no Biblical taboo against kinds evolving.

To argue that something can't be true because the Bible doesn't state it constitutes the fallacy *argumentum ad ignorantiam* — argument from ignorance. In other words, "Absence of evidence is not evidence of absence."

Paul also used broad terms to talk about *kinds* of animals; "All flesh is not the same flesh: but there is one *kind* of flesh of men, another flesh of beasts, another of fishes, and another of birds." (1 Cor 15:39) That would make four kinds of life with a wildebeest in the same *kind* as a duckbilled platypus. That leaves a lot of room for evolution within the beast *kind*.

Does Genesis support evolution?

An unbiased look at Genesis reveals that God commissioned the environment to bring forth life:

God does not say, "Let there be grass." Rather He says, "Let the earth bring forth grass."

God does not say, "Let there be marine life." Rather He says, "Let the waters bring forth abundantly,"

God does not say, "Let there be land creatures." Rather He says, "Let the earth bring forth the living creature."

By commanding the existing earth and the existing waters to do the job, God called upon natural processes in the earth and sea to produce life in a diversity of forms; fish, whales, fowls, etc. (Gn 1:11, 20, 24)

The same is true for humans. God did not say, "Let there be humans." Rather, He created humans "of the dust of the ground". (Gn 2:7) So the earth brought forth humans. God told Adam, "Out of [the ground] wast thou taken: for dust thou art, and unto dust shalt thou return." (Gn 3:19) How? Evolution!

Is it reasonable to think that God took dust, added some water to make clay, and lumped it together, patting it here and there as a potter would? If that were His method, why didn't He create a body for Jesus in the same way and avoid the awkwardness of Mary's mysterious pregnancy? Why couldn't Jesus have been created as an adult, ready to tackle His mission immediately instead of being delayed by a long childhood? That's not how God works. He used the same process to create Jesus as was used to create humans – He started with a predecessor, Mary, in this case.

The scriptural objections to evolution are based on misunderstandings. The most common goes something like this: "If evolution, then no Adam. If no Adam, then no Fall. If no Fall, no need for Atonement. If no Atonement, then no Gospel of Jesus Christ."

But this logic is faulty from the start. Adam and Eve could have come from earthly parents with an evolutionary ancestry. God could have raised them innocent, fed them Tree of Life fruit to keep them young and healthy, and placed them in the Garden of Eden as Adam and Eve. They fell. Jesus saved. And the gospel is true after all.

The Bible allows for the *physical* death requisite to evolution before Adam just as it allows for *spiritual* death necessitating a Savior before Christ. The atonement by Jesus is retroactive for both kinds of death. And the only death Adam and Eve brought about was their own!

Is forced evolution a sin?

Nature has been using various environments and ecosystems to force evolution on living organisms for more than four billion years. Humans have been at it for only about 10,000 years but have induced genetic changes much faster.

Archeology and genetic analysis prove that corn evolved from teosinte grass over thousands of years of selective breeding by humans. Carrots were developed from a tiny forked white root. The record-breaking 10-pound tomatoes grown nowadays are about a thousand times larger than the first tomatoes found in the fossil record. Similarly, domesticated animals are quite different than their wild ancestors. From the beginning of agriculture, gardeners and farmers have practiced the policy of "Save the best and eat the rest."

Scientists have now gone high-tech with genetic modifications. It's difficult to find a food item isn't genetically modified. As high as 75% of processed foods have at least some GMOs. In the USA, 94% of soybean acreage and 92% of corn acreage grow genetically modified plants.

What does God think of gene insertion? It seems to be part of His plan because He's doing the same thing in nature. One of the roles viruses play in evolution is to insert genes from one host organism into another. It is estimated that more than 100 genes in humans came from viruses and microbes. (sciencemag.org/news/2015/03/humans-may-harbor-more-100-genes-other-organisms, DL 24 Aug 2018)

Righteous Jacob used selective breeding to change his uncle's herd. The deal was that the striped, speckled, dark, and spotted goats would be Jacob's wages. Much to his uncle's bewilderment, Jacob's herd grew disproportionately. Unbeknownst to his uncle, Jacob was stanchioning his nanny goats and allowing only striped, speckled, dark, or spotted rams to breed. Thus, it was written that the Lord blessed Jacob with great herds. (Gn 30:25-31:16)

The whole point of sexual reproduction is to make the offspring *different* from their parents! The father's body randomly chooses from his genes to form each sperm cell and the mother's body randomly chooses from her genes to form each egg cell. The random selection of genes from both parents means the offspring will never be clones.

The only "kind" constraint God places on living things is that physical bodies must be suitable kinds for the spirits that ensoul them, though not necessarily perfect matches. Grass is in the image of grass spirits, trees are in the image of tree spirits, and so forth; "That which is temporal in the likeness of that which is spiritual… the spirit of the beast, and every other creature which God has created." (DC 77:2)

Does evolution violate a law?

Some antievolutionists claim that evolution violates the Second Law of Thermodynamics which says that the entropy of any *isolated* system always increases. (Wordnik) Entropy is defined as, *A measure of the disorder in a closed system.* The argument is that the highly complex life forms on Earth represent a great degree of order, a result that must have been caused by divine intervention since it violates natural law.

The more scientific definition of entropy is; *A quantitative measure of the amount of thermal energy not available to do work in a closed system.* (Wordnik)

The flaw in the antievolution argument is obvious. Earth is not a closed system. It receives a continual infusion of energy from the sun. On an average day, the earth gains millions of tons worth of energy from the sun and radiates almost the same back into space. So, Earth is a very open system. The sun's energy is what powered evolution and what sustains life.

Humans expend a lot of energy to build a skyscraper. But the construction project is not an isolated system. So, the local increase in organization as the building is completed in no way violates the 2nd Law of Thermodynamics.

The universe is a very large place. God uses natural mechanisms to transform vast amounts of energy into tiny bits of complexity here and there. "The theory of evolution has to include a notion that the dice have been loaded from the beginning in favor of more complex life forms." [*Reflections of a Scientist* pp61-62, Henry Eyring, 1983]

Is irreducible complexity valid?

Some religionists argue against evolution by claiming that certain species have *irreducible complexity*, that living organisms have biological systems or features that simply could not have evolved through small incremental changes to form and function. The argument in each example is that life simply can't get here (complex) from there (simplicity) via evolution, that the feature or function had to be created at once as part of the whole organism in order to serve its purpose. Among the examples used are the bombardier beetle's artillery, the human eye, and bacteria flagella.

However, almost as fast as such examples have been identified, scientists have identified the intermediate steps in nature — hence the probable improvement path toward full function, debunking the irreducible complexity argument in each case. Scientists are still working out the evolutionary pathways of certain phenotypic functions but just because complexity is not yet fully understood in a particular field of inquiry doesn't mean it's irreducible.

Where in scripture are T-rexes?

Paleontologists have named over 700 species of dinosaurs in the fossil record and are still finding new species. Does Genesis mention any of these? No. Biology and paleontology were not typical hobbies for the average OT prophet.

What about the beast mentioned in Job 40:15? "Behemoth, which I made as well as you; He eats grass like an ox." If he eats like an ox, it's likely a grazing animal, possibly a hippo. The Hebrew word (בְּהֵמוֹת) that the KJV translated as "behemoth" just means beast, animal, or cattle.

Could the *leviathan* mentioned in Job 41 mean dinosaurs? No. The text describes it as a sea monster that spits sparks, breathes fire, and boils the ocean. So, leviathan must be metaphorical.

There are no other plausible candidates in the Bible for dinosaurs. So, we must assume that they all died prior to Adam. God did not place their fossils in the earth just to fool scientists or to test the faith of believers. There is no vast global paleontological conspiracy. Scientists are simply reporting what they find.

How does evolution work?

Not even the staunchest fundamentalist anti-evolution zealot denies that viruses evolve. HIV is noted for being the virus that evolves the fastest. The human body detects the presence of the virus and begins producing antibodies tailored to attack it but, by the time the human body is able to deliver enough antibodies to kill the infection, the virus has evolved to look like a totally new intruder, and the cycle repeats itself.

This ability to evolve quickly is what keeps the host so busy fighting the HIV that it is unable to defend against some other infection and the patients die from other diseases.

Bacteria evolve quickly also. When dosed with insufficient antibiotics to wipe them out, new strains arise that are resistant to the drugs. Bacterial evolution has resulted in untreatable cases of tuberculosis, gonorrhea, and staph. Not only do bacteria mutate, they can also get large chunks of DNA from other types of bacteria to become new species and exploit a different niche or defend against new hazards.

The overwhelming evidence compels most creationists to accept microevolution but many still deny macroevolution despite the fact that macroevolution is essentially the same set of processes as microevolution and selective breeding, just too slow to be easily observed in the lab.

One of the most famous examples of adaptation over time is the giraffe. As competition for tree leaves increased, giraffes with longer necks survived at a higher rate than those with shorter necks. So, the long-necked ones more frequently passed their traits to the next generation. Scientists have traced the changes in neck lengths through the giraffe fossil record all the way back to their okapi-like ancestor.

Blue eyes are the result of a copy error in the OCA2 gene. The error bestowed no functional advantage on the first blue-eyed person and yet about 30% of Europeans have blue eyes nowadays. Why? The best guess is that blue-eyes increased an individual's chances of mating, thus spreading OCA2 across European populations.

Nature doesn't always select *for* a trait. Sometimes it selects *against* a trait. This can happen when a trait begins to be a disadvantage in a new or changing environment. If long legs begin to hamper a lizard population's ability to survive or reproduce, the percentage of shorter legged lizards will increase and eventually legs could disappear entirely. Legless lizard species are the result of selection against a trait. By far, most mutations are deletions.

Sometimes there is nothing in the environment to sustain a trait. For example, many species of spider, fish, salamander, and shrimp, after reproducing in darkness for millions of years, no longer have eyes. Kiwis have tiny nonfunctional invisible wings. Copy errors over the course of countless generations can result in genetic drift that corrupts the coding for a particular function rendering it useless or nonexistent.

Vestigial structures in humans suggest that unused traits are on their way out. Wisdom teeth are no longer needed for chewing roots and meat. Now that the human jaw is smaller, they are a liability. About 10% of humans don't have wisdom teeth at all. Most people still have a tail inside their body but some people have external tails, like Mauricio's in the movie *Shallow Hal*, replete with functional muscles for movement. The longest such tail recorded was 13 inches on a teen in East India. Most people still get goosebumps and spine chills despite the fact that human body hair is too sparse for such functions to provide warmth or intimidate enemies. Ear muscles allow most mammals to cock their ears toward noises but now only about 20% of people can even engage their ear muscles. About 6% of humans have an extra nipple somewhere in one of the mammary lines that extend from each armpit to the groin. Some people have up to eight nipples. Extra nipples are rarely functional in humans as they are almost evolved completely out of the genes.

Strands of DNA that are common among similar species are evidence of common ancestries. For example, deer share much DNA with moose and are therefore likely to have descended from common ancestors. The diversion of multiple species from a common ancestor is a result of changes in environmental or mating factors.

Over eons of evolution, almost all species of life accumulated a rich variety of genes, many of which lie dormant. So, populations that would have otherwise gone extinct for lack of an adaptive mutation have been able to activate latent adaptive genes or explode the number of individuals in the population that express the adaptive genes.

The following three mechanisms must be in play for evolution to work:

1. Variation – Recombinant DNA, genetic copying errors, infusions of genes by parasites, etc. If you are one of the 25% of adults in the world that can drink cow's milk without unpleasant side effects, you can thank a mutation in some farming community that occurred some 10,000 years ago.

2. Selection – Environmental conditions, competition, sexual advantage, etc. While visiting the Galapagos Islands, Darwin observed that natural selection in a dynamic environment played the same role as artificial selection by farmers in modifying heritable traits; "This preservation of favorable individual differences and variations, and the destruction of those which are injurious, I have called Natural Selection, or the Survival of the Fittest." (*Origin of Species*, Charles Darwin, 1859)

3. Inheritance – The most successful individuals pass their traits to their offspring. For example, people who inherited one copy of the sickle cell anemia gene had a survival advantage in areas of Africa hard-hit by malaria. So, even though people who inherit both copies of the gene usually died, the inheritance percentages favored carriers who were able to survive malaria and pass their genes on to their children.

Entirely new and unique species do not appear out of nowhere without antecedents. God doesn't just plop new animals on Earth whenever he has an itch to do so. Instead, He set the table with all of the requirements for life and commanded things to unfold and blossom according to the physical laws and principles.

Elder Parley P Pratt said that "spirit is eternal, uncreated... self-existing" and that when the earth and water were "filled with the quickening, or life-giving substance, which we call spirit, they produced living creatures". (*BYU Studies*, Charles R. Harrell, bk28, #2, p83) In Pratt's view, when God said, "Let the waters bring forth abundantly the moving creature that hath life," the spirit in the oceans responded and the evolution of marine life was the result. (Gn 1:20)

How does speciation occur?

Speciation has several modes:

♦ **Allopatric Speciation** occurs as a result of geographic isolation wherein a single group of organisms is split geographically (allo=other, patric=fatherland) due to changes in the environment such as a water barrier, mountain ridge, or some other type of habitat fragmentation such that mating between the separated groups stops or nearly stops. Over time, the characteristics of the two groups drift apart and the populations become two separate species.

♦ **Peripatric Speciation** is a special version of allopatric speciation wherein a subgroup carrying rare genes has very few individuals (peri=near). Over time, changes related to the rare genes eventually make the subgroup sufficiently different from its ancestral group that cross-mating stops.

♦ **Parapatric Speciation** occurs when individuals are more likely to mate with their close neighbors rather than randomly, resulting in restricted gene flows (para=beside). Over time, the geographic subgroup becomes genetically distinct and is considered a species apart from the original population.

♦ **Sympatric Speciation** happens to members in the same group living in the same place when some individuals change their behavior to exploit a different niche, such as eating a different type of fruit than before or when some individuals experience internal changes that significantly inhibit them from mating with peers that have not changed (sym=same, patric=fatherland). Over time, the subgroup that can exploit the niche becomes a distinct species from the group that cannot.

The *rate of evolution* can be estimated mathematically as (mutation rate) x (chance of being beneficial) x (chance of survival). It took billions of years for life to get where it is. But God is patient. Using evolution to create diverse species seems more ingenious than creating each of trillions of species from scratch! It allows God to appear quite hands-off.

What causes new traits in a species?

There are numerous mechanisms for genetic modification:

♦ **Epigenetic** changes occur when genes get activated or deactivated. Sometimes this occurs in response to environmental factors. Epigenetic changes can be heritable.

♦ **Alternative splicing** is the capability of a gene to code for more than one protein. It plays a major role in gene expression.

♦ **Recombination** is where DNA strands are exchanged to produce new nucleotide sequences as in *meiosis*. All sexual reproduction involves recombination. But even organisms as simple as yeast, when conditions are difficult, use recombination to generate new and genetically unique reproductive spores.

♦ **Hybridization** occurs naturally in the wild and occasionally produces offspring that become new species. In Arizona, a species of bird nesting in an invasive species of tamarisk tree was thought to be endangered by government conservationists and so efforts to eradicate the trees were stopped. Scientists later learned that the species was actually a hybrid of two other plentiful species. Conservationists now consider it a species in its own right, even though it could be reproduced by crossbreeding.

♦ **Migration** introduces a new species into an area to breed there, often resulting in a new indigenous species.

♦ **Lateral Gene Transfer** is where genetic material moves from one species to another. This occurs most commonly between species of microbes but transfers also occur between microbes and complex animals, allowing for the acquisition of new traits sufficiently different to constitute a new species.

♦ **Endosymbiotic Gene Transfer** is where cell organelles pass genetic material to the nucleus or vice versa.

♦ **Mutations** happen due to genetic copy errors during reproduction. For example, if a member of a white species of moth has a mutation that expresses as brown in a brown forest, it may be more likely to escape being eaten by birds and live to reproduce. Over time, a higher percentage of white moths will perish than of brown moths and eventually the entire population will have changed color.

♦ **Gene flow** is the movement of genes from one population to another population. For example, bees carry pollen from one population of flowers to another population.

♦ **Genetic Drift** happens when the natural selection mechanism that reinforces a trait is absent, allowing copy errors to eventually obliterate the trait. The most famous examples are the many species of cave fish that, after long separation from an environment where sight was needed to survive, have nothing left of their eyes except a slight anomaly in the scale pattern where the eyes would normally be. Other examples are the blindness in certain species of bats and moles. And, in the case of Ostriches, they survived so well by foraging on the ground that the traits for flight faded. If they don't need to use it, they eventually lose it.

♦ **Bottlenecks** happen when environmental events such as earthquakes, floods, fires, disease, droughts, volcanism, or asteroid impacts reduce the size and thus the gene pool of a population.

♦ **Founder effects** limit genetic variety when a very small number of individuals start a new population somewhere that consequently differs from the original population.

♦ **Adaptability** is a result of wide DNA variability or genetic richness that allows members of a population to express traits that facilitate survival as conditions change. The most germane example for humans began about 7 MYA in East Africa's rift zone where, as the tectonic plates pulled forested lands rich with arboreal primates apart, climate change gradually reduced some habitat types, forcing primates to travel on the ground. As the distance between trees increased, it became advantageous for hominids to cross the gaps between trees with their heads high to better scan for predators. Those primates that were best at walking and running on their hind legs were able to survive long enough to pass their genes on to the next generation. Most famously, 3.6 MYA, two early humans walked through wet volcanic ash in Laetoli, Tanzania, leaving footprints that could be precisely dated.

How did interdependency evolve?

If plants and animals had been created in six periods of a thousand or millions of years each, all the plants created in Epoch Three that rely on animals for survival would have died because the animals they needed would not have been created until Epoch Five. For example, all pitcher plants and bladderworts would have starved long before the arrival of the insects and other tiny animals they need for food.

Myriad plants would have gone extinct waiting until Epoch Five for the birds, bats, and bees to pollinate them. There are thousands of plant species in various sorts of symbiotic relationships with animals, proving that the sequence of creative periods in Genesis Chapter One cannot possibly represent the physical creation. But it's no problem if it's understood to be a spiritual planning and designing sequence.

Plants and animals that are in symbiotic relationships evolved together over many millions of years. This is coevolution. Individuals with traits that bestow cross-species survival advantages live to pass those traits to subsequent generations. It's a win-win and sometimes a win-win-win for symbionts.

Was it all Darwin?

Although the name of Darwin is synonymous with evolution for most people, he was certainly not the first to study it.

In the 1700's, French naturalist Count de Buffon published a 44-volume natural history in which he suggested that the environment alters animals.

In 1800, Jean-Baptiste Lamarck gave a lecture at the French National Museum of Natural History in which he proposed his ideas about evolution. He was the first to publish a fully formulated theory of transmutation in species.

In 1829, English author Patrick Matthew clearly explained the principle of natural selection in the appendix of his book *Naval Timber and Arboriculture*. He even had "Discoverer of Principle of Natural Selection" printed on his cards. (*Evolution and Human Nature*, Seaview/Putnam, NY, p31)

So, why don't we refer to evolutionary processes as Buffonian or Lamarckian? Because, when Charles Darwin published *On the Origin of Species* on the 24th of November 1859, the book became an international sensation and opened the door for other scientists to discuss their ideas and make contributions to the theory. Since then, thousands of modern scientists have discovered an overwhelming body of evidence supporting evolution. And, over time, the concepts of evolution were included in schools and popularized via mass communications.

Can evolutionists be Christians?

About 66% of all adults in the US believe in evolution while 82% believe in God. (Pew Research Center) So, there's a lot of people who believe in both. But belief in God is losing ground while acceptance of evolution is increasing.

Only about 33% of scientists believe in God and, since almost all scientists believe in evolution (97%), there is hope that young people of faith can choose a career in science without compromising their beliefs.

Darwin himself went out of his way to dispel the notion that evolution eliminated God; "It seems to me absurd to doubt that a man may be an ardent theist and an evolutionist." (theguardian.com/commentisfree/belief/2009/sep/17/darwin-evolution-religion, DL 1 Aug 2021)

Is evolution still happening?

Which is easier to prove; some frogs are pink or no frogs are pink? Finding a single pink frog proves the first but finding a million non-pink frogs doesn't prove the second. Hence the saying, "You can't prove a negative." Well, you can, if you can prove evidence of absence. For example, you could prove there are no pink frogs in your bathtub.

But in large complicated systems, such as planet Earth, it's much easier for scientists to prove that at least some evolution occurs. Only one case is needed and there are many thousands of cases. For example, in the age of pest extermination, there are numerous cases of bugs and rodents developing heritable immunity to poisons.

Anti-evolutionists, on the other hand, would need to have observed the entire lives of every living thing, generation after generation since the world began, to prove that no evolution occurs.

What about changes in form? In medieval times, during the sea battle of Dan-no-ura, the Tairi Samurai were slaughtered by rival warriors and cast into the sea. The surviving Taira women started a legend that their warriors' spirits walked the ocean floor. So, when fishermen caught crabs that had any markings on their backs that resembled a samurai face, they tossed them back into the water. Over the centuries, the backs of the Heike crab population in the area acquired faces that looked, not just like humans, not just like Japanese, but like Samurai. (*Cosmos*, Carl Sagan, 2013, p20)

The Heike crab story is an example of anthropogenic evolution that was *unintentional*. Thousands of years of selective breeding have caused dramatic evolution in domestic plants and animals that was *intended*. But nature doesn't need humans for evolution to happen. And, since nature is still with us, so is natural evolution.

Are humans still evolving?

In ancient times, humans depended on their wits for survival. Individuals who were better at making tools, crafting weapons, outsmarting wild animals, protecting the family, managing domesticated animals, raising plentiful crops, or evading enemies were more likely to have thriving descendants than less clever people. Nowadays, society helps the disadvantaged. So, it should be expected, with all the social safety nets that have been put in place, that there has been a recent downward trend in performance on measurements of intelligence.

And sure enough, global mean Intelligence Quotient (IQ) scores have fallen from 91.64 to 87.20 in the last half of the 20th century and are expected to fall another three points by 2050. (fourmilab.ch/documents/IQ/1950-2050, DL 23 Jun 2018) Some studies show IQ scores in Europe falling by 14 points on average since the Victorian Era.

Sexual selection is also at work downgrading humans. It is becoming more common that intelligent people mate with other intelligent people and then choose to have fewer children or none. So, the genes of intelligent people are not being passed to the next generation as prolifically as was the case in the past. Inverse natural selection is the basis for the plot in the movie *Idiocracy* in which an average guy wakes up from a long-term suspended animation experiment to find that he is the cleverest person on the planet; "Deficiencies in intellect haven't made it impossible for reproduction." (*Dumb and Dumber: Study Says Humans Are Slowly Losing Their Smarts*, USNews, 13 Nov 2012)

There is hope that as information technology continues to blossom exponentially, a boost will be seen in the Flynn effect, which is that as more people achieve greater prosperity and education, IQ scores rise. However, due to the recent easing of evolutionary pressures on intellect, innate human intelligence is tapering down even as growing information and technology boosts intellectual achievements.

Are there any transitional forms?

The unscientific pejorative term "missing link" was used in times past to highlight the evolutionary gap in the fossil record between humans and their ape-like ancestors. In 1912, Charles Dawson (not Darwin) claimed to have discovered the "missing link" in a layer of Pleistocene gravel. The Geological Society named the specimen *Eoanthropus dawsoni* but the news media called it *Piltdown Man* after a gravel pit near the site of the find. The first paleontologist to see it estimated it at 500,000 years old. But not everyone was convinced and, in 1953, the specimen was proven conclusively to be a combination of a small-brained man's skull with an altered jawbone and some orangutan teeth.

Not hoaxes are the hundreds of fossils of Homo neanderthalensis and Denisova hominins from 300,000 to 500,000 years ago whose DNA has been found in trace amounts in modern humans. Homo erectus is thought to be another ancestor of humans. Whereas these species are considered intermediate forms between ancestors and descendants, their bones would qualify as *transitional fossils*.

In the 1860s, Thomas Huxley suggested that birds were descendants of dinosaurs. So, from then until the 1990s, people wondered why lots of feathered flying dinosaur fossils hadn't been found. Several fossils of *Archaeopteryx* were found in the late 19th century, but feathers had not survived well on any of the early discoveries and the stark differences from birds left the general public skeptical that it was an avian ancestor. *Archaeopteryx* had a long tail full of vertebrae and a mouth full of teeth. But Huxley insisted that *Archaeopteryx* was a wonderful example of a transitional form between dinosaurs and birds, but that did little to reassure lay people.

As time went by, paleontologists continued to find more species of birdlike dinosaurs. *Protarchaeopteryx* looked a lot like Archaeopteryx except that its feathers were somewhat less conducive to flight. *Sinosauropteryx* had feathery filaments that may have served as insulation. *Caudipteryx* had feathers, a beak, and claws. *Confuciusornis* had large curved claws on its wrists but a toothless beak. *Eoalulavis* is the earliest bird found to have an alula, a tuft of feathers attached to the thumb that greatly enhances maneuverability. *Fukuipteryx prima* looks like a chicken with an oversized head. *Microraptor* had four feathered wings. Over 50 ornithic dinosaurs have been found as of late 2017. (*Jurassic Flight School*, John Pickrell, Cosmos Magazine, Issue 74)

Many feathered dinosaur species, like *Velociraptor mongoliensis*, were fairly flightless but the plumage was useful for regulating body temperature, attracting mates, and incubating eggs. Flight also turned out to be a very useful way to escape earthbound predators.

More birdlike species are being found. One of the most famous dinosaurs was recently proven to have had feathers, the vicious velociraptor of *Jurassic Park* movie fame. (sciencedaily.com/releases/2007/09/070920145402.htm, DL 23 Jun 2018) Dinosaur feathers have even been discovered in amber in a number of cases, one of which included enough bones for scientists to identify the species. Perhaps the weirdest birdlike dinosaur walked like an ostrich, swam like a penguin, had a bill like a duck, had teeth like a crocodile, had a neck like a swan, and had killer claws on its wingtips. (foxnews.com/science/2017/12/06/duck-dinosaur-hybrid-baffles-scientists-with-mixed-up-body.html, DL 6 Dec 2017)

A classic experiment with chicken embryos shows that chickens can still express the genes for teeth, a trait not expressed in birds since dinosaur days. Latent DNA has a greater effect on evolution than do mutations. (*Transposons*, Discover Magazine, Ayala Ochert, Dec 1999, p59) Since these so-called junk genes are not required for functionality, how did they find their way into the genetic code unless acquired during the evolutionary past?

It suffices to say that anti-evolutionists' cries for transitional forms from dinosaur to bird are heard less and less often as time goes by. Hundreds of other types of fossils have also been found that fill transitional form requirements. But, the problem with holding up a new fossil as transitional is that antievolutionists will always point to a gap between the new fossil and its predecessor or its successor no matter how insignificant it may be. Just as there are an infinite number of gaps between 0.9 and 1.0, there will always be gaps in the biological tree of life to which evolution deniers can point. Paleontologists have no control over how well a particular fossil find "fills a gap".

In 1991, the fossil of a whale with external vestigial legs was discovered in Egypt at Tunis. They called it *Basilosaurus*. Antievolutionists claimed that the tiny legs were for grasping during mating, thereby serving a function and therefore not transitional. But there is no rule that transitional features can't be functional. In this case however, since none of the grasper-less whale species have had trouble mating successfully, it would be difficult to prove that the vestigial legs provided a useful function. Modern whales still have internal vestigial hind leg bones. So, today's whales are, in themselves, excellent examples of transitional forms.

A 95-million-year-old snake fossil was found to have two tiny *external* leg bones attached to its pelvis. Many snake species alive today have *internal* vestigial leg bones.

Tyrannosaurus rex is famous for its ridiculously small forearms. But, Carnotaurus sastrei holds the record for the proportionately tiniest forelimbs among the predatory giants at about a third shorter than the length of its head. Like T-rex, Carnotaurus sastrei had evolved a head and jaw so large and powerful for the prey of the day that the forelimbs went unused for tens of millions of years, resulting in vestigialization. Its wrist bones are gone, its fingers are fused, and its elbows are frozen. Creationists are hard-pressed to even speculate what the function of such vestigial arms might be.

The Giant Legless Lizard from Russia blinks and has earholes but looks like a snake. Snakes can't blink and don't have ear holes. The internal vestigial leg bones at the points where lizard hips should be are palpable, even though there is no sign of them on the outside. What function could internal vestigial leg bones serve on a snake-like creature? Technically, humans are transitional forms — Wisdom teeth, appendixes, tailbones, goosebump muscles and ear-moving muscles are left over from millennia gone by. Only about 10 to 20 percent of adult humans can wiggle their ears voluntarily.

Life is rich with transitional forms. Since all complex species have vestigial structures, the proper premise should be that all life forms are transitional because all are continuing to evolve.

What's sex got to do with it?

Sexual reproduction is an ingenious way to ensure genetic diversity in populations. There are more potential DNA permutations from a man and a woman for a gamete than the human mind can imagine. The calculation produces a number so large it dwarfs the estimated number of atoms in the observable universe.

Sexual preferences can drive evolution. For example, ancient Irish elk bulls developed 12-foot antler spreads. It is postulated that the bulls with the largest antlers more frequently won access to the females and were able to pass on their genes for hugeness.

Sexual selection is a greater force in polygamous populations than in monogamous ones. For the most part, only the alpha gorilla male has access to the females in his band. The other males are cut from the group when they reach breeding age and must wander alone or with other bachelors until they can attract females of their own or challenge a silverback for an existing harem. Thus, the genes of males with inferior strength and intelligence are propagated only when they can sneak a liaison with a female before the silverback chases them off. If weaker males were *never* able to reproduce, there would be an even greater sexual dimorphism among gorillas.

Where is God's hand in evolution?

Did life on Earth arise by chance or by divine will? The answer is both! Chance is merely a word that reflects mankind's inability to determine outcomes with unknowable causes, usually the future. But if God's knowledge of the future is absolute to the minutest detail, he sees no chance.

Computer programmers create massive fantasy universes inside their video games. The best programmers are the ones who craft the most popular video games, such as *World of Warcraft, Skyrim,* and *Halo.* But a programmer who could create a program that creates the best games would immediately be hailed the best game programmer on Earth. Likewise, God coded the programs that produce the grandest of all games – reality! Therefore, He is the Master Programmer for the universe.

God's programs have been running in the universe for at least 13.8 billion years. The evolution of galaxies, star systems, and the evolution of diverse life play out as per the rules designed by God on a game platform built by Him. But instead of binary bits, He codes with intelligence bits.

Francis Collins, the geneticist who led the National Human Genome Project to fame, said, "If God, who is all-powerful and who is not limited by space and time, chose to use the mechanism of evolution to create you and me, who are we to say that wasn't an absolutely elegant plan?" (thetheisticevolutionist.blogspot.com, DL 23 Jun 2018)

Theistic evolution makes a lot of sense. Fighting a growing body of evidence for evolution does not. Science, the pursuit of more knowledge about the creation, is a very good way to praise the Creator!

15: Creation of Man

Male and Female only?

God created mankind "male and female". (Gn 1:27) Does that mean everyone has either two X chromosomes or an XY pair? Hardly! There are people born with odd chromosome configurations, such as XXYY, XXXXY, XYY, etc. There are even people with nullisomes, neither X nor Y, for example in Turner's syndrome, the chromosome configuration is designated as X0 or just X.

The Y chromosome primarily carries genes related to maleness and lacks too many of the genes necessary for general development. Therefore no combination of Y and 0 is viable.

Thus, if a Y chromosome is present, the person is technically male, even though there are cases where, due to mutations, testes do not develop, causing a lack of testosterone that leaves the individual with a vagina, a uterus, and female external genitalia, but no ovaries. Individuals with Swyer syndrome are raised as females, identify as female, and are typically not diagnosed until adolescence (14 to 23 years) when puberty misses the normal milestones. With hormone replacement therapy, a patient can develop secondary sexual characteristics.

The term "intersex" refers to a range of variations in sex characteristics from birth, such as chromosomes, genitals, reproductive organs, hormones, or secondary sex traits that do not fit the typical male or female pattern. Some intersex people can become biological fathers or get pregnant, depending on their specific anatomy and reproductive capabilities but assisted reproductive technologies (like IVF, surrogacy, or sperm/egg donation) are often needed to achieve parenthood.

Several cases of dizygotic chimeras (people with two complete sets of DNA, one XX and the other XY) have been documented. Most do not discover their condition until their

DNA is tested for paternity, maternity, or donor compatibility checks. They are truly both "male and female".

Well, wasn't that interesting? But hold on — God created male and female in Genesis chapter One! Isn't that still the spiritual creation? Yes. God assigned "male and female" genders without mistakes when creating the spirits of men. "Gender is an essential characteristic of individual premortal, mortal, and eternal identity and purpose." (Family Proclamation)

Are we born just once?

Ancient Jewish mystics, Hindus, Buddhists, Jainists, Sikhs, and others believed that reincarnation is the process by which *every* soul gets a chance to cycle forward toward enlightenment. The principle is that people who behave and learn well in a prior incarnation can return to Earth as more advanced individuals. A spirit improved by a tour on Earth as a tarantula could return as a reptile, fish, bird, or mammal. No improvement might result in a return to the same species or to some other comparable species. Spirits that backslide might return as a lesser life form, such as slug or a dung beetle. The ultimate goal is for one's spirit to reach a final state of enlightenment, oneness with God, and freedom from reincarnation, temporal constraints and unpleasantness.

Reincarnation also exacts justice for otherwise unpunished offenses against others. For example, people who kill unborn babies might be returned to a baby body that is to be aborted.

Bible passages cited by reincarnation enthusiasts that hint at a prior life are the same ones cited by others as evidence that everyone existed as spirits prior to mortality. Some reincarnation enthusiasts cite the teachings of the prominent early Christian Church Father, Origen of Adamantius (184-253AD) but, upon inspection, those citations turn out to be writings by somebody else misattributed to him or instances where he was quoting someone else.

So, there is no evidence that any form of Christianity ever espoused reincarnation. It's incompatible with scripture; "It is appointed unto men *once* to die." (Hb 9:27) Joseph Smith called it "the transmigration of souls" and soundly rejected it. (TPJS, pp104-105) But there is one possible exception. A spirit who enters an unborn body that is then miscarried or aborted might get a chance to enter another fetal body.

NDE'rs and people who believe they have memories of past lives might be recalling scenes from watching earthlings or experiences from their own preparation for mortality or serving in the premortal war against evil or their participation in creation or they may be remembering glimpses of their future given to them before they were born.

Did you choose your parents?

There are cases where children remember choosing their parents. "A Dutch boy called Thom was four years old when he told his mother: 'Mum, when I hadn't been born yet, I wished myself [good] parents and I have got them.'" (prebirthmemories.com/Thom.htm, DL 24 May 2018)

There are cases in which a baby's spirit communicated with the expectant mother; "When I was pregnant with my daughter, she communicated with me in dreams… She told me what kind of person she was and what she needed from a mother." (cosmiccradle.com/category/choosing-parents, DL 1 Jun 2018)

What if vast numbers of children want the same wonderful parents? The angels who work with spirits about to descend to Earth mediate decisions acceptable to all parties.

What about all the undesirable fetuses – diseased, deformed, or conceived in terrible circumstances? There are more spirits wanting any kind of situation in mortality than baby bodies to supply them. And, although it's hard to imagine anyone wanting to be born disabled, or sick, or to a drug-addicted mother, or to a destitute family of primitive circumstances in a remote jungle, such situations build character. The more difficult life is, the more opportunity for spiritual progress. Spirits want that, some more than others.

Did you choose your situation?

Did people who are born into abject poverty, or oppressive circumstances or with birth defects, or to abusive parents do something in their premortal lives that landed them there? Did people who are born into wonderful circumstances do something to merit those blessings?

President Hinckley mused, "I do not know what we did in the preexistence to merit the wonderful blessings we enjoy." (*CR*, Oct 2001) His listeners were born into a world more wealthy, less hungry, more free, more educated, and more exposed to the gospel truth than at any time in history. If he considered they have had something to do with timing of their birth or with the place they were born, then it's probable they did.

Some may have chosen difficult circumstances for spiritual growth. Others may have chosen blessed circumstances for the opportunity to serve. We'll know for sure when spirit memories are restored to us in the afterlife.

Why is man like the animals?

Surely God in His omnipotence, omniscience and creativity could have uniquely designed each species from scratch, perhaps some being based on silicon, perhaps with traits coded in a triple helix. Instead, He created all life based on the carbon molecule and double helix DNA.

All life forms are genetically and morphologically similar to millions of other species. So similar is DNA across species that human genes that control the differentiation of embryonic cells into head, thorax, and limbs can be transplanted into fertilized fruit fly eggs and produce perfectly normal flies!

Humans have historically fancied themselves so far above the animals as to defy comparison, claiming certain abilities unique to humans such as tool usage, art, reason, language, love, etc. But, studies of animals have recently proven those assertions wrong.

Chimpanzees strip leaves off twigs for "fishing poles" to draw termites from mounds. They select appropriate rocks to crush nuts. Crows drop nuts in the path of cars for cracking. New Caledonian crows not only use tiny sticks to fish bugs out of holes in tree bark but also carve them into hooks with their beaks to double the harvest. Such skills are shared with other individuals and imitated by offspring, just as in humans.

Language was once thought to distinguish humans from animals but apes can learn sign language and monkeys, dolphins, birds, and other animals have distinct vocalizations with specific meanings.

The ability to plan and reason was once thought to set humans apart from animals. Chimps can plan an attack, pull off a deception, and start fads. Monkeys can reason through analogy. Some African Grey parrots can solve various math, comparison, and categorization problems as well as three-year-old children. Marine animals such as dolphins and octopi use learning and logic to solve problems.

The use of fire was once thought to distinguish humans from animals. Not so! Certain birds of prey in Australia have been observed carrying burning twigs to set new fires to flush out prey.

Animals and humans have many psychological traits in common. Dogs dream and elephants mourn the deaths of dear ones. Pigeons, like human gamblers, will choose to push a button that gives them a big, rare payout rather than one that offers a small reward every time.

The list of human-animal commonalities is long. The explanation is organic evolution. Humans have much evolutionary history in common with other animal species.

Most religious leaders had a negative reaction when Darwin's Origin of Species was published in 1859. They were children of their time. But there were some exceptions. For example, modern Apostle and scientist John Widstoe wrote, "It is remarkable that Joseph Smith taught the law of evolution as an eternal truth, twenty or more years before Darwin published his views." (*Joseph Smith as Scientist: A Contribution to Mormon Philosophy*, 1908, p154)

Renowned chemist Henry Eyring (1901-1981AD) wrote, "Some people object to the slightest hint of being related to the rest of the animal kingdom, particularly the hairy apes…. I've kind of enjoyed what little I've seen of them…. Animals seem pretty wonderful to me. I'd be content to discover that I share a common heritage with them, so long as God is at the controls." (*Reflections of a Scientist*, 1983, p60)

Are you an ape?

Italian philosopher Lucilio Vanini taught that humans and apes had common ancestry. His ideas were considered atheistic and the French eventually arrested him in 1619, cut out his tongue, strangled him to death, and burned his body. (religion.fandom.com/wiki/Lucilio_Vanini, DL 11 May 2022)

Theistic evolutionism proposes that "God formed man of the dust of the ground" through gradual increases in the complexity of life. (Gn 2:7) Saying "Ape thou art" credits a person with more brains than "Dust thou art". (Gn 3:19)

Darwin was not a demon. He prayed of his own volition as a child and studied theology for three years at Christ's College, Cambridge. He had every intention of becoming an Anglican minister. But, as his scientific studies progressed, he was unable to reconcile the prevailing Genesis interpretations with his observations of Nature. He tried not to offend anyone as his views evolved. But finally, in 1880, he replied to a direct question from Francis McDermott by letter, "I do not believe in the Bible as divine revelation, & therefore not in Jesus Christ as the Son of God." (christiananswers.net/q-aig/darwin.html, DL 5 Jul 2020) Darwin died two years later. The letter surfaced 100 years later. Perhaps Darwin would have remained true to Jesus if he'd had the correct interpretation of Genesis as laid out in this book.

We now live in an age where one can send a saliva sample to any one of a number of laboratories and, for a small fee, obtain results that include ethnicity percentages for British (Brit.), Irish, Western European, Native American, African, Asian, etc. For such DNA tests to be accurate, a databank of sample DNA must be loaded with samples from all over the world for legitimate comparisons.

What about our paleoanthropic ancestors who lived tens of thousands of years ago? Unfortunately, most fossils do not contain readable DNA. However, scientists have successfully extracted some from Neanderthal and Denisovans who went extinct 30 to 40 thousand years ago. The results say that Europeans have up to 4% of their DNA from Neanderthal ancestors. And, Melanesians, Papuans and aboriginal Australians have up to 5% of their DNA from Denisovans.

Evolution may sound crazy at first but, to those who are most familiar with it, there's no more brilliant scheme for creating the Earth's diversity of life. It's time for creationists to stop telling God what he didn't do. The overwhelming amount of continually growing evidence makes it a much more likely process than creation from nothing, creation directly from dust, or creation from a spare rib.

To answer the question of this section — yes, if you are reading this, science classifies you as an ape in the superfamily Hominoidea (gorillas, chimpanzees, bonobos, orangutans, and **humans**). Bonobo chimps have approximately 99% of their DNA in common with humans. (*Tiny Genetic Differences between Humans and Other Primates Pervade the Genome*, by Kate Wong, Scientific American, 1 Sep 2014) The simplest explanation is that humans and chimpanzees share common ancestors who lived six to eight MYA.

But don't be offended. As friend and evolutionary biologist Dr. David Ellis remarked that it bothers him more that he and all humans share 50% of their DNA with bananas. That's true. (popsci.com/humans-genetically-linked-to-bananas, DL 29 Oct 2025) But it is primarily *genes* that drive form, feature, and function, and they make up only about 2%

of our DNA — Humans and bananas have only 17-25% of their *genes* in common with bananas. (lightofevolution.org/en/banana-split, DL 29 Oct 2025) But hey, they may be our kith but not our kin — genetic divergence occurred 1.6 billion years ago.

Which churches embrace evolution?

Popes and cardinals are generally favorable toward evolution, although the Catholic Church has not published an official position on it, nor on pre-Adamites, nor on the age of the earth, leaving these questions for the individual members to decide.

The Southern Baptist Convention issued a resolution in 1982 rejecting evolutionary theory and has not changed its position since. Jehovah's Witnesses, Seventh-day Adventists, and Evangelical Lutheran's are also emphatically against it. Other Lutheran churches are of mixed opinions.

Mainstream Episcopalian, Presbyterian, and Methodist churches have no official position and their leaders are generally reticent to state opinions publicly. The same is true for most restorationist sects such as The Church of JESUS CHRIST of Latter-Day Saints; "Whether the mortal bodies of man evolved in natural processes to present perfection, through the direction and power of God... [or] whether they [Adam & Eve] were born here in mortality, as other mortals have been, are questions not fully answered in the revealed word of God." (*Improvement Era*, CJCLDS, Apr 1910)

When B. H. Roberts of the LDS Seventy's Quorum included some quasi-evolutionary stances in his writings, President Joseph Fielding Smith publicly opposed them in favor of a young-earth view. But, the Quorum of the Twelve Apostles conducted a debate on the issue and "declared a draw: Neither the existence nor the nonexistence of pre-Adamites would constitute church doctrine". (*The Creationists*, Ronald Numbers, 1992, p312)

However, apostle and geologist James Talmage later told an audience at the SLC tabernacle, "The whole series of chalk deposits and many of our deep-sea limestones contain the skeletal remains of animals. These lived and died, age after age, while the earth was yet unfit for human habitation." (*The Earth and Man*, Talmage, 9 Aug 1931, p3)

Poll questions are usually worded in ways that respondents perceive as a choice between God and science. So, none will be cited here. But, it is clear from those results that more than half of laypersons in the Catholic, Orthodox, and mainline Protestant religions accept the science of evolution irrespective of the official positions of their churches, and the percentages are growing.

Public schools teach evolution as fact but creationism cannot be taught. So, in a few generations, believers will either be willing to consider theistic evolution as a means to resolve scripture-science conflicts or they will abandon religion altogether. A lot of sermons will need to be changed!

Can human evolution be proven?

Solid evidence of human presence has come from more than sixty countries, findings that are tens of thousands of years older than Genesis. For example, the remains of more than 400 Homo neanderthalensis skeletons have been unearthed, some of which have yielded DNA found in modern humans.

Other specimens of bipedal hominids include: Sahelanthropus tchadensis, Orrorin tugenensis, Ardipithecus ramidus, Australopithecus anamensis, Kenyanthropus platyops, Australopithecus africanus, Australopithecus garhi, Australopithecus sediba, Australopithecus aethiopicus, Australopithecus robustus, Australopithecus boisei, Homo habilis, Homo georgicus, Homo erectus, Homo ergaster, Homo gautengensis, Homo antecessor, Homo heidelbergensis, Homo naledi, Homo neanderthalensis, and Denisova hominins.

There is also Australopithecus afarensis, the famous 3.2-million-year-old "Lucy." She stood only 3 feet 7 inches tall.

And we must add Homo floresiensis, which at the same height as Lucy, was dubbed the "Hobbit" by science fans.

Fossil evidence strongly suggests that the ancestral lines of Homo sapiens and related primates diverge about 4 MYA. The record from the earth illustrates an obvious progression from primitive humanoids to modern humans over time.

How long has man been modern?

The oldest fossils of modern humans date back to about 300 TYA. Scientists also look at morphology, technology, culture, and skills to gauge modernity:

- ◆ 4.4 MYA – Walked on two legs.

- ◆ 2.6 MYA – Butchered large animals with sharp tools.

- ◆ 1 MYA – Controlled and used fire.

- ◆ 500 TYA – Killed large animals with wooden spears.

- ◆ 170 TYA – Made clothes.

- ◆ 71 TYA – Painted with red ochre and used bows and arrows.

- ◆ 50 TYA – Developed civilizations and culture.

- ◆ 20 TYA – Manufactured pottery.

- ◆ 12 TYA – Domesticated plants and animals for food.

- ◆ 8 TYA – Wrote characters to represent ideas.

- ◆ 6 TYA – Manufactured pottery wheels.

- ◆ 2.5 TYA – Manufactured indoor plumbing.

- ◆ 1.1 TYA – Manufactured gunpowder.

- ◆ 200 YA – Invented compressor-condenser refrigeration.

- ◆ 150 YA – Invented modern electric generator.

- ◆ 100 YA – Popularized the use of power in homes and businesses.

Are you a mutant?

Evolution comes in handy for answering questions from curious children or others:

Question: Why do humans have armpit and groin hair?

Answer: Hair retains pheromones from sweat that signal genetic and health information to potential mates, improving the likelihood of offspring. The hair and sweat in these regions also reduces chafing when one moves one's arms and legs a lot. Such activity was critical to primitive survival.

Question: Why are people different colors?

Answer: Dark skin in early humans reduced damage from the sun in Africa, where modern humans first appeared. Lighter-skins appeared in some groups that migrated to cooler climates where more clothes were worn to keep warm and less sunlight made the ability to synthesize vitamin D more important for survival. "Highly pigmented people will need to stay in the sun around 6 times longer than light people in order to synthesize the same amount of vitamin D." (livescience.com/7863-people-white.html, DL 11 May 2022)

Question: Why do people yawn?

Answer: They inherited the tendency from their animal ancestors who have been yawning for millions of years. Unborn babies have been caught on video yawning at 30 weeks. Horses, dogs, mice, kittens, foxes, hedgehogs, walruses, elephants, and birds are just a few of the many species that yawn. Darwin noted, "Seeing a dog & horse & man yawn, makes me feel how much all animals are built on one structure." (*1838 notebook*)

Question: Why do humans have long head hair?

Answer: Head hair protected early humans from heat stroke and cancer, extending their reproductive years. Hair also promotes blood clotting after injury. And, hair may have played a role in displaying for the opposite sex.

Does DNA prove Adam and Eve?

A huge sensation swept across the educated world when a group of scientists in the late 1970's announced the discovery of "mitochondrial Eve," a female ancestor common to all humans on earth. But theologians could not jump on the story as proof of Adam's wife without agreeing with the scientists that Eve lived at least 100 TYA – far too early for Bible chronology.

In a nutshell, mitochondrial DNA (MDNA) is passed from mother to child, normally without changes. But, every once in a long while, a mutation happens. So, over a long period of time, mitochondrial DNA diversified across human populations. The greater the MDNA difference between any two individuals, the more generations have passed since they shared a common grandmother. Using an estimated rate of mutation and a survey of MDNA differences across populations, scientists determined that Earth's human mitochondrial DNA converges on a single female ancestor somewhere between 100 and 230 TYA, the most recent common mother of all living.

But what about the most recent common father of all living? Well, since the Y chromosome passes from father to son, normally without changes, but suffers from mutations every once in a while, scientists did the same type of worldwide survey of men everywhere, applied a similar methodology as for MDNA and determined the "Y-chromosomal Adam" lived around 200 to 300 TYA. So, the likelihood that the Y-chromosomal Adam and mitochondrial Eve were even on the earth at the same time is infinitesimally small. The DNA lines leading to the last universal mother and last universal father are totally independent of each other.

When scriptures say Eve was "the mother of all living", we must exclude bacteria, fungi, plants, animals, and Homo sapiens not descended from Adam and Eve.

Is Adam figurative?

The Hebrew word *adam* (אָדָם), meaning mankind, was translated as *man* in Genesis Chapter One. Not until Genesis 2:19 is it translated as *Adam*. So, some scholars have suggested that *adam* should be understood as representing *humankind* rather than the distinct individual described in Genesis Chapter Two.

However, nobody suggests that Eve is a metaphor for womankind. Scriptures elsewhere in the Bible make it clear that the prophets and apostles understood Adam to be a real individual, married to a real woman named Eve and that they actually lived in a physical Garden in a region called Eden and that they had children from whom many nations came and from whom the Jews descended. In fact, the Jews tallied their genealogies back to Adam. For example, "Enoch also, the seventh from Adam." (Jude 14) The gospels of Matthew and Luke traced the genealogy of Jesus back to Adam. Metaphorical people are never included in genealogies!

Who is Adam really?

Jesus said, "The hour cometh that I will drink of the fruit of the vine with [all the prophets and] Michael, or Adam, the father of all, the prince of all, and the ancient of days." (DC 27:5-11) So, Adam is not only the chief archangel, he's also the "Ancient of Days" spoken of by Daniel. (Dn 7:9-22)

Christian clergy generally teach that the Ancient of Days is Jesus but Daniel's vision distinguishes between the two; "The *Son of man* came with the clouds of heaven, and came to the *Ancient of Days*." (Dn 7:13) Daniel's vision is about thrones and Judgment. The Ancient of Days sitting on a throne means that father Adam receives a throne as one of the "joint-heirs with Christ". (Rm 8:17) Adam will receive the "crown of glory that fadeth not away". (1 Pt 5:4)

Enoch saw the Ancient of Days and the Son of Man as two individuals; "I inquired of one of the angels, who went with me, and who showed me every secret thing, concerning this

Son of man; who he was; whence he was; and why he accompanied the Ancient of days." (*Book of Enoch*, 46:1)

Adam will sit on a throne and assist in the final judgment; "Judgment was set and the books were opened... the Ancient of days came, and judgment was given to the saints of the most High." (Dn 7:10, 22) Prophets and apostles from all dispensations will help Jesus with the judging of humanity. As Jesus told his apostles, "Ye also shall sit upon twelve thrones, judging the twelve tribes of Israel." (Mt 19:28)

Jesus told Joseph Smith, "The Holy One of Zion... hath appointed Michael your prince, and established his feet, and set him upon high, and given unto him the keys of salvation." (DC 78:15-16) In the end, all those prophets and apostles to whom priesthood keys have been given over the centuries will give an accounting of their stewardships to Adam who will give an accounting of them to Christ when He returns to Earth in glory. (*TPJS*, pp122, 157, 167)

Adam was the great and noble Archangel Michael who led the forces against the dragon and helped create the earth. (*The Man Adam*, Ensign, Jan 1994) He continues to watch over the human family and will again serve under Christ as the commander of God's forces in the final battle against evil at the end of His 1,000-year reign of peace. (DC 88:111-115)

Is Adam the first man or not?

If the Earth was peopled before Adam, why does scripture call him the first man? "The *first man* Adam was made a living soul... The *first man* is of the earth." (1 Cor 14:45, 47) And why did Moses call him "the first flesh upon the earth, the first man also"? (*PGP*, Moses 3:7)

It should be noted that, because animals came first, Adam was not "the first flesh" by any interpretation of scripture or science, suggesting that the reader be open to the idea that ancient prophets wrote from a very different perspective than is common today.

Adam was the first human to be termed a "man" made from the earth by God in scripture. He was also the first man to be ordained to the office to which Abraham was later ordained, "a High Priest… [with] the right of the *firstborn*, or the first man… or first father… [which is] the appointment of God unto the fathers concerning the seed". (*PGP*, Abr 1:2-4) Adam was the first man in the line of prophets who would carry His message to the world and the first patriarch of the covenant people, the Hebrews. He was the first man in God's history of dealing with men. And, as the first prophet of the first dispensation of the gospel, Adam is the "father" to the House of Israel.

So, Adam was the first man in lots of ways but, to be the first man *anthropologically*, he would have had to have been born more than 300,000 years earlier when modern humans appeared on planet Earth. The scriptures don't say how long Adam was in the Garden of Eden. So, God had at least 300,000 years of human families from which to choose a male and female and declare them to be "the first man" and "the first woman".

One good piece of evidence of humans before Adam is a set of 6 parallel lines scratched on a bovine bone recently found in Israel. They were made by a right-handed person with the same stone tool in one sitting about 120,000 years ago. That the markings had meaning is obvious. (ancient-origins.net/news-history-archaeology/symbols-0014889, DL 10 Feb 2021)

Why doesn't the Bible mention people who are not descended from Adam and Eve? Firstly, most authors of biblical books probably believed that everyone on earth descended from Adam. They did not know about the peoples like the Australian aborigines and the North Sentinel Islanders who were contemporaries of Adam and Eve.

Secondly, the Hebrew keepers of sacred records tended to focus on their own history to the exclusion of matters outside their circle of engagement, just as the Book of Mormon writers ignored the aboriginal tribes around them who had been living in the Americas for thousands of years.

Thirdly, the Bible *implies* that non-Adamites existed. For example, Cain protested that God's punishment for killing his brother was too harsh saying, "Every one that findeth me shall slay me." (Gn 4:14) Who was *everyone*? Was he talking about siblings the Bible failed to mention or about people outside Adam's family? Then, "Cain went out from the presence of the LORD, and dwelt in the land of Nod, on the east of Eden. And… his wife… conceived, and bare Enoch: and he builded a city." (Gn 4:16-17) So, either it was a very small city with just Cain and his family, or Enoch took several generations to build and populate the city, or *other people were already living in the land of Nod and joined the city*.

Where was Adam created?

"God had planted a garden in the east, in Eden; and *there he put the man*." (Gn 2:7-8) Adam was created elsewhere. Whether the garden was east of Adam's birthplace or east of the Genesis author, since he was created first and *then* put in Eden, he could have been born anywhere on Earth.

Per one ancient Jewish, "God took the man from the mountain of worship, where he had been created, and made him dwell in the Garden of Eden." (*Targum of Palestine*, Gn 2) Jewish writers often use "mountain" to reference the Lord's House; "As it is said: 'And the Lord, [God], took the man. (Gn 2:13)' From which place did [God] take him? From the place of the Temple." (*Midrash Tehillim* 92:6)

So, Hebrew tradition narrows Adam's birthplace down to the Jewish Temple Mount (Moriah) in Jerusalem or, if the Samaritans are correct, the Samaritan Temple Mount (Gerizim) 27 miles to the north toward Nazareth.

A Jewish text from the 1st-century BC, still considered canon by Ethiopian Christians and Jews, says that "Adam and his wife went forth from the Garden of Eden, and they dwelt in the land of *Elda* in *the land of their creation*." (Jub 3:32) But the location of Elda is unknown. An ancient Christian document says Adam was created where Christ was crucified. (sacred-texts.com/chr/bct/bct04.htm, DL 4 Jun 2025)

When was Adam created?

Adam was created no later than about 4,000BC, according to Bible chronologies. If we knew how much time he spent in the Garden of Eden prior to being ejected from it, we would know the approximate time of his birth.

Traditional telling of the story portrays Adam and Eve as being kicked out before they even had time to enjoy sex, an act that the Bible first mentions *after* their banishment from the garden. But the likely reason coitus is only mentioned later is because Eve did not conceive until later. (Gn 4:1)

But it would be a harsh indictment of their character if Adam and Eve transgressed their first day in the Garden of Eden. God chose His chief archangel to be Adam and a most noble spirit daughter to be Eve. One would expect such souls to be strong enough to avoid expulsion from the garden for at least a few days, maybe months, years, centuries, or even millennia. Satan was very persuasive but, as the story shows, Adam was more inclined to listen to his wife than to Satan, and she probably remained obedient for quite some time.

The ancient book of Jubilees says, "Adam and his wife were in the Garden of Eden for seven years tilling and keeping it." (Jub 3:15) But, Jewish writers put sevens on almost every finished thing as a sign of completeness.

If the traditional interpretation is correct that Adam named all the animals, at the rate of 1,000 per day (with each Sabbath off), he would have had to have been created 15 to 30 years prior to the fall to name the 10 million species of animals on the planet. (Gn 2:20) But, since it's likely that Adam only named the few animals that were in the Garden of Eden, he finished in one day.

According to Timothy, the archbishop of Alexandria in the fourth century, "Adam lived in the Paradise of Delight, he and Eve his wife, for two hundred years." (*Discourse on Abbatôn*, Brit. Mus. Ms. Oriental, #7025) But, there's no evidence that he had an inspired or historical source for that number.

If Adam is the first man in the anthropological sense, then he must have been born some 300,000 years ago. Homo sapiens (humans) have only been around for about that long.

Moses said Adam was the "first flesh upon the earth". (*PGP*, Mo 3:7) Was that human or animal flesh? If *human flesh*, then the answer is again 300,000 years. If *animal flesh*, then Adam had to have been born 640 million years ago before the first sponge arrived. (allthatsinteresting.com/first-animal-on-earth, DL 20 Jun 2020)

God is patient. He waited 13.7 billion years for His physical creation to blossom and produce humans. The Garden of Eden protected Adam and Eve from all hazards, potentially including ice ages, ferocious hominids, venomous snakes, voracious predators, etc. But, barring an answer revealed directly from God, we would only know Adam's date of birth if his Book of Remembrance still existed, and if it included the time before his transgression, and if he had transformed his reckoning of time into Earth years.

However, it may be that Adam would have been unable to do that because, at the time he was placed in the garden without a wife, Adam was not tracking days and years, for "the Gods had not appointed unto Adam his reckoning". (*PGP*, Abr 5:13) For all practical purposes, while not aging, Adam and Eve remained zero years old while in Eden. After the fall, he lived 930 more years. (Gn 5:5)

Did Adam have a belly button?

A navel is an umbilical exit scar, aka umbilicus. Adam would have had one *only* if He had been created by placental birth. It would be somewhat superfluous for God to have created him directly from clay and Eve from a rib and then to have given them navels.

Luke traced the genealogy of Jesus back to "Adam, which was the *son of God*". (Lk 3:38) If Adam was fathered by God Almighty, Jesus would not be the Only Begotten Son that scripture says He is. (Jn 3:16) So, that's probably not what Luke meant. In fact, Joseph Smith's translation corrected "the son of God" to "formed of God". (*JST (Joseph Smith Translation)*, Lk 3:38)

The original Greek used the word *tou,* meaning "of," rather than "son" (*tou* Enos, *tou* Seth, *tou* Adam, *tou* Theou). So, the best translation is "of Enos, of Seth, of Adam, of God," as some Bibles, such as the New Heart English Bible (NHEB) and others render it. Some Bibles use brackets or italics to indicate the word "son" was inserted at the discretion of the translators. (e.g., NASB, KJV)

In the Hebrew language, the name Adam (אדם) is related to the words *dam* (דם) meaning "blood" and *adamah* (אדמה) meaning "earth" or "ground". God told Adam, "As ye were born into the world by water, and *blood,* and the spirit… and *so became of dust a living soul,* even so ye must be born again into the kingdom of heaven, of water, and of the Spirit, and be cleansed by *blood*… Adam cried unto the Lord, and he was caught away by the Spirit of the Lord, and was carried down into the water… and was brought forth out of the water. And thus, he was baptized, and the Spirit of God descended upon him, and thus he was born of the Spirit." (*PGP*, Mo 6:58-65)

Joseph Fielding Smith confirmed, "Adam, our earthly parent, was also born of woman into this world, the same as Jesus and you and I." (*Deseret News*, 19 Sep 1936, pp2-8) Brigham Young said that God "created man, as we create our children; for *there is no other process of creation*". (*JD*, bk11, p122)

Under the direction of Joseph Fielding Smith, son of Joseph F. Smith, the First Presidency of The Church of JESUS CHRIST of LDS issued this statement about Adam : "He took upon him an appropriate body, the body of a man, and so became a 'living soul'… All who have inhabited the earth since Adam have taken bodies and become souls *in like manner." (Man: His Origin and Destiny*, 1954, p354)

So Adam's birth was like that of every human. His mother's water broke and the blood of her placenta accompanied his birth. In similitude of such birth, brought the ordinance on Adam that included three corresponding components, 1) baptism by water 2) Christ's atoning blood, and 3) reception of the Holy Spirit. (*PGP*, Moses 6:51–68)

The OT considers every person to have come from dust as is obvious from the statement that "man shall turn *again* unto dust". (Jb 34:15) The term for "mankind" in Hebrew is *benei Adam (*אדם בני*)*, meaning "the children of Adam". So, Adam is a son of God in the same sense that all men are His sons. (Gn 6:2; Jb 1:6; Jb 38:7; Jn 1:12; Rm 8:15, etc.)

"The Lord God planted a garden eastward in Eden; and *there he put the man whom he had formed.*" (Gn 2:8) Either God took charge of a couple of orphans or He had an arrangement with Adam's and Eve's biological parents, as he had with Hannah who gave her son to God. (1 Sm 1:11)

Could God have created Adam on some other world and then brought him here as Brigham Young seemed to think? (*JD*, bk2, p6) No. Scripture says Adam and Eve were created from *this* earth's dust. As God said to them, "In the sweat of thy face shalt thou eat bread, till thou *return unto the ground*; for *out of it wast thou taken*: for dust thou art, and unto dust shalt thou return." (Gn 3:19) They couldn't *return* to the ground of *this* earth if they were aliens. "Adam was made a living soul… *of the earth, earthy.*" (1 Cor 15:45-47)

Of course, if God had wanted to, He could have created Adam directly from raw materials. That is largely what He will have to do when resurrecting cremated bodies.

Why Eve before the fall?

Adam and Eve had no children before the fall. So, why did God give him a wife in paradise? Adam gets all the blame anyway, so why not let him be the sole transgressor? Because Adam needed Eve to persuade him to eat the forbidden fruit!

God had pronounced everything "good" in Genesis Chapter One. But after creating Adam, there was something that was not good; "It is not good for man to be alone." (Gn 2:18) God brought animals to Adam "and Adam gave names to all… but for Adam there was not found an help meet for him". (Gn 2:20) So God created one. (Gn 2:18)

What kind of help was meet for Adam? Not man's best friend nor a beast of burden nor any of the animals known to Adam. Some say God was checking whether Adam would settle for an animal companion. But no! He was giving Adam the privilege of naming Eden's animals – male and female animals – before bringing a wife to him.

Why the rib story?

One of the silly questions still often seen on the internet is, "Do men have one less rib than women?" If a surgical alteration were heritable, the demand for plastic surgery would be even greater than it already is! Who doesn't want all their children to look beautiful? But no, it's ridiculous to suppose that only men would inherit the rib deficit!

Even if that were true, men would weirdly be missing only *one rib*, not *a pair of ribs*! "The Gods caused a deep sleep to fall upon Adam; and he slept, and they took *one of his ribs…* and of *the rib* which the Gods had taken from man, formed they a woman, and brought her unto the man." (*PGP*, Abr 5:14-16)

Humans normally have 12 rib pairs, as did Neanderthals and Denovisans, our closest ancestors. Our closest relatives, chimps, usually have 13, as do gorillas. Any human ancestry with 13 ribs would have lived hundreds of millennia ago.

Some men have one less rib or pair of ribs but so do some women. 1 in 200 people have more than 24 ribs. Between .2% and .6% of humans have an unusual rib count. Most discover it from an x-ray taken for some other reason. It's not impossible that Adam had an extra rib on one side and God balanced him by using it in the creation of his wife, Eve.

If Adam's rib was the only ingredient for making Eve, she would have been his exact clone, right down to the Y chromosome, and would have needed transgender surgery, gene enhancement, and hormone therapy before becoming a mother.

The rib story must be symbolic of something the author of Genesis was reticent to spell out explicitly. Placental birth, already common for all animals except marsupials and monotremes, was simpler than a ribectomy on Adam and full-body-construction surgery for Eve.

Professor Ziony Zevit incurred the ire of a broad spectrum of Jewish and Christian readers in 2001 by suggesting that the bone taken from Adam was actually his baculum and that's why every human male lacks one. But Zevit's theory does nothing to avoid the clone problem. Nor does it explain why God wouldn't start from scratch. Furthermore, human males are not the only mammals lacking bacula, yet nobody claims God took the baculum of males of these other species to create their mates.

A very meaningful answer for the symbolism comes with a correct translation. The primary meaning in Strong's Concordance for *tsela* (צֵלָעֹת) is *side* rather than rib. (Str#6763) The Jubilee Bible 2000 got it right. Then, as many have taught, it would represent the side-by-side partnership of man and wife.

After finding no match for himself among all the animals God showed him, Adam was pleased to see Eve; "This *at last* is bone of my bones and flesh of my flesh." She came from man kind or "out of man she was taken". (*ESV*, Gn 2:23)

If Eve had been her husband's biological daughter in any way, it would have made the marriage a bad example for posterity. "Bone" is from the Hebrew *'etsem* (עצם) which also means *substance*. (Str#6101) So, Adam recognized that Eve was made of mankind flesh and bones, like himself, and that they were "one flesh" as a married couple, for the next thing out of Adam's mouth was, "Therefore shall a man leave his father and his mother, and shall cleave unto his wife: and they shall be one flesh." (Gn 2:24) Adam had left his father and mother to become *one flesh* with Eve.

Church President Spencer Kimball taught, "The story of the rib, of course, is figurative." (*Doctrines of the Gospel Student Manual*, CJCLDS, p18) This quote was used in the Ensign (Mar 1976, p71) and in the text for the college course Religion 327. (*The Pearl of Great Price Student Manual*, CJCLDS, p11)

What did Adam's sleep signify?

If God created Eve without doing a ribectomy on Adam, then there was no need to render the man unconscious for surgery.

Was God doing a close-your-eyes-I've-got-a-surprise thing on Adam? Why didn't God just put Eve into the Garden of Eden at the same time as Adam and dispense with the drama of bringing her later? Like the rib, there must be some significance attached to the nap God caused him to take, but what? (Gn 2:21)

Some have speculated that Adam's nap represents amnesia, his forgetting of premortal memories, after all, there is no mention of him awakening. It sounds like a good idea but Eve lost her memories of premortality *without* a nap, like virtually everyone else. So, Adam's sleep must denote something else, perhaps a dream coming true when Eve showed up.

Without offering an opinion of its own, BYU Religious Studies Center cites several early Christian church fathers as having argued that "the symbolic message in the 'deep sleep' that came upon Adam and the creation of Eve through that sleep is that Christ's death on the cross is the event that gave birth to his Church". (rsc.byu.edu/gospel-jesus-christ-old-testament/types-shadows-symbols-christ-seen-church-fathers#_edn10, DL 12 May 2022)

In other words, had Jesus not "slept" in the tomb, His bride, the Church, would not exist.

Was Eve Adam's equal?

Paul said, "Adam was not deceived, but the woman being deceived was in the transgression." (1 Tim. 2:14) By partaking of the fruit, Eve put Adam in the difficult position of having to choose between avoiding transgression and accompanying wife into the world outside it. In the end, both were equally transgressive, though for different reasons. Still, it's interesting and seemingly misandrist that nobody calls the event "the fall of Eve".

Those who believe the fall was necessary for the progression of humanity are happy that Adam is not still languishing in Edenic naiveté and laud Eve for her desire to acquire knowledge and for persuading Adam to partake.

In sentencing Eve for her role in the transgression, God told her, "Thy desire shall be to thy husband, and he shall rule over thee." (Gn 3:16) The "rule over thee" part falls harshly upon the ears of most modern women. Was that her punishment for being the first to partake of forbidden fruit? Was God giving Adam dominion because he had held out against temptation longer than Eve?

Church president Spencer Kimball said this about the word *rule*; "I would prefer to use the word *preside* because that's what he does. A righteous husband presides over his wife and family." (*The Blessings and Responsibilities of Womanhood*, Ensign, Mar 1976, p72) As Brigham Young allegedly said to the woman who complained her husband had told her to go to hell, "Don't do it." Likewise, if Adam failed to lead in a godly way, Eve would not have been obligated to follow.

Eve's family needed strong leadership. For the primary leadership responsibility to be placed on Adam was a blessing to Eve rather than a curse. Her curse would be the labor pains she would experience giving birth to children. (Gn 3:16)

Eve was similar to Adam physically as well, according to one vision shown to Joseph Smith and several other men; "The heavens gradually opened, and they saw a golden throne... and on the throne were two aged personages, having white hair, and clothed in white garments... the two most beautiful and perfect specimens of mankind he ever saw. Joseph said, 'They are our first parents, Adam and Eve.' Adam was a large, broad-shouldered man, and <u>Eve as a woman, was as large in proportion.</u>" (*Minutes of School of the Prophets*, Zebedee Coltrin, SLC, UT, 11 Oct 1883)

In God's eyes, men and women are equal but, since they have different aptitudes, and for practical division of labor reasons, each has different but equally important roles. But the way Adam and Eve divided roles is not law for people in other circumstances except that women must be the mothers. Couples should do what is necessary irrespective of roles that tradition has defined. Many Old Testament women worked outside the home.

Hebrew women were greatly esteemed compared to other cultures in Old Testament times. Some were revered as prophetesses such as Deborah, Hannah, Huldah, Miriam, Rachel, Rebekah, Ruth and Sarah. When Deborah was chief judge, she led a successful counterattack against the armies of Canaan's king Jabin. Even today, Israel stands above all other countries in the Mideast at bestowing equal rights upon women. Golda Meir, the fourth Prime Minister of Israel, was a very strong female leader.

Valiant and godly people of both genders have equal opportunity to be exalted as priests and priestesses, joint heirs with Christ. Eve was different but equal from Adam like a pound of wheat is equal to but different from a pound of barley.

What was Adam's language?

Hebrew tradition says that Adam and Eve obviously spoke Hebrew or God wouldn't have given them Hebrew names; *Adam* means "man" (Str#129) and *Eve* means "life" (Str#2332). The Hebrew Midrash says that God called the language of their first parents the "Adamic language."

An early version of the CJCLDS temple ritual stated that Adam used the words "pe le El" to begin his prayers to God. This transliteration of the Hebrew (לאל פה), means literally "mouth to God" or "God, hear my mouth" lends credence to the Jewish claim.

Apostle Orson Pratt said that "Ahman" in the place name "Adam-ondi-Ahman" is God's name in the Adamic language. (*JD*, bk2, p342) An 1832 handwritten page from the Joseph Smith Papers titled, "A Sample of the Pure Language," reportedly dictated by the Prophet to "Br. Johnson," asserts that the name of God is "Awmen". (*A Sample of the Pure Language*, Joseph Smith Papers, 1832, as dictated by Smith to Br. Johnson) One definition of the Hebrew word *awman* is "foster-father". (Str#539)

So, if Hebrew was the language of Adam, why wasn't it replaced by the confusion of tongues at the Tower of Babel? There is a lot of confusion about the tongues being confused. (Heb *mingle/mix* Str#1101) True, Genesis says, "The whole earth was of one language, and of one speech." (Gn 11:1) But "earth" was translated from the Hebrew word *erets* which more commonly means "land" or "nation." Old Testament writers had no global perspective, living as they did without faster-than-foot communication. The family of Noah and his righteous descendants were not on the tower project and were not living among its wicked builders. So, God had no reason to garble their language by mixing it with the languages of those working on the tower.

Adam, who lived until Noah's father was 65 years old, continued to speak whatever language he learned in the Garden of Eden. And Noah's father would have taught the same language to his own kids. They kept a Book of Remembrance in the original tongue. (*PGP*, Mo 6:5-6) So, all who remained under the influence of Adam's patriarchal line were taught generation after generation from those language-stabilizing records. Hebrew is the best candidate for Adamic; "Just as the Torah was given in Hebrew, so was the world created with Hebrew." (*Midrash Rabbah*, Gn 31:8)

Adam and Eve were educated in language by God for it was "by them their children were taught to read and write, having a language which was pure and undefiled". (*PGP*, Mo 6:6) Despite being dead as a spoken language from the

Middle Ages to the 19th century, and despite being degraded by thousands of years of use and abuse, Hebrew is still extremely rich and nuanced. And, if the language of Adam had been lost to the earth at Babel's confusion of tongues, his book of remembrance would have been unreadable to his posterity.

What color was Adam's skin?

If Adam and Eve had white skins, narrow noses, and brown or blond hair, as they are traditionally pictured, why does over 90% of the world's population have olive to black skin and black hair? How could Adam and Eve possibly be the sole parents of *all* humanity?

A good number of Bible apologists have provided explanations of the genetics of melanin production in an attempt to show how the diversity of skin colors could arise from a single set of monogamous parents in just 6,000 short years. But apologetic explanations, despite efforts to sound scientific, are extremely oversimplified and fall dismally short. In reality, there are about eight different genes that determine skin color.

Using Occam's Razor – The answer with the fewest assumptions is that that Adam and Eve were not the progenitors of all modern humans.

Scientists have found compelling archaeological evidence of multiple human migrations out of Africa starting about 1.6 MYA. People arriving in Europe around 1 MYA still had moderately dark skins. Europeans acquired the genes for lighter skin, SLC24A5 and SLC45A2, from mingling with populations from the Mideast around 8,500 years ago. (smithsonianmag.com/smart-news/heres-how-europeans-quickly-evolved-lighter-skin-180954874, DL 16 July 2016) So, the idea that all the light-skinned Europeans and Mideasterners could have descended from one mother only 6,000 years ago strains credulity.

Recent research suggests that the more ancient ancestors of humans likely had pale skin similar to chimpanzees. Later, as additional sweat glands displaced hair follicles to improve cooling in hot climates, higher melanin levels in skin became an advantage.

Some human populations in darker climates expressed genes for white skin again. This helped them avoid vitamin D deficiency. But many dark-skinned peoples also managed to survive in cold climates, such as the Eskimos. Sunlight reflecting from snow can help raise vitamin D levels along with a diet of fish rich in vitamin D.

So, dark skin predated Cain by many thousands of years. The stories that his curse was dark skin and that it survived the flood through Ham's descendants are total fabrications by Protestant slave-holders in the southern states of the USA, an example of gross twisting of scripture for self-justification. As ancient scripture professor Hugh Nibley asserted, "There is no exclusive equation between Ham and Pharaoh, or between Ham and the Egyptians, or between the Egyptians and the blacks, or between any of the above and any particular curse." (*Abraham in Egypt*, pp 219-220)

Some people have pointed out that the name *Adam* is a variation on the Hebrew word *adom*, meaning red, and that it's also related to *adamah*, which is associated with the reddish-brown soil from which the man was created. But red or brown skin is no better than white skin for explaining how diverse skin colors evolved in a mere 6,000 years. A first-century Jewish interpretation of the Genesis creation of man states that God "created him red, black, and white". (*Targum of Palestine*, Gn 1) But three colors are too many colors for a man with only one wife. So, the text refers to Day Six of the Divine Council *planning*.

The ancient Hebrew word for light, רא "Ore" (Str#216) is spelled similarly and pronounced identically to the word for skin, רע "Ore" (Str#5785). This might suggest that Adam and Eve had lighter skin than today's Jews whose ancestors mingled with darker skinned neighbors and conquerors for thousands of years.

Were cavemen children of God?

Science has extracted DNA from the fossils of Denisovans who last walked the earth during the most recent ice age, and from Neanderthals who were buff and brainy cave dwellers about 30 TYA. Some humans of Melanesian descent have 5% of their DNA from Denisovans and some Eurasians have 4% of their DNA from Neanderthals, to name just two examples.

So, if only real humans can be children of God, what criteria should be used for distinguishing the first human from his or her nonhuman parents? Ancient hominids crafted tools, invented weapons, controlled fire, created art, wore clothes, wore jewelry, buried their dead, performed funerary rituals, played musical instruments, and nursed their sick very much like humans! Where in God's plan of happiness do they fit?

Professor Hugh Nibley wrote, "Do not begrudge existence to creatures that looked like men long, long ago, nor deny them a place in God's affection or even a right to exaltation--for our scriptures allow them such." (*CWHN*, bk1, ch4, pp82-83) God will provide for "the happiness of man, and of beasts, and of creeping things, and of the fowls of the air". (DC 77:2) Adam was the lone man in the Garden of Eden for a spell, but a preponderance of evidence shows that he was not the only human on the planet.

A just and merciful God will not deny happiness to those of His creatures who died without the gospel due to factors beyond their control. Neither the incapacity to understand gospel principles nor the lack of an introduction to Jesus will prevent any creature from receiving desired blessings in the long run. Pre-adamites who died without knowing about the Word will get, or have gotten, the chance to accept Him in the afterlife, the same as any such who die nowadays. That's good news for the 100 or so currently isolated and uncontacted hunter-gatherer tribes around the world. (survivalinternational.org/uncontactedtribes, DL 31 Aug 2018)

16: Eden

Why a Garden?

"God planted a garden eastward in Eden; and… every tree that is pleasant to the sight, and good for food; the tree of life also in the midst of the garden, and the tree of knowledge of good and evil." (Gn 2:8-9)

Why did God go out of His way to plant the garden when there were thousands of naturally occurring paradises in places like the Bahamas, the Caymans, the Fiji Islands, and Costa Rica? A balmy tropical island would have made a nice honeymoon spot for Adam and Eve and effectively isolated them from all the unpleasantness elsewhere in the world.

Surely the Tree of Knowledge, the Tree of Life, and all the other trees in the garden orchard would have thrived in a place like Tahiti where the fruits could have looked and tasted delicious, especially compared to the wild fruits of 6,000 years ago, which were mostly small and tangy or bitter.

God's garden was better than all other earthly paradises, not just in natural beauty, but also in supernatural beauty. It was supra-earthly, radiant with a degree of heavenly glory suitable for in-person visits by the Almighty. The memory of paradise would serve forever as a symbol of the heavenly home for which Adam and his posterity were to strive.

Only after Adam and Eve were banished from the Garden of Eden did they see the contrast with real world conditions. Even the best of places in the ordinary world were lone and dreary by comparison, indicative of the increased separation of mankind from God in consequence of their transgression.

The garden was also a place to learn agriculture; "God took the man and put him in the Garden of Eden to work it and take care of it." (*NIV*, Gn 2:15) It would have meant starvation if they had been ejected from paradise without first learning how to provide food for themselves.

The garden also isolated Adam and Eve from the dangers that lurked in the world outside. Free of worries about food, shelter, clothing, safety, and healthcare, they were free to feel an appetite for a more meaningful existence and whether the Tree of Knowledge was the key to satisfying that hunger.

The Garden of Eden also served as a symbol of holy temples for the next 6,000 years, especially in association with the Garden of Gethsemane where Jesus tackled spiritual death, the Garden of the Empty Tomb where Jesus conquered physical death, and the Edenic earth that will be after the return of Christ.

Where was Eden?

The "garden of God" was part of a much larger region called Eden. (Ezek 28:13) But where was Eden? The Urantia Book, a series of channelings from the paranormal realm, claims the garden is now on the floor of the Mediterranean. Jewish tradition puts it east of Jerusalem. A lot of Latter-day Saints think it's in Missouri, USA.

But most Christians and Bible scholars have good reasons to think it's somewhere in Iraq near the convergence of the Tigris and Euphrates rivers, in present-day Iraq:

♦ The name Eden is related to the Assyrian *Edin* meaning "Plain". (biblestudytools.com/dictionary/eden, DL 20 Feb 2019) The vast floors of the Tigris-Euphrates valley qualify.

♦ Genesis says that Eden was "eastward". (Gn 2:8) Iraq is east of Israel and the wilderness where Moses wrote Genesis.

♦ Tablets found in Iraq describe the Sumerians as keepers of the *immortal garden*, a probable allusion to the garden and part of the larger set of god garden mythology from ancient Mesopotamia. (uoomiacu.blog.free.fr, DL 12 Dec 2018)

♦ While laying siege to Jerusalem in 701BC, the Assyrians bragged about having conquered the children of Eden about 100 years prior: "Have the gods of the nations delivered them which my fathers have destroyed, as Gozan, and Haran, and Rezeph, and the *children of Eden* which were in Thelasar?"

(2 Kg 19:12) Thelasar was situated midway up the Euphrates. (*Asimov's Guide to the Bible,* Isaac Asimov, bk1)

♦ The locations of the four rivers feeding Eden are all in the area of Iraq and surrounding country; "A river went out of Eden to water the garden; and from thence it was parted, and became into four heads." (Gn 2:10) "From" is not in the original Hebrew text and "thence" should be translated as "there". Rivers don't usually split into four going downstream unless it's a delta situation. And the word "heads" always refers to upstream tributaries that converge. The four tributaries converging and flowing into the garden were:

1) "The name of the first is **Pison**… which <u>compasseth the whole land of *Havilah*</u>, where there is gold… bdellium and the onyx stone." (Gn 2:11-12) Where is *Havilah*? The Ishmaelites, who occupied the Arabian Desert, were spread "from *Havilah* unto Shur, that is before Egypt, as thou goest toward Assyria". (Gn 25:18) From Egypt toward Assyria points to Arabia. So, the Pison flowed from Arabia. "Saul smote the Amalekites from *Havilah* until you come to Shur, that is over against Egypt." (1 Sm 15:7) *Havilah* means "sandy land" in Hebrew. Arabia is famous for sand and gold. One of the oldest gold mines in the world is located in the Hijaz Mountains where the area is named, "The Cradle of God". Arabia is also a source of bdellium, a fragrant gum resin similar to myrrh. (Wordnik) So, why doesn't the Pison show on the map? Because it is dry most of the year and is classified as a wash; *Wadi al-Rummah*. Landsat photos in the 1990's revealed that it once flowed continuously 530 miles from the glaciers of the gold-rich Hijaz Mountains to the Euphrates via its extension, the *Wadi al-Batin*, aka the *Kuwait River*, until about 5,000 years ago when it was several miles wide in places and carried pebbles all the way to Kuwait that match geology in the Hijaz Mountains. (downtoearth.org.in/news/river-of-long-ago-31026, DL 21 Jun 2020) Much of river bed can be seen using Google Earth.

2) "The name of the second river is the **Gihon**; it winds through the entire land of *Cush.*" (*NIV*, Gn 2:13) The Gihon River is most likely the Karun River that flows generally southward through Iran and empties into the Shatt al-Arab downstream from the confluence of the Tigris and Euphrates. *Gihon* means "a bursting forth" in Hebrew and, indeed, the Karun is more prone to violent and irregular flooding of its marshlands than the Tigris or the Euphrates. (iranicaonline.org/articles/karun_1_2, DL 8 Nov 2021) Cush may have been a region named after Ham's grandson after the family migrated southeasterly from the Ararat Mountains to climes more familiar and conducive to their pre-flood lifestyle.

3) "The third river is **Hiddekel**… which goeth toward the east of Assyria." (Gn 2:14) Hiddekel comes from the Assyrian *idiiklat*, meaning *east of Assyria.* (*In the Beginning*, Isaac Asimov, 1981) It is translated as *Tigris* in most modern translations. Daniel placed the Hiddekel River in Iraq when he said, during his Babylonian captivity, "I was by the side of the great river, which is Hiddekel." (Dn 10:4) The Tigris of today still flows eastward from Assyria and joins the Euphrates.

4) "The fourth river is **Euphrates**." (Gn 2:14) The name of the Euphrates has not changed much in thousands of years and can be found easily on modern maps.

All four rivers that watered Eden converged where the Tigris and Euphrates rivers still join to become the Shatt-al-Arab River 120 miles upstream from the Persian Gulf. Heavy rainfall and melting mountain ice caused the rivers to flood Noah's homeland repeatedly, as evidenced by ancient layer upon ancient layer of mud in the area. Jubilees says, God sent "the waters of the flood upon all the land of Eden". (Jub 4:24)

Science is in general agreement that agriculture began in the "Fertile Crescent," which is the lands between the Mediterranean Sea and the Persian Gulf in Iraq, the perfect place for the most dazzling garden ever to grace planet Earth!

Why not in America?

Many Latter-day Saints believe Eden was in Missouri. But, the scripture they most often cite to support that theory actually doesn't; "Spring Hill is named by the Lord Adam-ondi-Ahman, because, said he, it is the place where Adam *shall* come to visit his people, or the Ancient of Days *shall* sit, as spoken of by Daniel the prophet." (DC 116)

God said that Adam *shall* come and *shall* sit at Spring Hill in Missouri, not that he *had* sat there. So, Spring Hill is named Adam-ondi-Ahman in honor of Adam's ancient meeting place because it is where he will hold his future meeting.

Most of the stories in which the Prophet Joseph Smith identified one or more altars or towers in Missouri as having been built by Adam have been conflated with similar stories that tell of altars built by Nephites. None are first-hand accounts. Most were recorded years after Smith's death:

♦ Per Abraham Smoot: Twenty-two miles from Jackson County, Missouri, Smoot and Alanson Ripley discovered a *wall* 30 feet long, 4 feet high and 3 feet thick, laid in mortar... When Joseph Smith visited the ruins, he said it was an *altar* built by Adam upon which he offered sacrifices after he was cast out of the garden. (*BYU Studies* bk13, #4, p565)

♦ Per Oliver B. Huntington: His family lived about 250 yards from Adam's *altar*. There was also *Adam's tower*. The wall was 6 feet long and 13 inches high, and disappeared back into the hill. There were no tool marks on the stones. About a quarter mile to a half mile away, Joseph pointed out where *Adam built a tower*. (*Adam's Altar and Tower*, Juvenile Instructor, Oliver B. Huntington, 30:700-701)

♦ Per Benjamin F. Johnson: He owned the lot on which *Adam's altar* stood. The stones were so scattered that not one remained upon another. (Letter of Benjamin F. Johnson to the *Deseret Evening News*, 5 Nov 1895, published 9 Nov 1895)

♦ Per Chapman Duncan: His house was 40 rods south of *Adam's altar*. Joseph Smith dug with a shovel to uncover the altar, which was three layers deep of dressed stone, each stone

10 inches thick by about 18 inches long. (*Autobiography of Chapman Duncan*, Special Collections Library, BYU)

♦ Per Orson Whitney: Heber Kimball said that Joseph showed him, Brigham Young and others the ruins of *three altars* of stone, one above and back of the other and said, "There is the place where Adam offered up sacrifice after he was cast out of the garden." The altar stood at the *highest* point of the bluff. (*Life of Heber C. Kimball*, Orson F. Whitney, 1988, pp209-210)

♦ Per other early Latter-day Saints: The arrangement of stones was "marking the site of Adam-ondi-Ahman" and/or "the *grave of Adam*" and was one of "a number of such stone mounds or ruins in that vicinity". (*HC*, bk3, pp35-37)

♦ Per Joseph Fielding Smith, Jr: The ruins were Nephitish. "When the Prophet Joseph Smith first visited Spring Hill he called it '*Tower Hill*, a name I gave the place in consequence of the remains of an old *Nephite altar or tower.*'" (*The Way to Perfection,* Joseph Fielding Smith, Jr., p287)

♦ Per later on-site investigators: *No identifiable altars* or trace of ruins or of ancient human artifacts were found there.

Historians have researched and attempted to craft a single truth from the disparate stories but, after more than a century of trying, have been unable to do so. Even the simplest questions remain unanswered. Were Tower Hill, Spring Hill, and Adam-ondi-Ahman all the same place? Who built the wall, or altars, or tower, Adam or Nephites? What shape and size were the stones? Did Adam know how to build with perfectly dressed brick-shaped stones or cement? Were there three altars or just one? How could all signs of the exposed, or partially exposed, or totally buried ruins in the area suddenly change and then vanish? Why would Adam have a 30-foot-long alter unless he was sacrificing dinosaurs?

Adding to the difficulty of finding reliable answers is the fact that much of *The History of the Church* attributed to Joseph Smith or his journal was not written by Joseph himself. He employed scribes who often made entries that were not reviewed by him. The church at the time was suffering persecutions and hardships that frequently prevented the prophet from looking at everything his scribes recorded. Furthermore, there exists no Joseph Smith journal for the year 1838 in which the ruins were discovered. (scholarsarchive.byu.edu/cgi/viewcontent.cgi?article=1611 &context=byusq, DL 6 Jul 2018)

Most references to ancient relics in Missouri came from weary members of Zion's Camp during a time of illness and death on their miserable trek to Missouri to reclaim lands from which they'd been driven by mobs and tyrants. So, it was not unusual for some bedraggled person, in the bleakness of an evening, to speak inspirational but uninspired words to others in the camp as they huddled around the fire. Speculation by Joseph Smith intended to boost morale may also have been written down as gospel. Being human, he had his own conjectures and opinions.

Nowhere in the revelations in which God told Joseph Smith that "Independence is the center place" was any mention made of Eden. (DC 57; DC 78:15; DC 107:53-57) In calling several of the brethren to dispose of Kirtland properties and come to Missouri, God asked, "Is there not room enough on the mountains of Adam-ondi-Ahman, and on the plains of Olaha Shinehah, or the land where Adam dwelt, that you should covet that which is but the drop, and neglect the more weighty matters?" (DC 117:8-11)

Adam-ondi-Ahman in Missouri has no "mountains." The plaque on the hill overlooking it says, "The Valley of Adam-ondi-Ahman." So, God was speaking poetically and metaphorically. The plains of Olaha Shinehah mean the plains where the Moon and the Sun shine. (See *Olea* and *Shinehah*; Abr 3:13) Since they shine virtually everywhere, their mention is not meant to pinpoint Adam's dwelling place.

The use of old names for new places is common. The future Zion in Missouri will be called New Jerusalem, not because that is where Jesus *was*, but because that is where he *will* be. The Jordan River in Utah is named after the one in Israel because both drain from a freshwater lake into a salty lake, not because it's where Jesus was baptized.

If Adam had lived in North America, his posterity would have had to migrate across the ocean to the Mideast. Some have suggested that Noah floated the family 7,000 miles to the Mideast on the ark. (*MD*, "Flood of Noah")

But the geological evidence is overwhelming that the hemispheres separated 175 MYA. Furthermore, all flood-related theories for getting Adam's posterity to the eastern hemisphere are contradicted by bullet-proof scientific evidence that the Noahic flood was not possibly global.

Finally, Missouri's archeology lacks any trace of biblical populations while the Tigris-Euphrates plains are so chock-full of ruins dating to Biblical times as far back as Enoch that it was difficult for Desert Storm allies to pick a target that would not damage an ancient archeological site.

What is "Adam-ondi-Ahman"?

The dashes in Adam-ondi-Ahman are word separators. *Adam* means *man* and *Ahman* is a name for God. Ondi is related to the Coptic *ente* and to the Egyptian *nty* and to the Book of Mormon *anti*, meaning "of" as in Anti-Nephi-Lehies — "of Nephi-Lehi people" So, Adam-ondi-Ahman means "Man of God". (thelunchisfree.com/2017/05/28/the-meaning-of-adam-ondi-ahman, DL 23 May 2019)

Was Eden a temple?

The Garden of Eden was where God personally visited Adam and Eve. It was the holiest place on Earth. The Holy of Holies in Solomon's temple represented the Garden of Eden and included garden symbolism, such as the golden lampstand. (Ex 25:31-40) "The seven candle holders and three joints where the branches meet the central column represent the ten sefirot of the Tree of Life." (byzant.com/Mystical/Kabbalah/Menorah.aspx, DL 11 Jun 2017) Other garden elements in Solomon's temple included cedar gourds, flowers, palm trees, and cherubim. (1 Kg 6:18, 29, 32, 35) Temple washbasins were decorated with lions, oxen, and cherubim. (1 Kg 7:29) Bronze pillars included artwork of pomegranates and lilies. (1 Kg 7:18, 22) There were wooden carvings of palm trees and flowers. All these decorations are reflections of the verdant Garden of Eden.

The Contemporary English Version of the Bible says that after Adam and Eve were cast out of the garden, God posted cherubim "at the *entrance* to the garden". (*CEV (Contemporary English Version)*, Gn 3:24) Likewise, the doors to Solomon's and Ezekiel's temples had cherubim statues guarding the holy of holies. (Ex 25:18-22; 1 Kg 6:23–28; 1 Kg 8:6-7; 6:29; Ezek 41:18)

The temple's outer court represented the world, the inner court Eden, and the Holy of Holies the Garden of Eden, as that was the most likely place God might be found. (Gn 3:8)

An angel in one ancient Hebrew text said, "[Enoch] was taken from amongst the children of men, and we conducted him into the Garden of Eden in majesty and honor… And he burnt the incense of the sanctuary." (Jub 4:23-25) Enoch served as High Priest in the Garden of Eden for others like himself who had been translated up. (Hb 11:5) Anciently, it was called "the Garden of the Righteous". (1 Enoch 60:23)

"Melchizedek was a priest of this order… And his people wrought righteousness, and obtained heaven, and sought for the city of Enoch which God had before taken." (*JST (Joseph Smith Translation)*, Gn 14:32-33) So, Melchizedek's holy city of Salem ascended to the same heaven as Enoch's city to dwell in the Garden of Eden.

Another ancient Jewish work says, "[Noah] knew that the Garden of Eden was the holy of holies and the dwelling of the Lord." (Jub 8:18-19) As it will be when the Paradise of God returns to the new glorified earth, the presence of God is what makes it a temple, its dimensions being described as a cube to signify Holy of Holies. (Rv 21:16-22; 1 Kg 6:19-20)

So, although Eden could not have been in Jackson County, Missouri, the temple lot for the New Jerusalem, Zion's center place, is there. Christ's visit to Zion at the beginning of His millennial reign will usher in the return of the Tree of Life and the garden that contains it.

Why tend the garden?

Many people think of the Garden of Eden as being maintenance-free. But if that were true, God wouldn't have told Adam "to dress it and to keep it". (Gn 2:15) Adam and Eve could destroy plant cells by eating "of every tree of the garden". (Gn 2:16) Therefore, the plants were not immortal and could die if not tended. God didn't give Adam unnecessary work just to keep them busy.

The garden also needed to be watered. If not so, God wouldn't have made a river run through it. (Gn 2:10) Natural sub-irrigation can serve plants with sufficiently deep roots, but water doesn't just jump out of a river to water shallow-root plants. During rainless times, Adam probably brought water some of the plants he tended.

A plant is deemed to be a "weed" only if it is unwanted where it grows. Seeds that God deemed wonderful were distributed throughout the garden by wind, water, and winged creatures. There was nothing to stop them from falling and germinating amid Adam's vegetables. So, he had to tend to them. Likewise, he had to protect his plants from rodents, bugs, and birds.

Indeed, the historical Hebrew perspective was that Adam and Eve had to work to keep the garden productive. The Gods said, "We gave him work and we instructed him to do everything that is suitable for tillage. And he tilled... and he protected the garden from the birds and beasts and cattle, and gathered its fruit, and eat, and put aside the residue for himself and for his wife." (Jub 3:15-16)

So, even before their transgression, Adam and Eve were farmers. Their experience tilling, planting, and harvesting in the Garden of Eden undoubtedly helped them later when they were forced to survive outside Eden where "he tilled the land as he had been instructed in the Garden of Eden". (Jub 4:35) "And Eve, also, his wife, did labor with him." (*PGP*, Mo 5:1)

If the Garden of Eden needed fertilizing, cultivating, weeding, pruning, mulching, and thinning, how was it any different from other gardens? Fruits and vegetables 6,000 years ago were much smaller than they are today because they were still wild and had not yet benefitted from agricultural methods that selected and bred the large produce now sold in grocery stores.

So, crops during Christ's millennial reign will likely be larger and tastier on average. Furthermore, there is nothing in scripture to suggest that farmers will be able to forget about seeding, feeding, weeding, and watering during the millennium of peace.

Could the animals talk?

One might expect that mankind's closest animal relatives would be best at speech but, biologically, none of the non-human primates even have the throat equipment for it. But Koko the gorilla understood about 2,000 spoken words and had an active vocabulary of about 1,000 hand signs. Koko's male companion knew about 500 hand signs.

Alex, a parrot, could speak about 150 words and was doing far more than just mimicry! When shown a dozen or so blocks and balls of different colors, he correctly answered, for example, how many green blocks there were or how many red balls were there. He could also read small numbers and a few short words. When shown three items he could say which one was different and why, such as shape, color, or size.

Ravens, starlings, orcas, beluga whales, elephants, and various pets have been observed trying to mimic human speech without success. But many kinds of animals converse with each other in their own ways.

Interspecies communication is easy for animal spirits. They use telepathy. Such was the case in the Garden of Eden where the elevated state of glory and holiness that allowed Adam to talk face-to-face with God without withering in His presence also enabled spirit-to-spirit communication among all creatures.

Serpents lack the type of tongue, teeth, vocal equipment and brain circuitry necessary for speech. So, Satan could not have talked through its physical body. At best he could only use the serpent as a visual, like a puppet or ventriloquist dummy, while communicating with Eve telepathically.

Eve didn't seem surprised that the serpent communicated with her. Rather than being shocked and repulsed by him, she allowed him to seduce her into eating the forbidden fruit. So, interspecies communication was not a strange thing to her.

Satan used the same trick that God's angel used to help Balaam's donkey speak except that the donkey's words were its own while the serpent's words were Satan's. (Nm 22:26-31) Joseph Smith's expansion on Genesis indicates that Satan *"spake by the mouth of the serpent"*. (*PGP*, Mo 4:5-7)

Why did it have to be a snake? Perhaps snakes were easier to manipulate than other animals. Or, maybe God limited Satan's choice of animal to less attractive, less cuddly ones so that Eve's shame would be full.

So, when did animals stop communicating with humans? The Book of Jubilees, sometimes called Lesser Genesis, says, "On that day on which Adam went forth from the Garden… was closed the mouth of all beasts, and of cattle, and of birds, and of whatever walks, and of whatever moves, so that they could no longer speak." (Jub 3:27-28)

Did Adam name each species?

Sunday school children have historically been given the impression that Adam named the animals with the labels a visitor to a zoo might see on the creature enclosures. The text is sufficiently imprecise to allow that Adam gave his animal friends in Eden names like Makaiah, Imlah, Neriah, and Keziah, rather naming the animals taxonomically.

"Adam gave names to all cattle, and to the fowl of the air, and to every beast of the field." (Gn 2:20)

Most theologians agree that the text means he named "kinds" (major groups of animals) rather than every macro-

and microscopic organism that is able to move itself. And some theologians say the naming was limited to "kinds" present in Eden and excluded animals that did not qualify as "cattle", "fowl", or "beast of the field".

What purpose did the naming of animals serve anyway? Was it so that Adam and Eve could talk about them without having to describe them or point a finger at them? The naming was God's way of confirming the fact of Adam's dominion over the animal kingdom and all creation.

According to Jewish commentary and Kabbalah, rather than pull names randomly out of his imagination, Adam perceived the spiritual nature of each "kind" and gave it a Hebrew name that reflected its spiritual characteristics. (chabad.org/kabbalah/article_cdo/aid/565870/jewish/Namin g-with-Divine-Inspiration.htm, DL 4 Jun 2025)

Where is Adam's scripture?

There is a *Book of Adam*, but it is not even written from Adam's point of view and is an apocryphal work by an unknown author thousands of years after the fall from paradise. It contains spurious details about Adam and Eve that are not found in the Bible. There are also other spurious ancient texts bearing Adam's name, such as the *First Book of Adam and Eve* and the *Second Book of Adam and Eve*, but none seem to be based remotely on anything Adam wrote.

Adam and Eve were literate as attested to by the fact that their children "were taught to read and write, having a language that was pure and undefiled" and were "called upon by God to write by the spirit of inspiration" and a "book of remembrance was kept". (*PGP*, Mo 6:5-6) Unfortunately, the whereabouts of Adam's book of remembrance is unknown.

Perhaps Adam's book survived in some form and was among the records Abraham had from "the patriarchs" containing a knowledge of "the beginning of creation" as "made known unto the fathers". (*PGP*, Abr 1:31) But Abraham didn't say whether Adam had authored any of those records.

Moses gets credit for Genesis and for compiling the rest of the Pentateuch. Although the creation story was had by Abraham, God revealed it again to Moses, "I reveal unto you concerning this heaven, and this earth; write the words which I speak." (*PGP*, Mo 2:1) So, if a version of the creation saga had been written by Adam, it was either lost or inaccessible to Moses.

Were Adam and Eve bald?

Since Adam and Eve could not die, they must have bodies with cells that could not die. But hair and fingernails are made of dead cells. So, did Adam and Eve lack hair, eyelashes, eyebrows, and nails on their fingers and toes?

Did lions, tigers, and bears have fur and claws? All creatures not partaking of the fruit had cell death cycles commensurate with the special ecological and spiritual circumstances in the Garden of Eden. But monkeys and squirrels might have been helping themselves to the fruit. Were they furless? What about birds? Feathers are made of dead cells. Did the birds that perched in the Tree of Life and ate of its fruit have feathers?

The Tree of Life fruit was not a wonder drug. In symmetry with the Tree of Knowledge, Adam's and Eve's use of its fruit signaled a choice that invoked the power of God upon them. In fantasy fiction it would be said that an enchantment boosted the health of Adam and Eve when they chose Tree of Life fruit. The spell had no effect on anyone or anything else.

So, an ant eating the fruit of the tree still died of old age, squirrels still had hair, and birds still had feathers. And, the trees in the Garden of Eden likely had bark and heartwood, both of which are comprised primarily of dead cells.

It's difficult to picture Adam and Eve without hair. So, it's best to think that they had hair and nails before being placed in the garden and were preserved in that condition from the moment they began eating the fruit of the Tree of Life.

How was Adam married?

God made Eve "and brought her unto the man". (Gn 2:22) That's it! We know nothing of any marriage ritual – no music, no vows, and no rice – at least not that are in the record.

That Adam knew right then he was married is obvious from his advice to his posterity; "They shall be one flesh." (Gn 2:24) The next verse cinches it; "[they] were both naked, the man and his wife, and were not ashamed." (Gn 2:25)

Orson Pratt said, "Says one—'I never knew before that immortal beings were to be connected as husbands and wives, I thought marriage pertained to mortality, and until death should us part...' Two immortal beings--Adam the bridegroom, Eve the bride, stood up together, and the Lord gave the bride to the bridegroom." (*JD*, bk18, 1875, p48)

According to the book of Jubilees, Adam and Eve had sexual relations as soon as God united them. (Jub 3:6) By placing Eve and Adam together and pronouncing them man and wife, God married them for eternity. Whatever else He may have said is apparently none of anyone else's business. Like many Bible marriages, the union was arranged by their Father but it seems both parties were quite amenable to it.

Was Eve Adam's only Wife?

Adam's first wife was Lilith according to ancient Hebrew mythology. She fought with Adam and split. God created Eve to replace her. The idea for Lilith was probably inspired by the Sumerian *Lillu* or the Mesopotamian night demons called *lillin* but was used to explain why an unnamed female was created on Day Six ("male and female created he them") and another named Eve on Day Seven. (Gn 1:27; 2:22)

Lilith was born of a misinterpretation of the Bible's first two chapters. An elaborate fable developed over time. Lilith dumped Adam because he tried to force himself on her in a moment and manner unacceptable to her. When confronted by God for leaving Adam, she protested that she should not have to be submissive, arguing that she had been created from the same dust as Adam and was therefore his equal.

When God did not solve her problems, she rebelled and began consorting with demons, which resulted in her giving birth to countless minions who have been haunting babies and men while they sleep ever since, doing unspeakable immoral things to them. As the queen succubus, she is blamed for nocturnal emissions, stolen babies, crib injuries, and sudden infant deaths, to name a few. To protect their babies, many Jews placed garlic, red ribbons, and amulets with warding scriptures or images of angels in baby cribs, sometimes around the infants' necks.

This "female demon of the night" became associated with the Hebrew word for "screech owl' or 'night bird' or 'night creature' (Heb לִילִית) and is transliterated to *liyliyth*. (Str#3917) This became *Lilith* in modern times and is a common name for witches, tramps, crazies, seductresses, dominatrixes, porn actresses, strident feminists, strong dark females, or the daughters of parents weird or naïve enough to give their baby that name.

Hebrew moms used Lilith as a bogeywoman to scare their children into obedience, telling them she would steal or eat naughty children. But there is no mention of her in the Bible or revered Hebrew literature. Scholars generally consider her to be a myth that was likely borrowed from the Babylonians and tacked onto the creation tradition.

If there had been a wife prior to Eve, Adam would have given her that name "because she would become the mother of all the living." (*NIV*, Gn 3:20) "The Hebrew name חַוָּה (chava) has a root connection… to words such as חַי (chai) and חַיִּים (chayim) communicating the idea of "life". (lp.eteacherhebrew.com/lp_et_modern_hebrew_eve-en.html, DL 24 Aug 2024)

Why no kids in the garden?

In the most common telling of the story, Adam and Eve got themselves kicked out of the garden before they had time to produce any children. Perfect bodies would normally result in pregnancy within one month of sexual intimacy. So, if the common view of the story is correct, Eve was so gullible that she succumbed to temptation in less than a month!

Was their relationship platonic until after their banishment from paradise? Did Adam and Eve lack libido until their bodies became completely mortal? Or were they too naïve to discover the mechanics of coitus before they fell? The Bible doesn't mention intimacy between Adam and Eve until Cain was conceived *after* they were expelled. (Gn 4:1) But that doesn't mean there was none. Nonetheless, an ancient Armenian text says that the couple were virgins when they left the garden. (*Armenian Apocrypha Relating to Adam and Eve*, Michael Stone, "The Sethites and the Cainites," p203)

On the other hand, it was one of Adam's first declarations that "a man... cleave unto his wife". (Gn 2:24) So, it's unlikely that intimacy never crossed *his* mind after receiving his perfectly beautiful wife. He never questioned whether she was "suitable for him". (*NIV*, Gn 2:18)

Surely God would not have commanded them to "be fruitful and multiply" without their knowing what that entailed! (Gn 1:28) The prophet Nephi said, "the Lord giveth no commandments unto the children of men, save he shall prepare a way for them that they may accomplish the thing which he commandeth them." (*BOM*, 1 Ne 3:7) How did God expect Adam and Eve to multiply if they were incapable of reproducing before the fall? The answer is in the question; *the fall was the path to progeny*!

God told Eve, "With painful labor you will give birth to children." (*NIV*, Gn 3:16) This was a warning and a promise. Eve would start bearing children *and* it would be painful!

Nephi's brother wrote, "If Adam had not transgressed he would not have fallen... and *they would have had no children.*" (*BOM*, 2 Ne 2:23) Eve said, *"Were it not for our transgression we never should have had seed."* (*PGP*, Mo 5:10-11) Leaving paradise was a prerequisite to childbearing.

Fourth-century Archbishop Timothy of Alexandria wrote, "[Adam and Eve] were even as the angels of God." (*Discourse on Abbatôn*, Brit. Mus. Ms. Oriental, #7025)

Another text says that when God created Adam, the spirits in heaven expressed awe of his splendiferous and glorious appearance. His face and eyes were described as shining like the sun and his body sparkling like crystal. They were greatly moved by his beauty and the divine honor given to him, highlighting the moment's cosmic significance and the special place of Adam in creation. (sacred-texts.com/chr/bct/bct04.htm, DL 4 Jun 2025) It was their unmortal angelic nature that prevented conception.

So, their bodies were in a translated-like state and, as Elder Bruce McConkie wrote, "Translated beings cannot be killed nor can they procreate." (*DNTC*, bk4, p189, 1994)

Their cells did not age, sicken, die or divide. Placental tissues *must* die and apoptosis, or programmed cell death, is mandatory for embryonic development. If cells couldn't die while others divide and differentiate, an embryo would not morph into a baby. Apoptosis is an important part of development on into adulthood.

So, angels just don't get pregnant. After being ejected from Eden, Adam and Eve reproduced as soon the effects of their angel-like condition wore off. This may have taken months or years, during which time the couple could get their farming going without having to lug babies around the fields.

Could ants be crushed?

The no-death-before-the-fall idea has raised a few questions. For example, didn't insects that Adam accidently trampled get crushed? One response has been to question whether small bugs are considered to be "alive" by the Bible.

Somehow Adam and Eve knew what death was. When God warned, "thou shalt surely die," they didn't have to ask, "What is death?" Unless there was some kind of bubble barrier around and over the garden to keep birds, beasts, and bugs out, Adam and Eve would have noticed the cycle of life and death firsthand.

They would also have noticed plant death. Eating the fruits of the garden killed cells. Picking a flower to put in Eve's hair doomed that bloom to die.

A staunch defender and general authority of The Church of JESUS CHRIST of LDS wrote, "To limit… the whole of life and death to this side of Adam's advent… some six or eight thousand years ago… is to fly in the face of the facts so indisputably brought to light by the researcher of science in modern times." (*The Truth, the Way, the Life*, B.H. Roberts, 1995, from 1920s manuscript)

So, sufficient archaeological evidence existed in the early 1900's to convince Roberts that living things were dying prior to Adam's transgression. In fact, millions of years of dying happened millions of years ago to create the vast oil reserves beneath the nations of Iraq and Kuwait where Eden was.

Similarly, for those who supposed that Eden was in Jackson County, Missouri, it sits on hundreds of feet of sedimentary rock millions of years old - shale and sandstone containing fossils of crinoids, corals, brachiopods, and other aquatic creatures along with limestone composed mainly of skeletal fragments and calcium from marine organisms such as coral, sponges and mollusks.

Yes, there was death before the fall, both inside and outside the Garden of Eden. So, Adam and Eve did not kill everything and everybody, just the ants they stepped on and the flowers and fruit they picked. But, by losing access to the Tree of Life fruit, they killed themselves, though very slowly.

Which tree had bitter fruit?

The ancient American prophet, Lehi, spoke about "the forbidden fruit in opposition to the tree of life; the *one being sweet and the other bitter*". (*BOM*, 2 Ne 2:15) So, which fruit was bitter and which was sweet?

Eve thought the forbidden fruit was delicious, and that is enshrined in temple ritual. She "saw that the tree was good for food". (Gn 3:6) Brigham Young recounted, "[Eve] tastes of it, and does not die, and likes it so well that when Adam comes along she says, 'Husband, this fruit is delightful; I have tasted it, and it is desirable to make one wise; take some.'" (*JD*, bk16, pp40-41)

So, if the Tree of Knowledge fruit was delicious, then the Tree of Life fruit must have been bitter – right? No, Lehi himself declared its fruit to be "most sweet". (*BOM*, 1 Ne 8:11) Later, the prophet Alma spoke of eternal life as "the fruit of the tree of life... which is sweet above all that is sweet". (*BOM*, Alma 32:42) So, how can either fruit be bitter as Lehi claimed?

Lehi was speaking metaphorically. Both types of fruit were delicious to the palate, but the forbidden fruit was bitter to the soul because partaking of it brought banishment from paradise, separation from God, aging, injury, suffering, sickness, sweat, and eventually death. Hence God's warning.

But Christ would "drink the *bitter* cup" and save Adam and his posterity from the nasty consequences of eating the forbidden fruit. In opposition to the Tree of Knowledge stands the Tree of Life, a representation of Christ's atonement, bringing His sweet gift of salvation, "a representation of the love of God". (*BOM*, 1 Ne 11:25-27)

Why the trees' names?

God set two trees against each other, "The forbidden fruit in opposition to the tree of life". (*BOM*, 2 Ne 2:15) So, why didn't the names reflect that? Why not "The Tree of Death" versus "The Tree of Life" or perhaps "The Tree of Knowledge" versus "The Tree of Ignorance"?

The trees were in opposition to each other in the sense that they were mutually exclusive. The consequences of eating of them were not exactly opposite, rather were asymmetrical. Adam and Eve had God's permission to eat of the Tree of Life, along with all the trees except the Tree of Knowledge

— eating its fruit was a one-way path out of the garden. Only the Savior could provide a path back to a Tree of Life.

The one tree was a promise of sustained life. The other tree was a promise of new knowledge; "Wherefore, the Lord God gave unto man that he should act for himself. Wherefore, man could not act for himself save it should be that he was enticed by the one or the other." (*BOM*, 2 Ne 2:15)

Of course, Adam and Eve acted for themselves every day, choosing among many varieties of permitted fruits. God invited them to eat of all the trees save one. Such mundane choices had no impact on their eternal destinies, whereas the choice between the acquisition of great knowledge and status quo certainly did. Without knowledge, they would be damned, unable to become more like God.

To realize their potential, Adam and Eve had to disqualify themselves from God's presence and experience mortality. God warned them, "Thou mayest choose for thyself, for it is given unto thee; but, *remember that I forbid it*, for in the day thou eatest thereof thou shalt surely die." (*PGP*, Mo 3:17) In mortality they could "taste the bitter, that they may know to prize the good". (*PGP*, Mo 6:55)

In short, God set up two trees "to bring about his eternal purposes". (*BOM*, 2 Ne 2:15) What was His eternal purpose? "To bring to pass the immortality and eternal life of man." (*PGP*, Mo 1:39) Never dying is not eternal life. Everyone will resurrect. Eternal life is being saved by Christ to the kind of Life that God lives. Joseph Smith said, "A man is saved no faster than he gains knowledge." By choosing the Tree of Knowledge, Adam and Eve opened the gate leading back to the Tree of Life whose gift of immortality awaits His true followers.

What species were the trees?

Tree of Life

Joseph Smith dubbed one of his revelations "the 'Olive Leaf' … plucked from the Tree of Paradise". (DC 88 preface)

Poetic, but not proof that the Tree of Life was an olive tree. Millions of people have eaten olives regularly without having their mortality suspended. Smith was using the olive leaf metaphorically as did John when he wrote that "leaves of the tree [of life] were for the healing of the nations". (Rv 22:2)

The ancients used olive oil to promote health. The health benefits of it are now widely known. So, it has a health association to the Tree of Life and the oil still used today by priesthood bearers in administering to the sick and afflicted.

The Ethiopian Bible includes this description of the Tree of Life and its fruit; "Its leaves and bloom and wood wither not forever; its fruit is beautiful and resembles the dates of a palm." (*Ethiopic Book of Enoch* 24:4)

The term *species* is defined by scientists and has undergone many changes over time. The Tree of Life is *sui generis* (in a class by itself) so, by definition, it must be its own species, in which case, trying to identify it with a known species is folly.

Tree of Knowledge

The art of Western Europe often depicts the forbidden fruit as an apple, probably for the same reason an apple was used in the story of Snow White. It was a very common fruit in the region. The larynx in the human throat became known as the Adam's apple from the notion that the fruit stuck in Adam's throat. This myth would suggest that Eve took a smaller bite since women have a much smaller Adam's apple.

Certain Jewish rabbis have said that the forbidden fruit was a fig. Michelangelo Buonarroti even immortalized a fig tree as the forbidden fruit in his masterpiece fresco on the Sistine Chapel ceiling. "And the eyes of them both were opened, and they knew that they were naked; and they sewed fig leaves together, and made themselves aprons." (Gn 3:7) But it seems awfully silly for intelligent people to cover themselves with the leaves from the tree of their crimes.

Using any kind of leaves for clothing was such a bad idea that it could only have come from Satan. Surely Eve knew that leaves were not stylish and surely Adam must have known leaves were not durable. Furthermore, the scanty fig leaf aprons failed to mitigate their sense of nakedness when God arrived. Like with the bikini, Satan was having fun.

Midrash Rabbah opined that the fruit of the Tree of Knowledge was an etrog (a citron similar to a lemon). But anything that tasted like a lemon would have to be ruled out by the scriptures that describe the Tree of Knowledge as delicious.

Other traditions say grapes were the forbidden fruit. Some even say the grape vine was entwined around the Tree of Life. That would have enabled them to eat both types of fruit simultaneously, perhaps in a smoothie. The *Apocalypse of Abraham* says the tree was large enough for Adam and Eve to stand under it and that "the fruit of the tree was like the appearance of a bunch of grapes of the vine". (marquette.edu/maqom/box.pdf p44, DL 15 Feb 2020)

An ancient account of Enoch's visit to the Garden of Eden said, "The tree of knowledge also was there, of which if any one eats, he becomes endowed with great wisdom. It was like a species of the tamarind tree, bearing fruit which resembled grapes extremely fine; and its fragrance extended to a considerable distance." (*Book of Enoch* 31:3-4) The tamarind is a leguminous tree that matures to between 40 and 60 feet.

It would be poetic if olives were the forbidden fruit. Israel's economy at the time of Christ centered on olives. And, Israel itself was compared to an olive tree. (*BOM*, Jacob 5) But, the fact that olives are incredibly bitter fresh off the tree rules them out as the forbidden fruit. They are only edible after being soaked in salt brine for several months, a process that very hungry ancient people likely discovered after years of trial and error. Adding lye to the brine speeds up the process but they had to first learn how to extract lye from wood ash.

An opinion from the Midrash is that God withheld the identity of the forbidden fruit so as not to tarnish the reputation of a tree which itself was not at fault. There is no reason to believe the Tree of Knowledge was supernatural.

Any attractive and delicious fruit could have served God's purpose.

If Eve judged correctly "that the tree was good for food," then the tree wasn't inherently evil and it wasn't the nature of the fruit that caused the fall, rather the act of disobedience to God. The fruit was just as nutritious other trees in the garden. Any tree would have served to test obedience. The monkeys, squirrels, birds, and bees that partook of the forbidden fruit did not suddenly get wise, though they may have distributed its seeds. But they hadn't been commanded not to partake.

Were the aprons the 1st clothing?

The lice that have plagued human bodies below the neck over the millennia are, in reality, clothing lice. Head lice are a different species. Research into the evolution of body lice strongly suggests that humans began wearing clothes about 170,000 years ago, thousands of years before they first migrated from warm climates to cooler ones.

The oldest evidence of sewing is a 50,000 year old needle made from a bird bone found in Siberia's Denisova Cave. It is almost 3 inches long with an eye for thread carved at the top. Evidence from China indicates the use of sewing needles 45,000 years ago. Animal sinews or plant fiber were used as threads to stitch clothing components together from animal skins, furs, grass, leaves, or bark. (thoughtco.com/history-of-clothing-1991476, DL 9 Nov 2023)

However, early clothing probably did not need stitching. Humanity needed coverings far earlier than the invention of sewing needles. Copying animals by using their hides was the obvious think to do. Hence, aprons of leaves would likely not be the first apparel worn by humans living outside Eden.

17: The Fall

Did God expect Adam to fall?

If God didn't want Adam and Eve to partake of the fruit of the Tree of Knowledge, why did He plant it in the garden? Almost every answer includes the fact that God wanted Adam and Eve to have the choice to obey or disobey Him. Why did they need to make that choice? Because love is an action verb and Adam and Eve could not learn to love God without exercising their agency to submit to His will. To love God is a prerequisite to glory with Him in His kingdom; "This is the first and great commandment." (Mt 22:37-38)

That God knew Adam and Eve would fall is evident from the fact that He cast Satan to Earth where he could tempt Eve. God could have locked Satan in a phantom zone outside this universe, but then Eve might not have been persuaded that the fruit would make her wise "like God". (*NIV*, Gn 3:5)

That God planned that Adam would fall is evident from the fact that He had already commissioned the premortal Jesus to save Adam's family from the resulting spiritual and physical deaths. "He indeed was foreordained before the foundation of the world" to be that sacrifice. (1 Pt 1:20)

Timothy, archbishop of Alexandria, basing his comments on ancient texts he found in the famous library there, paraphrased the premortal conversation between the Father and the Son about Adam; "My Father said unto Me, 'If I put breath into him, My beloved Son, Thou wilt be obliged to go down into the world, and to suffer many pains for him before Thou shalt have redeemed him...' And I said unto My Father, 'Put breath into him; I will be his advocate, and I will go down into the world, and will fulfill Thy command.'" (*Discourse on Abbatôn*, Brit. Mus. MS. Oriental, #7025)

Did God's foreknowledge deprive Adam of his agency? No. Simply because parents can correctly predict their children's choices doesn't in the least rob them of agency. It would be unjust for parents to punish or reward children for errors the parents knew they *would have made* if they'd been allowed to choose. Likewise, God did not rob Adam and Eve of their agency by kicking them out of the garden before they actually ate the forbidden fruit. He let them choose so they could own the consequences.

Why were Adam and Eve naïve?

The fact that Adam and Eve failed to recognize Lucifer behind the serpent is evidence that they had lost all memory of their premortal existence, just like all humans. But, if they were functional adults in the Garden of Eden, they must have known something. The Bible doesn't say they were naïve. This assumption is made because they were unashamedly naked and sufficiently enough for Satan to manipulate them.

The prevailing belief is that God created Adam and Eve as fully-grown adults, skipping right over childhood. They were made with just enough information and cognitive abilities to enable them to walk, talk, eat, move, and think. But, that's not true. God said Adam was "born into the world by water, blood, and the spirit". (*PGP*, Mo 6:59) President Joseph F. Smith taught that Adam was "born of woman". (*Deseret News*, 27 Dec 1913)

So, who raised Adam and Eve? If their mortal parents had brought them up, their youth experiences would hardly have left them naïve. Only God could have raised them without exposing them to life's school of hard knocks. But, if they weren't placed in the garden until they were adults, where did they grow up?

God must have taken Adam and Eve as babies, while they were still innocent, transfigured them to compatibility with His glory, raised them in His own home, and moved them to the garden free from sin, and comfortable in their nakedness. There were no celebrities, peers, advertisers, or social media to create body shape standards, so their bodies were the most beautiful they knew. And they accepted the physical differences between them as normal. The animals didn't dress every morning, so they knew nothing of clothes. God hadn't told them to avoid nudity and Satan would wait until after he got them to transgress before body shaming them.

Why didn't God just teach them about good and evil instead of making disobedience a prerequisite to wisdom? Knowing *about* it does not equal *living* it. Even God the Son had to enter sojourn in a mortal body, "suffering pains and afflictions and temptations of every kind; and… take upon him the pains and the sicknesses of his people…And… take upon him death… and… take upon him their infirmities, that his bowels may be filled with mercy, according to the flesh, that he may know according to the flesh how to succor his people according to their infirmities". (*BOM*, Alma 7:11-12) Despite being the omniscient God, Jesus still needed to "know according to the flesh", as did Adam and Eve.

God taught them only what they needed to know to for the Garden of Eden scene. Since He did not give them any moral code, they remained as innocent as toddlers; "All little children are alive in Christ, and also all they that are without the law." (*BOM*, Moroni 8:22) Any mistakes they made would not be considered evil because Christ's "blood atoneth for the sins of those... who have ignorantly sinned". (*BOM*, Mosiah 3:11) So, Adam and Eve started their garden life with clean slates.

How naïve were they?

Adam and Eve were naïve only in that they lacked a knowledge of God's laws vis-à-vis good and evil; "If ye shall say there is no law, ye shall also say there is no sin. If ye shall say there is no sin, ye shall also say there is no righteousness." (*BOM*, 2 Ne 2:11) So theoretically, if Adam got annoyed at his wife for some small thing and said something unkind, this would not have counted as sin, for God had not yet shared with them His "law" regarding the treatment of others.

There are totally naïve people to this day. The Pirahã jungle tribe in Brazil make decisions solely based on anticipated short-term outcomes rather than any sense of right and wrong. For example, some of them once readily agreed with a travelling salesman to murder a visiting researcher in exchange for booze. Fortunately, the researcher, who understood the language, overheard discussion of the plot and escaped with his wife and three children. The salesman was guilty of conspiring to commit murder because of his knowledge of right and wrong. But the jungle native didn't know the difference.

The Pirahã have no names for colors, no words for numbers, and no past or future tense. There is no system of writing and they live and think only in the present. There is no oral tradition and no history passed from parents to children for entertainment, education, or preservation of heritage. Their view of creation is, "Everything is the same, and things always are." They do not believe in a higher being of any kind. (spiegel.de/international/spiegel/brazil-s-piraha-tribe-living-without-numbers-or-time-a-414291.html, DL 5 Aug 2016)

Adam and Eve were much better educated than the Pirahã. But, like the Pirahã of the Amazon, Adam and Eve knew nothing of right and wrong. Adam was nice to Eve because it was in his nature to please her and she reciprocated. They preferred each other's company to loneliness.

God told them that He was their Father. They were aware that he was wiser and yearned to be knowledgeable like Him. He had taught them a few essentials:

♦ how to talk.

♦ how to garden.

♦ that they should reproduce.

♦ which tree would result in death.

But they did not know what lay beyond death. They did not know how difficult life outside the garden would be. They did not know pain and sorrow. They did not have definitions for good and evil.

Primitive tribes robbed other tribes of their food, stole their women, and enslaved or sold their captors. Without knowing any absolute morality, were those tribes more wicked than other godless tribes that peacefully grew their own food and minded their own business? We cannot impose gospel morals on naïve peoples; "Where no law is, there is no transgression." (Rm 4:15)

Dictionaries define "good" as that which is morally right. And "morality" is defined as principles that distinguish between "right" and "wrong." Then "right" is defined as "being in accordance with that which is good." And it's back to the start. The circularity of these dictionary definitions stands in stark contrast with what God eventually gave Adam and Eve, namely that good is any deed or condition pleasing to Him without any circuitous definitions. To be wise is to know divine principles that inform good decisions without the necessity of a law for each situation.

Was it a sin or a transgression?

God commanded Adam and Eve to multiply, something that could not be done unless they became mortal. But they could only become mortal if they ate the forbidden fruit. (*BOM*, 2 Ne 2:23) Disobeying God is a sin. So, it seems God set up a catch-22 where they would sin by avoiding the fruit and not having children, or sin by eating the fruit so they could.

Eve's desire to gain knowledge was righteous and Adam's desire to remain with his wife after she partook was also righteous and their desire to have children was righteous. Saul imprisoned Christians with righteous intent and thereby got himself a visit from Jesus and a new lease on his soul! God looks on the heart to differentiate transgression from sin.

A literal translation of God's advice to Adam and Eve is, "And Jehovah God laid a charge upon the man saying, from all garden tree food eat, [but] from tree [of] judgment good and bad you no eat because on the day you eat from [it] by death you will perish." (Gn 2:16-17)

Did God's instruction to eat of *all* trees constitute a commandment? If they missed a couple, would that have been a sin as much as God's command to not eat of the one?

God gave the naïve couple two mutually exclusive options and said, "Thou mayest choose for thyself, for it is given unto thee." (*PGP*, Mo 3:17) God's warning was an explanation of consequences, some of which would be unpleasant; "The Lord said to Adam that if he wished to remain as he was in the garden, then he was not to eat the fruit, but if he desired to eat it and partake of death, he was at liberty to do so." (*AGQ*, bk4, p81)

Transgression is often conflated with *sin* because all sins are transgressions. But not all transgressions are sins. If repentance is needed to invoke Christ's atonement for cleansing, then it is a sin. If forgiveness is granted without repentance, then it is merely a transgression.

The Lord said that little children "cannot sin, for power is not given unto Satan to tempt little children, until they begin to become accountable before me." (DC 29:47) When one toddler hits another or disobeys a parent, it is not imputed to them as sin, rather as transgression. Adam and Eve, being naïve, were in the same class as little children. Likewise, mentally delayed people and people raised in isolation from spiritual light cannot sin in total naiveté.

If the fall had been a sin, Adam and Eve would have needed to repent. But they did not repent of their transgression. Instead, Eve rejoiced, "Were it not for our transgression we never should have had seed, and never should have known good and evil, and the joy of our redemption, and the eternal life which God giveth unto all the obedient." (*PGP*, Mo 1:11) But God forgave them anyway; "The Lord said unto Adam: Behold I have forgiven thee thy *transgression* in the Garden of Eden." (*PGP*, Mo 6:53)

With the transgression forgiven, there was no guilt to be passed to their posterity; "Men will be *punished* for their own *sins*, and not for Adam's *transgression*." (*PGP*, AOF #2) Sin is done with adequate knowledge. Adam and Eve did not know how to keep both commandments simultaneously. Furthermore, eating a fruit is not inherently wrong or offensive to God.

Sins offend God, but he was not offended. Joseph Fielding Smith said, "I never speak of the part Eve took in this fall as a sin, nor do I accuse Adam of a sin... This was a transgression... but not a sin... it was something that Adam and Eve had to do!" (*DOS*, bk1, pp114–15) Dallin Oaks wrote, "The act that produced the fall was not a sin – not inherently wrong – but a transgression – wrong because it was formally prohibited." (*The Great Plan of Happiness*, CR, Oct 1993)

Babies are often self-centered, gluttonous and injudicious. But they are forgiven because they are naïve, like Adam and Eve; "All little children are alive in Christ, and also all they that are without the law." (*BOM*, Moroni 8:22) Christ's atonement unconditionally covers the transgressions of children and childlike adults. All who have not learned about the laws of the gospel are in the childlike category; "Sin is not imputed when there is no law." (Rm 5:13) Sin is also not imputed when there is ignorance of the law; "Jesus said unto them, If ye were blind, ye should have no sin: but now ye say, We see; therefore your sin remaineth." (Jn 9:41) Speaking to the Greeks at Mars' hill, Paul said of their idolatry, "The times of this ignorance God winked at; but now commandeth all men everywhere to repent." (Acts 17:29-30)

Did eating the fruit bring sin?

Without knowledge of the Law of the Lord, no creature could be guilty of sin; "For until the law sin (transgression) was in the world, but sin is not imputed when there is no law." (NKJV Romans 5:13) But as God revealed His law through Adam, mankind became accountable before Him such that knowingly transgressing His law constituted sin.

So, eating the forbidden fruit did not bring sin into the world direction. Rather it opened the door for God to reveal His law to Adam and Eve and their posterity. At the center of the Law was the promise of a Savior to redeem all penitent sinners. Prior to the first prophet, Adam, sins were deemed transgressions in naiveté and were covered by Christ's redeeming sacrifice.

Were the trees allegorical?

Was the Tree of Knowledge merely a metaphor for something? Clement (2nd century) speculated that Adam and Eve's transgression was their rush into intimacy while still immature. (*Adam, Eve, and the Serpent*, Elaine Pagels, 1988, p28) Down those lines, other scholars and literary theorists have speculated that the Tree of Knowledge represented forbidden sexual intercourse with the serpent as a phallic symbol and they therefore covered themselves afterward in shame.

Some say the Tree of Knowledge represents the gaining of understanding from the school of life. Others say the forbidden fruit represents the one evil thing in a person's life that they are too weak to resist. Still others say it represents temptation in general. Tertullian (2nd century) used the story to warn against gluttony. (*Adam, Eve, and the Serpent*, Elaine Pagels, 1988, p63)

Closer to the truth, Jewish tradition says that the eating of the forbidden fruit represents the mixing of good and evil together for the first time.

As for the Tree of Life, it is a common element in many cultures around the world. Ancient Egyptians used the Tree of Life to represent the branching causal events in the sequence and method of creation. In the Kabbalah, the Tree of Life represents the nature of God and His creative processes. The former Pope Benedict XVI said the real Tree of Life is the cross. The Tree of Life in Lehi's dream symbolizes the "love of God" as manifested through the condescension of His Son to become human and save mankind. The fruit is therefore "the greatest of all the gifts of God". (*BOM*, 1 Ne 11:9-22; 1 Ne 15:36)

The Hebrew from which "evil" is translated in the KJV of Genesis 2:9 referring to the forbidden fruit is *ra*, primarily meaning "adversity". (רַע, Str#7451) Elsewhere in the KJV *ra* is rendered "evil" where it can't possibly mean that in the modern sense. When God said, "I make peace, and create evil" in the KJV, it is more accurately rendered "disaster", "bad times", "calamity", "woe", "trouble", and "sorrow". (biblehub.com/isaiah/45-7.htm, DL 24 Sep 2023)

In the same verse, the fruit of every tree but one is described as "good for food". The Hebrew from which "good" is translated in the KJV of Genesis 2:9 referring to the forbidden fruit is *towb*, primarily meaning "beauty" but also "pleasant" and "agreeable". (טוֹב, Str#2896) (biblehub.com/hebrew/2896.htm, DL 24 Sep 2023)

Thus we see that the original Hebrew has a likely breadth and richness of meaning that portended the experience Adam and Eve would have outside the garden after partaking of the fruit. Yes, both trees serve as symbols. But to say they are nothing more than symbolic implies that the entire Genesis story is merely symbolic – Light, Dark, Satan, Adam, Eve, God, Cain, Abel, Enoch, and Noah! At what point in scripture would reality start? Real things often serve as symbols too!

How did the fatal fruit work?

God drove Adam out of the garden "lest he put forth his hand, and take also of the tree of life, and eat, and live for ever". (Gn 3:22-24)

Would just one bite from the Tree of Life have canceled the effects of eating the forbidden fruit and allowed them to live forever? If that had been true, Adam and Eve would have run to the Tree of Life immediately after eating the forbidden fruit and saved themselves from death. God didn't show up right away, so they would have had time if the two trees were sufficiently close, especially if the Tree of Knowledge had been a grapevine wrapped around the trunk of the other tree, as some have supposed. Or, Eve could have told the Devil, "Hold that thought," while she fetched some live-forever fruit to eat right after tasting the forbidden fruit. Or, she could have dried some Tree of Life fruit in advance and hung it around her neck in case she happened to succumb to temptation.

If Adam and Eve were paying attention, they would have already eaten of the live-forever fruit. God said, "Of every tree of the garden thou mayest freely eat," except the Tree of Knowledge. (Gn 2:16) So, it's a good bet that Adam and Eve tried a wide variety of fruit including the fruit of the Tree of Life. Both trees were right "in the middle of the garden". (*NIV*, Gn 2:9)

If Tree of Life fruit had been a one-bite-live-forever deal, God would have separated the two trees to such a distance that Adam and Eve could not run from one to the other before He could intercept them. Since Adam and Eve didn't even try to run to the Tree of Life, it's a good bet that one measly bite of its fruit would not have immortalized them. No. They headed for a fig tree where they picked some leaves and painstakingly stitched some makeshift clothing, totally unconcerned that their window of opportunity to eat the antidote fruit was slowly closing while they sewed.

It's silly to think that God would create an antidote for the forbidden fruit and then block access to it as soon as someone needed it!

What makes the most sense is that Adam and Eve remained healthy and young as long as they ate regularly from the Tree of Life. The fact that they knew this is evidenced by the fact that they did not immediately rush to the Tree of Life after eating the forbidden fruit. And, even if they had gorged themselves on fruit from the Tree of Life after transgressing, the effects would have soon worn off after their banishment to the outside world where they could no longer feed on it. Regular doses were needed in sufficient amounts to keep them young and healthy.

Dr. Trent Stephens came to this same conclusion, "Adam and Eve were created as mortal beings, kept in a state of immortality by the Tree of Life, separation from that tree returned them to their original mortal condition." (fairmormon.org/conference/august-2003/evolution-and-latter-day-saint-theology-the-tree-of-life-and-dna, DL 10 Dec 2016)

So, the two trees were far too different to be polar opposites. A single bite of the Tree of Knowledge was a one-way ticket out of paradise because it was disobedience. On the other hand, Adam and Eve needed continued access to the Tree of Life to enjoy its blessings.

The Tree of Knowledge could have been any kind of fruit. As has been said in a homonym quip on what caused the fall, "It wasn't the *apple* in the tree, rather the *pair* on the ground."

Theologian and philosopher, St. Augustine (354-430AD), wrote, "Man was furnished with food against hunger, with drink against thirst, and with the tree of life against the ravages of old age." (*The City of God*, bk14, ch26) This idea is supported by details in the Apostle John's vision of the Tree of Life under the future reign of Christ on earth; "The leaves of the tree were for the healing of the nations." (Rv 22:2) So, the most sensible exegesis is that God created Adam and Eve mortal, the same way He creates every human, and then gave them access to the Tree of Life so they would stay young, be immunized against sickness, and acquire miraculous self-healing abilities to cover injuries of any kind.

The global anti-aging industry is about a 200-billion-dollar market. Supplements are being sold. Specialty clinics are helping their clientele look and feel younger. Thousands of scientists are researching causes of aging and advising people to avoid age accelerators, like ultraviolet rays, smoking, alcohol, stress, sleeplessness, poor diet, and inactivity. But there is still no cure for aging.

"All come from dust, and all return to dust." (Ec 3:20) This was true for Adam. Brigham Young said, "He was made as you and I are made, and no person was ever made upon any other principle." (JD, bk3, p319) Adam grew from a physical baby to an adult with the normal human body features of programed cell death (apoptosis) and senescence necessary for growth. But, the fruit of the Tree of Life suspended all such processes for Adam and Eve once they accessed the Tree of Life. It was their fountain of youth. As Trent Dee Stephens, Ph.D., Professor of Anatomy and Embryology said, "Adam and Eve were created as mortal beings, kept in a state of immortality by the tree of life, separation from that tree returned them to their original mortal condition."

(fairlatterdaysaints.org/conference/august-2003/evolution-and-latter-day-saint-theology-the-tree-of-life-and-dna)

The state of Adam and Eve before the fall was similar to that of the three translated Nephites; "That they might not taste of death there was a change wrought upon their bodies, that they might not suffer pain." (*BOM*, 3 Ne 28:38) The transformation of the three Nephites was done off-planet; "They were caught up into heaven, and saw and heard unspeakable things. And it was forbidden them that they should utter; neither was it given unto them power that they could utter the things which they saw and heard… for it did seem unto them like a transfiguration of them, that they were changed from this body of flesh into an immortal state, that they could behold the things of God." (*BOM*, 3 Ne 28:13-15)

Their glory was upgraded so that they could see with their spiritual eyes, just as Adam and Eve could before the fall as they talked with God. Mortality is suspended for translated people just as Adam's was before the fall, just as will be done to the bodies of the righteous during Christ's millennial reign. (Rv 22:2) Elijah was taken up to spare him from death. (2 Kg 2:11) Of John, Jesus said, "I will that he tarry till I come." (Jn 21:23) Translated women cannot bear children, as that would require cell death. Their state is similar Adam's in Eden. So, a Tree of Life was likely used for translating the bodies of Enoch, Elijah, John, and Enoch, who, as Apostle Paul put it, was "translated that he should not see death". (Hb 11:5)

When Enoch's body was translated, according to the Talmud, he and his city were taken up into heaven to their new home in the off-planet Garden of Eden. (Tract *Aboth,* Talmud, ch1) Enoch "was taken from amongst the children of men, and… into the Garden of Eden in majesty and honour". (Jub 4:23) Saint Irenaeus (AD 130-202) wrote, "The elders who were disciples of the apostles tell us that those who were translated were transferred to that place [the Garden]… and that there shall they who have been translated remain until the consummation [of all things], as a prelude to immortality." (*Against Heresies,* bk5, ch5)

Translated people go to the Garden of Eden so they have access to the fruit of the Tree of Life which keeps them healthy and whole until their missions in the flesh are complete, at which point they skip death and are glorified to immortality; "He that liveth in righteousness shall be changed in the twinkling of an eye." (DC 43:32)

That the state of Adam and Eve exempted them from the second law of thermodynamics (entropy) prior to the fall was confirmed by Abraham's statement on eating the forbidden fruit; "Of the tree of knowledge of good and evil, thou shalt not eat of it; for in the time that thou eatest thereof, thou shalt surely die… [time being highly durative] for as yet the Gods had not appointed unto Adam his reckoning." (Abr 5:13)

Was the fruit poison?

A fruit designed to trigger aging would have to reprogram almost every single system and cell in the body. No poison could do that without major negative side effects, none of which are mentioned in Genesis. So, how did the forbidden fruit inflict gradual aging and eventual death? Was there a virus loaded with deleterious designer genes in the fruit?

Eve quoted God as saying, "You shall not eat it, *nor shall you touch it*, lest you die." (Gn 3:3) Was the skin of the fruit so toxic that merely touching it would kill Eve? Merely touching the skin of even the most deadly poison dart frog won't kill a healthy person unless there's a break in the finger skin. The Emberá people eat what they kill with frog-poisoned darts with no side effects. No known plant or fruit is anywhere near the toxicity of the golden poison frog. And, after eating the fruit, it still took Adam 930 years to die.

An early edition of the LDS Bible Dictionary stated, "Before the fall, Adam and Eve had physical bodies *but no blood*." (*BD (Bible Dictionary)*, "Fall of Adam and Eve," pre-2013) If spiritual fluid had been circulating in their systems, no *physical* poison nor virus could have hurt Adam and Eve. God would have had to prepare a special fruit that would cause blood to begin circulating in their bodies or administer transfusions from compatible donors, if such could be found.

Fortunately, the Adam-had-no-blood idea was abandoned by The Church of JESUS CHRIST of LDS in the early 21st century with the 2013 republication of the scriptures in which Bible Dictionary entry was changed to "Before the Fall, there were no sin, no death, and no children." (*BD*, "Fall of Adam and Eve," post-2013)

There was nothing about the forbidden fruit that changed Adam and Eve. Its species didn't matter. Rather, by choosing to eat of it, Adam and Eve got themselves banned from the garden, cut off from regular access to the Tree of Life fruit.

After eating the forbidden fruit, Adam lived 930 more years, more than ten times today's life expectancy for men. That's not how poisoning works!

People who believe the forbidden fruit directly changed Adam's body *and* that there was no death in the world previously need to ponder how all the other life on Earth got poisoned without eating the fruit of the Tree of Knowledge.

If by one person death came upon all, then why did Eve have to persuade Adam to partake of the forbidden fruit? Just as she did not bring death to Adam by partaking, neither did they bring death to all other life on Earth by partaking. From this it's obvious that the poison theory is incompatible with the universal death theory, for there is no way all life on earth could have eaten of the Tree of Knowledge to acquire their mortality. And, if other life aged and died without eating the forbidden fruit, then the fruit was not what killed Adam!

Could the environment outside the garden have aged Adam and Eve? Unless there was a glass dome over Eden, its garden had the same atmosphere as the outside world. And unless God imported dirt from another planet, Adam and Eve walked on the same types of soils after they were exiled. And the fact that God made four rivers to flow in from outside the garden to water the garden is evidence that the fauna and flora inside and outside were very similar. So, the idea that the outside environment killed them makes no sense.

Invoking Occam's razor again, we see that the simplest explanation for aging is the most likely – the loss of access to the Tree of Life is what killed Adam and Eve! They and their posterity lived for centuries until the effects of the Tree of Life fruit in their bodies and their descendants' bodes waned.

As it has been said in a play on words, "The apple on the tree didn't cause the fall. It was the pair(pear) on the ground."

Why not let Adam live forever?

Why couldn't Adam and Eve have apologized to God, promised never to do it again, begged forgiveness, and returned to status quo in paradise?

One reason is that continuing to eat the Tree of Life fruit would have prevented Adam from having any descendants, among whom was Jesus, the One who would be sacrificed for his and everyone's transgressions. All other life would die without any possibility of being raised to true immortality.

Another reason is that Adam would have failed miserably as the first prophet to bring the gospel to the world as long as he was cloistered in paradise.

After ejecting them from the Garden of Eden, God posted "cherubim and a flaming sword… to guard the way to the tree of life." (*NIV*, Gn 3:24) Why? "Lest they partake of the tree of life and *live forever in their sins.*" (*The Healing Power of Christ*, J. Richard Clarke, BYU Devotional, 27 Mar 1984)

What sins? The fruit transgression did not rise to the level of sin. God was talking about the iniquities that Adam and his posterity would inevitably commit. Eating the forbidden fruit unleashed Satan upon the world and gave people sufficient knowledge to do things that were inherently wrong. This was apparent from the fact that Adam and Eve hid from God, an inherently wrong act, a sin. They would never have achieved the same level of comfort in God's presence as before.

So, if the Tree of Life had remained accessible to Adam's family, Cain might have helped himself to it after murdering his brother, and been quite content to live indefinitely in his sins. But it would have been a disaster for him and all the wicked people that followed, such as those in Noah's day, to be eating the fruit of the Tree of Life while committing atrocious turpitudes for millennia to come. Tribes would have been fighting each other over the limited fruit supply. Wars would have been waged to conquer the land of Eden for exclusive access to the tree and it would have fallen into the hands of whomever was the most powerful.

Finally, under no circumstances could Adam have literally lived *forever*. Unless God intervenes, the sun will eventually run out of fuel, expand into a red giant, and incinerate the earth and her wicked occupants. That would have destroyed all the wicked in their sins. The greatest tragedy would be sin for which repentance is never done. But, scientifically speaking, it would be impossible to live *forever*, let alone in sin. The Hebrew means *long duration* (Str#5769 עוֹלָם).

Why did Eve partake?

Is it a commentary on the innate nature of women that Eve was dissatisfied with the status quo while living in Paradise? Is it a commentary on men that Adam stubbornly refused to listen or consider his options? It's not in the Bible, but he may have bristled to discover that he was not the only sapient male creature in the Garden of Eden.

When God asked Eve what she had done, she said, "The serpent beguiled me." Or, as another translation puts it, "The serpent caused me to forget." (*YLT*, Gn 3:13) Forget that the fruit was forbidden? She partook of it just moments after reminding Satan, "Of the fruit of the tree which is in the midst of the garden, God hath said, Ye shall not eat of it, neither shall ye touch it, lest ye die." (Gn 3:3) If she really did forget, she did so willingly. People often push God out of their minds in the fog of making a forbidden choice.

The fact the fruit was forbidden may have been part of its appeal, but it was also naturally alluring; "The woman saw that the tree was good for food, and that it was pleasant to the eye." (Gn 3:6) She is not the only woman to have looked for new and interesting menu items. She likely had tried all the other varieties of fruit in the garden. So, much like Disney's Belle who was overcome by curiosity about the off-limits West Wing in the Beast's castle, Eve craved knowledge.

Satan's role in hyping the forbidden fruit cannot be overstated. He minimalized the negative consequences with a mixture of truth and lies. "Ye shall not surely die: For God doth know that in the day ye eat thereof, then your eyes shall be opened, and ye shall be as gods, knowing good and evil."

According to ancient texts, Satan also told Eve, "God was a man like you. When he ate of the fruit of this tree he became God of all." (*Adam and Eve and the Incarnation*, 4, M5913, as quoted in Michael Stone's, *Armenian Apocrypha Relating to Adam and Eve*, 1996, p25) Satan omitted a few details but he was telling the truth that partaking would make them more like the Gods, who said, "Behold, the man is become as one of us, to know good and evil." (Gn 3:22)

Eve "saw that the tree was... to be desired to make one wise". (Gn 3:6) What Satan told her was consistent with the name of the Tree of Knowledge. She wanted to become wise like her Heavenly Parent. Her name denoted motherhood and she yearned for it. The status quo did not satisfy her. She wanted progress. The key to realizing all her aspirations was the knowledge that the tree promised.

While previously pondering on the forbidden fruit, Eve must have thought about what would become of Adam if she partook and he refused. Would she survive all alone outside the garden? Would God would give Adam another wife? How could she fulfil her maternal destiny without him?

Despite all of these concerns, she must have eventually become confident that she could persuade Adam to join her in whatever destiny the forbidden fruit handed them and that their loving Heavenly Father would be merciful.

We should thank Eve. Without her determination to move things forward Adam might still be all alone in the Garden of Eden, doggedly eating of every edible fruit except one. God was counting on her and she came through.

Did Satan hand Eve the fruit?

Satan was a spirit back then and, as such, would have had difficulty manipulating physical objects. If it were not so, devils would be causing an unbearable amount of physical chaos in the world. So, he would not have plucked the forbidden fruit for Eve. Any representation that Satan pressed the fruit into Eve's hand is merely symbolic of his power to persuade.

Why did Adam partake?

"Adam was not deceived." (1 Tim 2:14) So, with a perfect idea of what was going on, why did Adam eat the fatal fruit?

When Eve brought some forbidden fruit to Adam and asked him to taste it, he asked, "Why have you eaten the fruit?" She replied, "The fruit is very sweet. Take and you taste, and notice the sweetness of this fruit." (*Transgression of Adam*, 6-9, M5913, as quoted in Michael Stone's, *Armenian Apocrypha Relating to Adam and Eve*, 1996, p25)

Determined to be obedient to God, come what may, Adam refused, "I cannot taste it." Eve began to cry and implored Adam, "*Do not separate me from you and do not abandon me in this nakedness...* Take and eat and because of love of you, God will turn and have mercy upon us." Adam thought about it for a while and reasoned, "It is better for me to die than to become separated and detached from this woman." And he partook. (ibid.)

Though the above conversation from ancient apocrypha was not available during Joseph Smith's day, he somehow managed to include all its elements in the temple script.

How did the fruit bestow wisdom?

Was the fruit of knowledge a wisdom pill? The fact that Adam's and Eve's next choices were quite unwise suggests otherwise. They stitched fragile fig leaves into woefully inadequate aprons and futilely tried to hide from an all-seeing God instead of gorging themselves on Tree of Life fruit or packing their bags for their trip to the outside world.

Neither did Adam and Eve realize they were naked all on their own. That they found out from Satan is obvious from God's question, "Who told thee thou wast naked?" (Gn 2:11) Harking to Satan's advice to eat the forbidden fruit invited him to further mess with their heads. The silly fig leaf aprons were obviously his idea also. If God had not previously limited him to just the forbidden fruit angle, he would have messed up all aspects of their paradisiacal lives much sooner.

The Tree of Knowledge had no magical properties to increase IQ or trigger a wisdom download from heaven. Adam and Eve had to acquire wisdom by making good and bad choices and reaping the results, the same way as everyone else. Solomon didn't need knowledge fruit to acquire his wisdom. Mozart didn't need music fruit. Einstein didn't need physics fruit. God lets people learn precept upon precept; line upon line; here a little and there a little. (Is 28:10)

When God confronted the couple about their transgression, Adam was extremely distraught. But God comforted him, "Adam, Adam, do not fear. *You wanted to be a god*; I will make you a god, not right now, but after the space of many years." (*Syriac Testament of Adam*)

In the ancient *Adam and Eve* books, God is quoted as saying, "You will work the earth and it will give you no fruit... You will be hungry and not sated. You will suffer from bitterness and not taste sweetness. You will be afflicted by heat and experience cold. You will poor and not rich. You will eat but remain malnourished." (*The Bible's Cutting Room Floor*, Joel M. Hoffman, 2014, p187)

The road to perfect wisdom is a long line-up of all the right experiences to maximize spiritual growth.

What if no fall had happened?

The bulk of Christianity believes that God's "Plan A" was for all humanity to thrive in a world-wide paradise forever and that, because Adam and Eve failed that plan, we are now living "Plan B" which necessitates a Savior. It approaches blasphemy to characterize our Savior as Plan B.

Nor does it make sense for an omniscient God to have a Plan B if He knows the future. To resolve the paradox, some say that God only knows what's knowable and that the outcome of placing the forbidden fruit tree in the Garden of Eden was unknowable.

Okay, let's assume for a moment that God did not foresee Adam's choice. He could have had the disobedient Adam and Eve die immediately and started Plan A over again with a new couple He deemed more likely to stick with Plan A. If the next Adam disobeyed, He could repeat that processes as often as necessary to get the desired outcome. The body count would probably have been lower than the carnage of the big flood or even the destruction of Sodom and Gomorrah.

Alternatively, God could have suggested to Adam that he kill the Tree of Knowledge to make sure his wife did not partake. Lacking an ax, he could have girdled it with a sharp rock (removed a ring of bark around the trunk). And, assuming God wanted Plan A, killing the tree would have prevented Adam's posterity from partaking of it.

But here's the nail in Plan A's coffin. If the entire population lived Plan A perfectly, humans would soon have overburdened Earth's capacity with a population of over a hundred trillion by now. And, if Plan A meant perpetual youth for animals and plants as well, the explosion of life would have destroyed earth's biosphere long ago. There would not be enough room to grow the food necessary to sustain life. Humanity would be dead or headed for Mars.

Plan A would also have left Adam and Eve and their posterity in state of naïveté. Trillions would be running around naked and naïve without any shame. And, lacking any knowledge of good and evil, society would be a dystopian culture of garden children.

At the latest, Plan A would end when the sun runs out of hydrogen fuel, turns into a red giant, and fries the earth unless God transplanted everyone to another star system to wait for the universe to die. Being ignorant of good and evil, none would be fit for heaven.

Playing out the scenarios illustrates the absurdity of the Plan A hypothesis that God didn't foresee everything. The godhead resides "on a globe like a sea of glass and fire, where all things for their glory are manifest, past, present, and future". (DC 130:7) God doesn't do guessing. What may look like a contingency plan to humans is actually *the* plan" because God knows beforehand whether it will be invoked and how it will play out.

Did all creation fall?

Many Christians believe in the "cosmic scope of the fall". (creation.com/the-fall-a-cosmic-catastrophe, DL 13 Jul 2018) The idea first became popular around the 16th and 17th centuries and has persisted in Christian circles until today. The LDS Bible Dictionary says; "After Adam fell, the *whole creation fell and became mortal*." (LDS.org/scriptures/bd/fall-of-adam-and-eve, DL 13 Jul 2018) The doctrine has even been sung in church; "This earth was once a garden place." (*Adam-ondi-Ahman*, Phelps, LDS Hymns, 1985, #49)

However, Genesis doesn't say Adam's transgression caused the entire universe to fall, including Earth, unleashing aging and death on all things. And Science proves just the opposite. There is overwhelming evidence that the creation is billions of years old. The question of how the forbidden fruit snack could have transformed the entire universe into a physical state will remain unanswered because it's not true.

If all life was cursed to die within one God day of 1,000 years, as many believe, then why have so many individual specimens lived far longer? One scientist coaxed a 34,000-

year-old salt-captured bacteria to reproduce. Another scientist planted seeds from a 32,000-year-old fruit found in an ancient squirrel cache and the resulting plant bore viable seeds. Pando, the clonal aspen in Utah, is more than 80,000 years old. And a Cal Poly biology professor was able to grow some yeast that had been trapped in amber for 45 million years. (sanluisobispo.com/news/local/article39142146.html, DL 20 Jul 2020) The immortal jellyfish (*Turritopsis dohrnii*) and the hydra never die of old age. So, if some organisms have escaped the curse of death for 6,000 years, perhaps it's time to reexamine the impact and scope of the Adam's fall.

If Adam and/or Eve imposed mortality on the world, then it was not *after* their choice, but merely *because* of it, meaning that God, knowing at least 13.8 billion years in advance that Adam would partake of the forbidden fruit, incorporated entropy into the universe from the get-go.

Similarly, penitent people received forgiveness and cleansing from sins *because* of Christ's sacrifice, *not only after* he suffered and died for them, but also throughout all the time prior to His death! However, in the case of physical death, the effects of the fall on all creation could be reversed *only after* Christ's resurrection.

The cosmos did not wait until Adam ate the forbidden fruit to fall. Eve ate it first! It would be sexist for the entire universe to wait until the *man* took a bite before falling! The Earth did nothing to incur punishment. Mortality was an opportunity and a blessing. Without temporality, Earth would never have attained celestial glory.

If the whole earth had been glorious and immortal – a garden place – before Adam fell, what made the beauty of Eden special? And why didn't the Garden of Eden immediately fall with the rest of creation? Adam and Eve were ejected into a world that was already lone and dreary, consistent with the temple narrative. When God said, "Cursed *is* the ground for thy sake... Thorns also and thistles shall it bring forth to thee," He wasn't cursing the ground at that moment. He was educating Adam as to the nature of the world that already existed outside the Garden of Eden. (Gn 3:17-18) When Adam fell, God simply "drove out the man". (Gn 3:24)

Some have repeated Brigham Young's opinion that the whole earth was originally "near the throne of our Father in heaven. And when man fell the earth fell into space and took up its abode in this planetary system, and the sun became our light". (*JD*, bk17, 1874, p143) If true, why do the results of tests done on meteors and moon rocks show that the planets were all created from the same swirling cloud of materials that collapsed to form the sun 4.5 BYA?

Furthermore, any classical journey through space from the vicinity of Kolob would have turned the earth into a snowball, stripped it of its atmosphere, and killed all life on Earth! And, it would have taken 27,000 years traveling at half the speed of light (310 million MPH). But Adam did not notice any acceleration of the planet nor the even the slightest change in climate after eating the fruit.

Did all the alien worlds lose their immortality *after* Adam ate the fruit? Did they wonder why they suddenly began to age, sicken, get injured, and die? Did God have to notify all the aliens in all the universes that someone far, far away on a world called Earth had just ruined life for everybody everywhere? Ridiculous! It's more likely that every world is given its own first prophet and prophetess and that their choices only affect their own world.

All geological and paleontological evidence is that Earth's rotation has been slowing and that life on its face has always been mortal! The fossil record contains hundreds of billions of years of history on thorny plants, weeds, predaceous animals, nasty parasites, slithery serpents, venomous spiders, and all kinds of biting bugs and stinging insects. And, nowhere in the fossil or geological record is there the slightest evidence of a 6,000BC catastrophic deterioration or change in Earth's climate, fauna, or flora. Earth's climate and biome has been incredibly stable for the last 8,000 years.

The entire evolution and character of the universe depends on 13.8 billion years of entropy with the full suite of physical laws active. So, it seems most sensible to believe the records

God wrote in the heavens and the earth. The creation itself tells us it has billions of years of natural law under its belt. Without entropy, fusion would be impossible and stars could not shine. Without entropy, stars could not explode and earth could not have been *"organized out of portions of other Globes* that has been Disorganized,"* as Joseph Smith said and with which science agrees. (*McIntire Minute Book*, WJS, 5 Jan 1841, pp60-61) The elements found on earth were all cooked up in the furnaces and explosions of stars.

The simplest assumption is that Adam and Eve shortened *only their own lives* circa 6000BC by eating forbidden fruit.

Why did Satan do it?

One way Satan could have foiled God's plan would have been not to tempt anyone. Satan remembered the premortal council and therefore knew that he was to play a key role in God's plan. So, why did Satan willingly go along with it?

Satan must have been kicked off the creation project before getting all the details! So, he planned "to beguile Eve, for *he knew not the mind of God*, wherefore he sought to destroy the world". (*PGP*, Mo 4:6) Perhaps he hoped to win the battle that would ensue between his forces and Jehovah's if Adam and Eve succumbed to his enticements. He also might have thought, if not him, some other devil would try it and he would lose his position as top devil! Or, perhaps he figured it would be worth it even if only few of souls who fought against him in premortality could be damned to hell. He especially hated Adam who, as Michael in the premortal realm, led the hosts of heaven in crushing him and casting him down.

Also, like a serial murderer already sentenced to death, Satan had nothing to lose. He could either stay in hell forever bored out his gourd doing nothing or delight in getting some vengeance for a few thousand years. Effort was not an issue as spirits never suffer fatigue.

Satan gloried in evil and relished the thought of recruiting mortal and deceased souls to his cause. And, he might have hoped to could get Adam and all of his posterity to "eat of the fruit of the Tree of Life and thereby live forever in their sins… in misery, therein being subject to the will of the devil". (*Commentary on the Book of Mormon*, Reynolds and Sjodahl, bk4, p222)

Being vainglorious, Satan wanted power and authority over as many individuals as possible. If he was weighing the decision whether to foil God's plan by not tempting anyone, he likely choose instead to maximize the number of spirits under his control. He may have even hoped to acquire sufficient power to take over the world, as evidenced by his efforts to conquer and eliminate the righteous entirely, even to the end of Christ's millennial reign. His position of authority had gone to his head. (DC 76:25)

The Talmud, written in the 1st and 2nd centuries, suggests that Satan wanted Adam to become mortal so he could kill him; "I will slay Adam and marry his wife, and I will be king of the whole world, I will walk erect, and will banquet on the best of the land." (Tract *Aboth*, ch1) Any intentions Satan had toward Eve were likely more malicious and vindictive than amorous. Perhaps he planned to possess a human from outside the garden to carry such a marriage.

Before their fall, Adam and Eve were immune to demonic possession. Satan hoped to possess a body or two from among their posterity after he had introduced sin into the world. And sure enough, he's had a few temporary successes. "The mortification of Satan consists in his not being permitted to take a body. He sometimes gets possession of a body but, when the proven authorities turn him out of doors, he finds it was not his but a stolen one." (*WJS*, 21 May 1843, p208)

If Satan stopped tempting, he'd be conceding that God's plan is perfect — and that's not in his playbook. The name "Satan" means "accuser" or "adversary" and it isn't just his job title — it's the identity he has adopted. His pride and rebellion would drive him to keep trying, even if he knew that failure is inevitable.

Does Satan tempt other worlds?

Brigham Young said, "Every world has had an Adam and an Eve... The first man is always called Adam and the first woman, Eve." (*Doctrine of the Priesthood*, 1854, bk2, #2, p13) He also said, "Every earth has its tempter." (*JD*, bk14, pp71-72)

If Lucifer were the tempter for all of God's planets, then his job would go on for forever, as worlds continue to be created ad infinitum. God has something else in store for Satan. After the final battle at the end of Christ's millennial reign, Satan will be "cast into the lake of fire and brimstone... tormented day and night for ever and ever". (Rv 20:10) Thus will end the career of Satan. Other worlds will have to be tempted by their own devils.

Brigham Young said, "There never was a time when there were not Gods and worlds, and men were not passing through the same ordeals that we are now passing through." (*DBY*, pp22–23) Ordeals include temptations. The scriptures say that Lucifer was cast down to *the earth*, not *earths* plural. (Rv 12:13)

Was the serpent metaphorical?

Snake? Why did it have to be a snake!? Because "the serpent was more subtle (crafty) than any beast of the field which the LORD God had made". (Gn 3:1) Hmm! The smartest snakes, like the king cobra, can remember basic things like a hiding or hunting place, and some successful fight moves, but they are bad at solving problems, acting mostly on instinct for complex behaviors.

Even if a snake were as smart as Einstein, it would be unable to talk as it lacks a larynx. It can't shape its lips or configure its tongue, teeth, and throat to form consonants and vowels. The best a snake can do is a hiss, huff, or wheeze – nothing that even sounds remotely like parseltongue.

So, it seems beyond impossible that a possessed snake could talk to Eve at all, let alone persuasively and seductively.

Eve would likely have recoiled at the sight and sound of such a serpent. So, why on earth did Satan choose a cold-blooded reptile instead of a cute and fuzzy mammal to entice Eve?

Is there a bad translation? No. The word translated as *subtle* comes from the Hebrew word *arum*, which also means *shrewd* or *prudent*. Also not true of snakes! But true of Satan!

What makes the most sense is that Genesis uses "serpent" as a metaphor to denote the shrewdness and slyness of Satan. He is called the dragon elsewhere in scripture, "that old serpent, called the Devil". (Rv 12:9) That's not to say there was no serpent there. In her naiveté, Eve may not have felt the natural revulsion to snakes that ordinary mortals feel. So, she may have had no problem exchanging thoughts telepathically with what appeared to be a harmless animal.

Did the serpent lose its legs?

God scolded the serpent for tempting Eve, "Upon thy belly shalt thou go, and dust shalt thou eat all the days of thy life." (Gn 3:14) A 1st-century translation from Hebrew to Aramaic says, "Upon thy belly thou shalt go, and thy feet shall be cut off." (*Targum of Palestine*, Gn 1) Midrash Aseret Malachim says, "When the snake is punished for leading them astray, it loses its legs forevermore."

Did just that one serpent lose its legs? Or did all populations of all 3,400 or so species of snakes in the world get a quadruple amputation along with a gene change that passed the curse on through all subsequent generations?

The archeological record in the earth shows that early snakes indeed had legs — about 150 MYA. (livescience.com/56573-mutation-caused-snakes-to-lose-legs.html, DL 20 Jul 2020) All snakes have vestigial leg bones. If God were to suddenly zap the genes for legs, wouldn't He do the job 100% and leave no vestiges of them? After all, when He changed the water to wine, it was excellent, not thin in the least!

Actually, the scriptures do not say that God removed the legs of any serpents. There are plenty of lizards with legs intact. There are several species of lizard that have no legs. Boas and constrictors have particularly conspicuous buds on their skin marking where the last vestiges of leg bones float in the muscles. So, although serpents evolved to slither instead of scamper, it's unlikely it was a result of God's pronouncement in Eden.

Besides, was it the snake's fault that Satan manipulated it? Can evil be imputed to such an ignorant and stupid creature? And, why would God curse all species of snake when only one was used by Satan? What could the poor snake be expected to do? What chance did it have against Beelzebub?

"Dust shalt thou eat…" This curse cannot be taken literally since snakes do not feed on dust. They stick out their tongue to taste the air. Only if the dust around them is disturbed do snakes "eat" dust to clean their tongues before tasting the next air sample. Snakes have been doing that for millions of years before the curse.

It's more likely that crawling on the belly and eating dust are metaphors for the degraded position of Satan. A serpent slithering on the ground is powerful imagery for his low predatory character. And God may have been providing an object lesson for humanity as He did when He cursed the fig tree that, having produced leaves without fruit, served as a metaphor for unproductive humans. (Mt 21:18-22)

What does the enmity mean?

"I will put enmity between thee and the seed of the woman… he shall bruise thy head, and thou shalt bruise his heel." (*PGP*, Mo 4:21)

Some say the enmity means the natural aversion to snakes that people have. But humans also have a natural aversion to bugs, spiders, and other creeping things that were not tools of Satan in Eden, that were not cursed by God. The enmity likely refers to the classic struggle between good and evil as Satan seeks the destruction of souls.

"Thou shalt bruise his heel" foretells the suffering Jesus. "He shall bruise thy head" refers to Christ's victory over death and hell. Christ shields us daily from Satan's full wrath.

Isn't it all God's fault?

Since Adam and Eve were created by a perfect God, shouldn't they have been perfect creatures, totally disinclined to make bad choices? Why isn't the Manufacturer to blame for their transgressions and for the sins of all his children?

The God of the OT appears to have taken credit for evil; "I form the light, and create darkness: I make peace, and create evil: I the Lord do all these things." (Is 45:7) Even though God is not responsible for the fundamental intelligence of which all things are made and is therefore not at fault for the bad choices they make, He did take responsibility for the sins of all mankind when He sent His Son to pay for them.

People are born with a vast variety of biological and situational potentialities, some of which facilitate righteous use of agency and some of which hamper it. Some people are born with bad circuitry in their heads. All animals are driven by survival instincts to one degree or another. Some people are born into situations that are abusive, lack parents, lack food, lack security, or lack opportunity. Is all of this God's fault? No. Mental, emotional, physical deficiencies and environmental challenges are opportunities for spirits to grow, no matter the species. Messy mortal experiences are part of God's perfect plan.

Every human has a history of choices that goes back to a point before the premortal spirit existence, before God began to mentor them. Every soul is the cumulative result of a chain of agency and accountability that began with the spirit essence or spark of light or intelligence that had no beginning. Spirits from imperfect stuff could not be totally perfect. And humans cannot be more perfect than the spirits that ensoul them. God does not force perfection on anything or anyone.

18: Mortality

Is nakedness a sin?

There is no sin in nakedness if you're a fish, cetacean, pig, naked mole rat, or any type of animal without a coat. There is also no sin in nakedness for babies. And, since Adam and Eve were as innocent as animals or babies, until they ate the forbidden fruit, their nudity was not considered evil!

Adam and Eve were the only humans in the entire garden, so they can't be accused of exhibitionism. They were married and the entire garden was their private living space. So, even by the highest of moral standards, their nudity in private was not sinful or unseemly.

So, why all the fuss about nakedness with nobody else present to see them? Well, God could see them, but that's true of anyone taking a bath or changing clothes. The same is true of Satan who may lurk and look at will.

Spiritually speaking, Adam was not naked. He and his wife were clothed in "divine light". (*Transgression of Adam*, 6-9, M5913, as quoted in Michael Stone's, *Armenian Apocrypha Relating to Adam and Eve*, 1996, p25) Another apocryphal work agrees, "Adam through this tree was condemned and was stripped of the glory of God." (3 Baruch 4:16) But, after partaking of the fruit, they lost their shroud of light. Ephrem, the 4th century Syrian Christian theologian, wrote, "It is because of the glory with which they were clothed that they were not ashamed. When it was taken away from them—after they had violated the commandment—they were indeed ashamed, because they were now naked." (*Commentary on Genesis*, 2:14)

Growing less naïve by the moment in the presence of Satan, Adam and Eve realized how glory-less they were compared to God, who always appeared in robes of light. They were no longer clothed in light. It was likely the Devil who told them they were naked, suggested the fig leaf clothing, and urged them to hide in shame at the voice of God.

Just hearing His voice overwhelmed them with guilt and shame; "Adam and his wife hid themselves from the presence of the LORD God amongst the trees of the garden." (Gn 3:8) God asked Adam an obviously rhetorical question, "Where are thou? And he said, I heard thy voice in the garden, and I was afraid, because I was naked; and I hid myself." (Gn 3:9)

Why clothes? Modesty rules were put in place after the fall so that Adam's posterity would be less likely to have inappropriate thoughts. Those parts of the human body that need to be covered are the same ones that serve as visual stimuli for mating responses. It's a simple and practical way to improve the odds of maintaining chastity. Children do best when raised by mature parents who are faithful to each other.

The descent in modern times from modest to immodest clothing, revealing dress, yoga pants, skimpy swimwear, etc., tracks with the increase in godlessness and denigrates individuals as sex objects. Besides the benefit of respect for body and self, since humans lack fur, clothes are also good protection from solar radiation, wind, cold, thorns, bugs, etc.

Why did God use animal skins?

The most obvious reason for God to make fur coats for Adam and Eve was because the fig leaf loin coverings were woefully inadequate. The risk of wardrobe failure was quite high. There is no record of Adam and Eve or their posterity ever reverting to fig leaf apparel.

Some modern pundits have suggested that God killed an animal, not only to clothe Adam's and Eve's physical nakedness, but also to serve as a symbol of the future salvific death that would cover their sins. Wearing the skin from a dead animal may also have served as a constant reminder to them that death was a consequence of their transgression.

An old Hebrew tradition says that the skin for the coats was from the serpent who tempted them. (3 Baruch) Of course, the serpent would have had to be quite large to make two coats that provided adequate coverage. Such garments would have been effective reminders to Adam and Eve of their transgression but much less comfortable than rabbit or chinchilla skins. Naturally-shed snake skin is far too brittle and gossamer to serve as robust and modest clothing material. Furthermore, if God had killed the serpent His prophecy that it would go on its belly and eat dust *all the days of its life* would have been nullified. Rabbi Johanan said, "They were like the fine linen… garments of skin meaning those that are nearest to the skin." (intertextual.bible/text/4q504-genesis-rabbah-20.12, DL 30 Mar 2025)

Per ancient Jewish and Christian texts, the coats of skins represented repentance and were substitutes for the cloaks of glory Adam and Eve lost when banished from the garden. Another old Jewish tradition says the coats God gave them were "garments of light". (*Midrash in Minhat Yehukh*, Gn 3:21) This tradition might have sprung from the conflation of the Hebrew words for skin and light that are pronounced identically (ôr). The "light" covering concept was reiterated in the NT by Paul; "Let us therefore cast off the works of darkness, and let us put on the armour of light." (Rm 13:12)

The King James version's use of plural "skins" seems to indicate that more than one animal got skinned. But the Hebrew word is actually singular as rendered by other later translations; "The LORD God made garments of *skin* for Adam and his wife and clothed them." (*NIV*, Gn 3:21)

Since there is nothing in the text to suggest the animals died from natural causes, it would seem that this is the first example of God intervening to take life. It would be a great foreshadowing of the Christ's sacrifice if it had been a male lamb without blemish! As the skins covered the fallen couple, the sacrifice of the Lamb of God would cover fallen mankind.

Why skins at all? Or, why clothing? Surely not to suggest their bodies were hideous! Perhaps God was teaching Adam and Eve; 1) how to respect the fleshly temples He had gifted to them, 2) how to protect their bodies from flies, thorns, sunburn, and cold, 3) how to make better clothing from more durable resources at hand — more primitive humans had been wearing fur and leather clothing for at least 40,000 years by then, 4) and, most importantly, that the blood of a lamb would be shed to restore them to the light — The garment of skins, like the temple veil, represent the Savior. Wearing the temple garment is deeply symbolic of taking Christ and His Atonement upon oneself to cover for sins. The Hebrew word for atonement (Str#3722, *kaphar*) also means covering.

Did Adam's coat have powers?

There is also the tradition regarding Adam's coat that it was associated with the law of primogeniture or birthright, which followed the male descendants and included priesthood presiding rights. Adam's coat was passed down from father to son until Noah from whom it was stolen by his son Ham because he thought it had special powers. Shem and Japheth allegedly used a woven copy of Adam's the coat to cover their father's nakedness when they found him drunk with wine and naked in his tent. (*CWHN (Collected Works of Hugh Nibley)*, bk5, pt2, ch1, pp169-171)

A Jewish midrashic work says that the coat endowed the wearer with the ability to command animals, it being the key to Noah's success in getting animals to enter the ark. (*Pirkei de-Rabbi Eliezer*, ch24) It goes on to say that the coat stolen by Ham was inherited by his great-grandson Nimrod who used it to attract animals and become a mighty hunter and king. (Gn 9:22-27; 10:9) So, if this legend is to be believed, the skins had something to do with the God-given right to have dominion over animals. (Gn 1:26)

Much of the legend is found in an 18th century pseudepigraphic text; "The garments of skin which God made for Adam… were given to Enoch, the son of Jared, and… he gave them to Methuselah, his son. And at the death of Methuselah, Noah took them and brought them to the ark, and… Ham stole those garments from Noah his father… And when Ham begat his first born Cush, he gave him the garments… and when Cush had begotten Nimrod, he gave him those garments… and Nimrod grew up, and when he was twenty years old he put on those garments. And Nimrod became strong when he put on the garments." (*Book of Jasher* 7:20-24)

However, since Nimrod was an evil man, it is unlikely that his power as a hunter and leader came from God or from a garment made by God. According to legend, the garment cost him his life when Esau, another great hunter, ambushed him in the wilderness, beheaded him, and took the garment that should have been passed down from father to eldest son anyway through the lineage of the patriarchs to Isaac. (fairlatterdaysaints.org/evidences/Source:Echoes:Ch12:19:G arment_of_Joseph, DL 14 May 2022)

Esau then sold the garment along with his birthright to Jacob for a bowl of pottage. The transfer of the garment represented Esau's sale of his birthright to Jacob. When Isaac, blind in his old age, determined to give the birthright blessing to his eldest son, it was the smell of the ancient garment, along with the feel of the goat hair on Jacob's hands and neck, that convinced Isaac to give the son at his knee the blessing. When Jacob passed the garment and the birthright on to Joseph, his other sons were jealous, kidnapped him, replaced the garment with a slave tunic, and sold him to an Egyptian. The original Hebrew should not have been translated "coat of many colors", rather "garment in which figures are woven." (ibid.)

What about the flaming sword?

God "drove out the man; and he placed at the *east* of the garden of Eden *Cherubims*, and a *flaming sword* which turned every way, to keep the way of the tree of life". (Gn 3:24) Placing angels only on one side of the garden would leave all other sides unguarded. And, since there was no shortage of angels anxious to help, why not have enough to guard all sides? The whole point of the sword of glory turning "every way" was to secure all sides of the garden!

The Tree of Life was "in the midst of the garden" rather than on any particular side. (Gn 2:9) Cain is the only one the Bible says settled east of the garden where the land of Nod was located. (Gn 4:16) The rest of Adam's family settled west of the garden where they settled in the cities and lands of Babel, Uruk, Acadia, Calneh, and Nineveh - all westerly from Eden. (Gn 10:10-30) That would place the "garden eastward in Eden" from the perspective of a scribe in the Holy Land. (Gn 2:8) Taking the shortest path from their post-exile home, the approach to the garden for Adam and Eve would have been on the west side of the garden.

So, with only Cain on the East and everyone else on the west, placing angels only on the east side as guards doesn't make sense. So, why does the text say angels were placed on the *east of the garden*?

A look at the Hebrew reveals that *east* is better translated as "before" or "in front of". (Strg#6924) So, angels were placed "*in front* of paradise" or "*before* the paradise," as the Douay-Rheims Bible renders it. (*DRB (Douay-Rheims Bible)*, Gn 3:24) Also, the Contemporary English Version says, "*at the entrance* to the garden." (*CEV (Contemporary English Version)*, Gn 3:24) If the garden was in the Tigris-Euphrates valley, it wouldn't make sense for the its entrance to be toward the Persian Gulf in view of the fact that the posterity of Adam's patriarchal migrated westward.

Had swords been invented at the time? Scientists think that swords were not invented until the Bronze Age, 3300–1200BC, the oldest sword found being from about 5,300 years old. (freerepublic.com/focus/f-news/880260/posts, DL 20 Feb 2019) How could Adam know what a sword was 700 years before they were invented? He probably didn't. But, the revelation to Moses that was the basis for the Genesis creation story was recorded centuries after swords were invented and was therefore in symbolic terms that the Hebrews of his day would understand. The word "sword" represented the power of God wielded by the angels. It would explain why there was only one sword – one power of God.

Swords are used elsewhere in scripture to represent divine force. For example, "David… saw the angel of the LORD standing between earth and heaven, with his drawn sword in his hand stretched out over Jerusalem. Then… David said to God… 'God, please let Your hand be against me and my father's household, but not against Your people that they should be plagued.'" (1 Chr 21:16) David knew that the angel did not intend to cut the people down with an actual sword, rather with a plague. The sword symbolized God's power to cut people down with whatever means He saw fit to use.

Why are the swords described as flaming? Unfamiliar with modern technology such as incandescent bulbs, light-emitting diodes, fluorescent tubes, et cetera, the ancients described glory using their terms for bright light – fire. Moses saw a "burning" bush without being consumed. No smoke was mentioned. Why? Because glory doesn't give off smoke.

The use of "fire" for glory is clear in the New Testament as well. For example, "Our God is a consuming fire." (Hb 12:29) God is not physical fire ($C+O_2 \rightarrow CO_2$). His glory and that of His angels is analogous to physical light which, in Bible times, humans could only produce using fire.

Where did Adam go after the fall?

An apocryphal work says that Adam returned to his birthplace to live after he was banished from paradise. There he offered prayers to God. And that is why Solomon chose that place as the site for the temple as directed by an angel. (*Life of Adam and Eve*, 51:6–7)

The 1st-century apocryphal *Life of Adam and Eve* says they lived near enough to the Tigris River to baptize themselves in it. (sacred-texts.com/chr/apo/adamnev.htm, DL 20 Jun 2020)

Where was Adam buried?

According to Jewish tradition, the burial place of biblical patriarchs and matriarchs is the Cave of the Patriarchs, aka Cave of Machpelah, an ancient Judean structure in modern-day Hebron on the Israeli West Bank about 30 miles south of the temple mount. The cave is Judaism's most ancient site, second only to the Temple Mount in sacredness.

But, St. Thomas Aquinas cited St. Jerome as having said, "Adam was buried near Ebron and Arbee." (ccel.org/ccel/aquinas/catena1/catena1.ii.xxvii.html, DL 16

Jul 2020) This is a reference to "the city of Arbah, which is Hebron". (Gn 35:27)

What are cherubim?

As for the angels, experts say cherubim are probably an order of angels somewhat lower than seraphim. Cherubim is plural as denoted by the suffix *im*, so there were multiple angels posted as guards, whether in combination or one after the other is not clear from the Bible text.

When angels showed Enoch the Tree of Life, he saw "three hundred angels very bright, who keep the garden". (*Secrets of Enoch* 8:9) Why more than one angel when one angel has more than enough power to keep any and all mortals from accessing the Tree of Life? Perhaps many angels volunteered and God granted the privilege to many. There might have been various shifts staffed by different groups of angels to give them all a chance to serve at the Tree of Life.

Why didn't Adam die that day?

God warned, "In the day that you eat of it you shall surely die." (Gn 2:17) And after they ate of the forbidden fruit, God confirmed, "Dust thou art, and unto dust shalt thou return. (Gn 3:19) So, how is it that Adam lived 930 years? (Gn 5:5) Here are some ways the statements from the Bible have been interpreted for harmony:

♦ The ratio could be used; "One day is with the Lord as a thousand years." (2 Pt 3:8) That would explain why Adam died before his thousandth birthday. The longest-lived biblical character was Methuselah who died only 31 years shy of that limit.

♦ The Hebrew word from which day is translated (*yowm*) is used to refer to the entire six-period creation epoch; "These are the generations of the heavens and of the earth when they were created, in the *day* that the LORD God made the earth and the heavens." (Gn 2:4) Using that undefined but finite period of time for *yowm*, God may have been warning Adam and Eve, "in the *age* that you eat thereof you shall surely die."

♦ One version of the Bible says, "If you eat its fruit, you are sure to die." (*NLT (New Living Translation)*. Gn 2:17) So, God may have meant, "As of the day you eat it, dying will become a certainty."

♦ If God was referring to spiritual death, aka separation from God, then Adam and Eve indeed died the same day they were ejected from the garden. Adam lamented; "God, when we dwelt in the garden, and our hearts were lifted up, we saw the angels that sang praises in heaven, but now we do not see… [And God answered], Thou hadst a bright nature within thee, and for that reason couldst thou see things afar off. But after thy transgression thy bright nature was withdrawn from thee." (*Conflict of Adam and Eve I*, 8:1-2)

Which view is true? They are not mutually exclusive interpretations and there is likely truth in all of them. Joseph Smith made a change to Adam's age in his second set of notes while making inspired correction to Genesis, increasing it 70 years to a round 1,000. But, Joseph was more precise in a sermon; "[Adam] was within 6 months of 1000 years old, which is one day with the Lord's time, thus fulfilling the Lord's decree, in the day thou eatest of the fruit of that tree thou shalt surely die, and he did 6 months before the day was out." (rsc.byu.edu/archived/selected-articles/ages-patriarchs-joseph-smith-translation, DL 5 Apr 2019, commas added)

What happened to the Garden?

Without God's presence to glorify the Garden of Eden, and without Adam and Eve there to "dress and keep it," all the plants in the garden would have eventually died unless God intervened. (*PGP*, Moses 3:15)

Nobody has discovered any 6,000-year-old tree in the area where the Garden of Eden is thought to have been. Neither does any remnant of the Garden of Eden show on satellite or aerial photos. Nor has anyone seen the angels that guard the Tree of Life nor found a place on earth fenced by invisible and inexplicable powers. So, we must conclude that the Garden of Eden either died or that God removed it.

If, as some believe, the whole earth had been paradisiacal before the fall, then the garden should have fallen with everything else and would therefore not be distinguishable from the surrounding countryside. It would have been obliterated by "the waters of the flood upon all the land of Eden". (Jub 4:24)

Might God have taken the Garden of Eden off-planet? "According to the early church Fathers, the paradise in which our first parents dwelt before the fall still exists, neither on the earth nor in the heavens, but above and beyond the world." (biblehub.com/str/greek/3857.htm, DL 19 Nov 2018)

Eliza R. Snow, one of Joseph Smith's wives, wrote a poem that was included in an early church hymnal that spoke to the Earth as follows, "When Enoch could no longer stay amid corruption here, part of thyself was borne away to form another sphere… and nearer to the throne of God his planet upward moved." (*Sacred Hymns and Spiritual Songs*, CJCLDS, ed14, #322a, 1871)

Some have speculated that the removed chunk included the Garden of Eden which served thereafter as the place where translated beings go with room for entire cities such as Enoch's Zion. According to one ancient Hebrew text, Enoch "was taken from amongst the children of men, and we conducted him into the Garden of Eden in majesty and honor… And he burnt the incense of the sanctuary." (Jub 4:23-25) So, Enoch was called to serve as a High Priest in the Garden of Eden which, according the Slavic book of Enoch, God placed into the third of at least ten heavens, as shown to him by angels prior to him being translated. (pseudepigrapha.com/pseudepigrapha/enochs2.htm#Ch8, DL 2 Mar 2019)

But why would God delay removal of the garden when He could just as easily have removed it immediately and saved the cherubs the trouble of guarding it? Perhaps God kept the garden around for a little while to remind Adam's family of the lessons of paradise lost. Or, perhaps He kept it around as a residence for himself when He came to visit earth and talk to Adam. After being ejected from the Garden, "Adam and Eve, his wife, called upon the name of the Lord, and they heard the voice of the Lord from the way toward the Garden of Eden, speaking unto them". (*PGP*, Mo 5:4)

Perhaps God maintained the garden on Earth until the city of Enoch was righteous enough to be taken up to heaven along with it. If Hebrew tradition is correct, Enoch is in the Garden of Eden to this day and the place serves as a home for translated persons. According to an ancient Hebrew text, there are other righteous people besides Enoch dwelling there also. (*Secrets of Enoch* 9:1) And it still existed in John's day; "To him that overcometh will I give to eat of the tree of life, which is in the midst of the paradise of God." (Rv 2:7)

If the Tree of Life mentioned in John's prophecies is the same as the one from the Garden of Eden, it had to be removed off-planet to preserve it until Christ ushers in His millennial reign. (Rv 22:2)

Is paradise lost forever?

The Tree of Life and the City of Enoch will both return to earth after Earth's glory is upgraded. (*PGP*, Mo 7:62-64; Rv 22:2, 14) For those who are righteous "is paradise opened, The Tree of Life is planted, the age to come is prepared, plenteousness is made ready, a city is built, and rest is allowed". (4 Ezra 8:52)

Until then, the Tree of Life remains off-planet until the prophecy is fulfilled that says, "As for this *fragrant tree* no mortal is permitted to touch it till the great judgment... It shall then be given to the righteous and holy. Its fruit shall be for food to the elect: it shall be transplanted to the holy place, to the temple of the Lord, the Eternal King." (1 Enoch 25)

After ejecting Adam and Eve from the garden, God "gave unto them commandments, that they should... offer the firstlings of their flocks, for an offering unto the Lord". (*PGP*, Mo 5:4–5) So Adam obeyed; "He built an altar and sacrificed on it as a burnt offering an ox." (*Babylonian Talmud*, Tract

Aboth, ch1) "Then Adam and Eve stood under the altar and wept, thus entreating God, 'Forgive us our trespass and our sin, and look upon us with Thine eye of mercy.'" (*The Forgotten Books of Eden,* ch23)

Adam and Eve first offered sacrifices out of blind obedience without understanding its symbolism. But then, "after many days an angel of the Lord appeared unto Adam, saying: Why dost thou offer sacrifices unto the Lord? And Adam said unto him: I know not, save the Lord commanded me. And then the angel spake, saying: This thing is a similitude of the sacrifice of the Only Begotten of the Father". (*PGP*, Mo 5:6–7)

Thus, by obediently looking to Jesus for redemption, Adam and Eve were born again to innocence and requalified to enter paradise. Through the atonement of Christ, an even better paradise would be possible, celestial glory, "through the merits, and mercy, and grace of the Holy Messiah". (*BOM*, 2 Ne 2:8)

Why punish Adam's posterity?

"The son shall *not* suffer for the iniquity of the father." (Ezek 18:20) That seems quite fair. Yet Christian churches persist in teaching the doctrine of Original Sin, i.e. "The sins of the forefathers leading to punishment of their descendants." (infogalactic.com/info/Original_sin, DL 20 Feb 2019)

At worst, God is "visiting the iniquity of the fathers on the children and the children's children, to the third and the fourth generation". (Ex 34:7) It's understandable that the *consequences* of sin reverberate through several subsequent generations, but guilt should not. A child is not guilty because his father robbed a bank, though he may suffer if his father is in prison. Likewise, nobody is guilty because Adam ate some fruit, though everybody inherits the legacy of his choice.

God does not use the sins of the parents as an excuse to punish people. "God is love" and "His mercy is great". (1 Jn 4:8; 2 Sam 24:14)

"Wherefore, as by one man sin entered into the world, and death by sin; and so *death passed upon all men, for that all have sinned.*" (Rm 5:12) Everyone dies physically whether or not they have sinned, babies die, animals die, plants die, etc. So, the type of death Adam brought to himself and his posterity as mentioned in scripture was spiritual death, aka a greater degree of separation from God than was the case in paradise. Mortality is not a punishment. But spiritual death is indeed the consequence of sin!

Many people further distance themselves from God through their own bad choices. But penitent people can be reunited with God through the merits of Christ and live again spiritually.

Adam's babies were innocent and God could have brought them back into His presence in the Garden of Eden. But that would have been a step backwards. Adam fell downward but forward. (lds.org/manual/true-to-the-faith/fall, DL 19 Nov 2018)

The spirit of every baby chose mortality to progress forward. The consequences of the fall experienced by all are not punishment for Adam's transgression, rather an opportunity to grow spiritually, the reward for choosing righteously in premortality.

Another reason that Adam's and Eve's transgression could not be passed down to their posterity is that God forgave them 6,000 years ago. He told them, *"I have forgiven thee thy transgression in the Garden of Eden.* Hence came the saying abroad among the people, that the <u>Son of God hath atoned for original guilt</u>, wherein the sins of the parents cannot be answered upon the heads of the children." (*PGP*, Mo 6:53-54) When forgiveness of sins is achieved through the grace of Jesus, He promised, "I, the Lord, remember them no more." (DC 58:42) "Though they be red like crimson, they shall be as wool." (Is 1:18) It is because God remembers the first transgression no more that "Men will be punished for their own sins, and not for Adam's transgression". (*PGP*, AOF #2)

Epilogue

Almost half of weekly churchgoers, 46%, believe that science sometimes conflicts with their religious beliefs. (*Faith Exploration*, Deseret News, 15 Dec 2013) Hopefully, rather than losing faith, people can resolve those conflicts while reading this book and then help others find resolutions also. If stumbling blocks still persist, there is hope that inspiration will light the way. Perhaps other people or sources have answers that make more sense to the reader than those offered in this book.

This book as been valuable as a reference for the author who has often copied sections from it and posted them on social media as answers to gospel questions. It has also been handy when discussing topics with friends and family.

Even if some of this book's ideas turn out to be incorrect, presenting a cohesive view of what is taught in the revealed word will at least prove that a person can cobble together an internally consistent theory of creation that is also consistent with science. And if one person can, many people can.

Bibliography

The following reference materials are key sources for this book. The abbreviations in parentheses are used in source citations throughout the book:

♦ <u>Analytical Concordance to the Bible</u>. Robert Young. New York NY, FUNK & WAGNALLS, 1955. (Young)

♦ <u>Answers to Gospel Questions</u>, Joseph Fielding Smith, 1957. (AGQ)

♦ <u>Articles of Faith, Joseph Smith</u>, Pearl of Great Price, CJCLDS, 1842. (AOF)

♦ <u>Collected Works of Hugh Nibley</u>, (CWHN)

♦ <u>Comprehensive History of the Church</u>, BYU, 1965. (CHC)

♦ <u>Discourses of Brigham Young</u>, (DBY)

♦ <u>Doctrine and Covenants</u>. SLC UT, The Church of JESUS CHRIST of Latter-day Saints, 2013. (DC)

♦ <u>Doctrinal New Testament Commentary</u>. Bruce R. McConkie, Deseret Book, 2002. (DNTC)

♦ <u>Doctrines of Salvation</u>, Joseph Fielding Smith Jr., 1956. (DOS)

♦ <u>Egyptian Alphabet and Grammar</u>, early to mid-1830s. (EAG)

♦ <u>Encyclopedia of Mormonism</u>, New York NY, Macmillan Publishing Company, 1992. (EOM)

♦ <u>History of the Church</u>, commissioned by Joseph Smith, 7 volumes, CJCLDS, 1858. (HC)

♦ <u>Journal of Discourses</u>, SLC UT, Compiled by George Watt, Deseret Book Company, 1886. (JD)

♦ <u>Lectures on Faith</u>, Joseph Smith, et al, 1835. (LOF)

♦ <u>Mormon Doctrine</u>, Bruce R. McConkie, 1958. (MD)

♦ <u>New Testament</u>. SLC UT, The Church of JESUS CHRIST of Latter-day Saints, 2013. KJV. (NT)

♦ <u>Old Testament</u>. SLC UT, The Church of JESUS CHRIST of Latter-day Saints, 2013. KJV. (OT)

♦ <u>Pearl of Great Price</u>. SLC UT, The Church of JESUS CHRIST of Latter-day Saints, 2013. (PGP)

♦ <u>Strong's Concordance</u>, James Strong's Exhaustive Concordance of the Bible. Thomas Nelson, 2001. (Str)

♦ <u>Teachings of Spencer W. Kimball</u>, by Edward Kimball, 2006. (TSWK)

♦ <u>Teachings of the Prophet Joseph Smith</u>, SLC UT, Compiled by Joseph Fielding Smith, Deseret Book Company, 1976. (TPJS)

♦ <u>Times and Seasons</u>, Nauvoo CJCLDS newspaper. (T&S)

♦ <u>Words of Joseph Smith</u>, Ehat & Cook, Gandin Book Company, 1991. (WJS)

Abbreviations, Acronyms & Initialisms

\# - number

2D – Two Dimensional

3D – Three Dimensional

4D – Four Dimensional

a – answer

Abr – Abraham, a PGP book brought forth by Joseph Smith.

AD – Anno Domino (Latin: Year of the Lord)

AGQ – Answers to Gospel Questions, by Joseph Fielding Smith

aka – Also Known As

AMS – Accelerator Mass Spectrometry

AOF – Articles of Faith, published in the Pearl of Great Price.

Apr – April

Ar – Argon

ASV – American Standard Version

Aug – August

BC – Before Christ

BD – CJCLDS Bible Dictionary

bk – Book or volume

BOM – The Book of Mormon

Brit - British

BYA – Billion Years Ago

BYU – Brigham Young University

C – Carbon

c – Celeritas

C12 – Carbon-12

C14 – Carbon-14

ca – Circa, meaning about, as in approximately

CCC – Catechism of the Catholic Church

CEV – Contemporary English Version of the Bible

ch – Chapter

CHC – Comprehensive History of the Church, by Roberts

CJCLDS – The Church of JESUS CHRIST of Latter-day Saints

CMB – Cosmic Microwave Background

Col – Colossians

COM – Communications

Col – Corinthians

CR – Conference Report (held in April and October by The Church of JESUS CHRIST of LDS)

CWHN – Collected Works of Hugh Nibley

DBY – Discourses of Brigham Young

DC – Doctrine and Covenants

Dec – December

DL – Downloaded

Dn - Daniel

DNA – Deoxyribonucleic Acid

DOS – Doctrines of Salvation, by Joseph Fielding Smith

DRB – Douay-Rheims Bible

Dt – Deuteronomy, a book in the OT

E – Energy

EAG – Egyptian Alphabet and Grammar (created by Joseph Smith's scribes during studies of papyrus)

Ec – Ecclesiastes, an OT book

Ed – Edition

EEG – Electroencephalogram

EMF – Electromagnetic Field

En – Enoch, the antediluvian prophet

EOM – Encyclopedia of Mormonism

Eph – Ephesians

ERV – English Revised Version of the Bible

ESR – Electron Spin Resonance

ESV – English Standard Version

ET – Extraterrestrial

ETI – Extraterrestrial Intelligence

Ex – Exodus

Ez – Ezra

Ezek – Ezekiel

F - Fahrenheit

Facs – facsimile

Feb – February

Fig – figure

fn – Footnote(s)

GMO – Genetically Modified Organism

Gn – Genesis

GR – General Relativity

Gr – Greek

GUT – Grand Unified Theory

Hb – Hebrews, a NT book

HC – History of the Church, a publication of The Church of JESUS CHRIST of Latter-day Saints

Heb – Hebrew, the language

Hel – Helaman, a book in The Book of Mormon.

i – Imaginary number square root of minus 1

i.e. – id est (Latin: that is)

Ibid – Latin, short for ibidem (in the same place)

IQ – Intelligence Quotient

Is – Isaiah

ISV – International Standard Version of the Bible

Jan – January

Jb – Job

Jdg – Judges, an OT book

Jer – Jeremiah

Jn – John

Js – James

Jsh – Joshua

JD – Journal of Discourses, 26-volumes of sermons by early leaders of The Church of Jesus Christ of Latter-day Saints

JPS – Jewish Publication Society

JST – Joseph Smith Translation (KJV Bible with inspired edits)

Jub – Jubilees

Jul – July

Jun – June

JWST – James Webb Space Telescope

K – Potassium

Kg – Kings, one of the Old Testament Books

KJV – King James Version (default for Bible quotes)

K-T – Cretaceous-Tertiary (geologic boundary)

L – Latin

LDS – Church of JESUS CHRIST of Latter-day Saints

lect – Lecture

Lk – Luke

Lv – Leviticus

LHC – Large Hadron Collider

LIGO – Laser Interferometer Gravitational-wave Observatory

LOF – Lectures on Faith, notes taken from Joseph Smith's School of the Prophets activities.

LUCA – Last Universal Common Ancestor

LXX – Septuagint (earliest extant Greek translation of the OT from original Hebrew)

m – Mass

Mal – Malachi

Mar – March

MD – Mormon Doctrine

MDNA – Mitochondrial DNA

MeV – Mega-Electronvolts

MIT – Massachusetts Institute of Technology

Mk – Mark, a NT book

Mo – Moses, a book in the *Pearl of Great Price*

MPH – Miles per Hour

Ms – Manuscript

MT – Matthew

M-Theory – Magic, Mystery, or Membrane Theory

Mus - Museum

MYA – Million Years Ago

n – Algebraic variable for number of something

NDE – Near Death Experience

Ne – Nephi, the first author in the Book of Mormon

NHEB – New Heart English Bible

NIV – New International Version (Bible)

Nm – Numbers, the OT book

Note - nt

Nov – November

NT – New Testament

Oct – October

OT – Old Testament

p – Page number

par – Paragraph

PBE – Pre-Birth experience

PGP – Pearl of Great Price (one of The Church of JESUS CHRIST of LDS standard works of scripture)

pH – Power of Hydrogen, a measure of acidity/alkalinity

PLOS – Public Library of Science (peer-reviewed journal)

p – Page

pp – Pages, describes a range or set of page numbers

Prv – Proverbs, an OT book

pt – part

Pt – Peter

q – question

QM – Quantum Mechanics

Rm – Romans

RNA – Ribonucleic Acid

RSV – Revised Standard Version of the Bible

Rv – Revelation (last book of the Bible)

sec – Section

Sep – September

SETI – Search for Extraterrestrial Intelligence

Sgr – Sagittarius

SLC – Salt Lake City

Sm – Samuel, one of the Old Testament Books

SMBH – Supermassive black hole

Str – James Strong's Concordance of the Bible

T&S – Times and Seasons, an early LDS periodical

Tim – Timothy

TOE – Theory of Everything

TPJS – Teachings of the Prophet Joseph Smith, selected quotes by Joseph Fielding Smith

TSWK – Teachings of Spencer W. Kimball, Deseret Book

TYA – Thousand Years Ago

UFT – Unified Field Theory

US – United States of America (I prefer USA, but some spellcheckers don't like it)

vlm – Volume

v – Verse

vv – Verses

WIS – The Wisdom of Solomon (a book of the Apocrypha in the Catholic Bible)

WJS – The Words of Joseph Smith, compiled and edited by Andrew F. Ehat & Lyndon W. Cook

YLT – Young's Literal Translation (of Bible)

Book Versions

Version number interval magnitudes reflect the extensiveness of the changes for the republication.

Version 1.0	1 December 2018 (initial publication)
Version 2.0	1 January 2019 (formatting corrections)
Version 3.0	5 June 2019 (proofreading corrections)
Version 4.0	28 June 2020 (wordsmithing, etc.)
Version 5.0	27 October 2020 (reorganized chapters, etc.)
Version 6.0	02 August 2021 (incorporated AW feedback)
Version 7.0	16 May 2022 (incorporated DE feedback)
Version 8.0	27 March 2023 (temple sequence update)
Version 9.0	04 January 2024 (miscellaneous additions)
Version 9.1	24 March 2024 (miscellaneous adds/edits)
Version 9.2	10 June 2025 (miscellaneous adds/edits)
Version 9.3	6 September 2025 (miscellaneous adds/edits)
Version 9.4	27 February 2026 (miscellaneous adds/edits)
Version 10.0	28 June 2026 (miscellaneous adds/edits)

Acknowledgements

I thank God for blessing me with the circumstances and inspired materials and tools that made this book possible. I apologize to Him for what He certainly views as a clumsy attempt to piece together the puzzle of creation in this book.

Thanks also to Aristarchus, Eratosthenes, Copernicus, Newton, Leibniz, Darwin, Curie, Maxwell, Lorentz, Einstein, Bohr, Feynman, Hawking, Penrose, and other great scientists whose contributions to the vast and wonderful library of scientific knowledge made this book possible.

I owe a lot to the teachers, preachers, authors, and fellow gospel enthusiasts whose efforts and ideas have shaped many of the views in this book, even when I disagreed with them. The positions taken in this book are solely my responsibility and do not necessarily represent the views of my family, my church, or anyone else, and sometimes not even my own views. I wrote what I thought was necessary to make the puzzle pieces fit.

I'm indebted to James Strong for his Bible concordance, to the purveyors of biblehub.com, and to the King James Version translators whose Bible is my primary source, and to Joseph Smith for volumes of enlightening revelation.

Thanks to my wonderful wife, Lavita, for patiently listening to me ponder possibilities and for leading the way by writing and publishing books of her own. Thanks to my son, Adam, for encouraging me, pushing back on ideas that seemed implausible, suggesting great ideas for inclusion, and reviewing a proof copy. Thanks also to my brother Bob and my friend David Ellis, Ph.D. Behavioral Ecology (evolutionary basis of animal behavior in response to ecological pressures) for their proofreading work!

Finally, thanks in advance to each of you who has read this book with an open and analytical mind.

Made in the USA
Monee, IL
07 July 2026

56550375R00260